Selected Papers from 27th European Biomass Conference & Exhibition (EUBCE 2019)

Selected Papers from 27th European Biomass Conference & Exhibition (EUBCE 2019)

Editors

David Baxter
Solange I. Mussatto

MDPI • Basel • Beijing • Wuhan • Barcelona • Belgrade • Manchester • Tokyo • Cluj • Tianjin

Editors
David Baxter
Former European Commission, Joint Research Centre
UK

Solange I. Mussatto
Department of Biotechnology and Biomedicine,
Technical University of Denmark
Denmark

Editorial Office
MDPI
St. Alban-Anlage 66
4052 Basel, Switzerland

This is a reprint of articles from the Special Issue published online in the open access journal *Energies* (ISSN 1996-1073) (available at: https://www.mdpi.com/journal/energies/special_issues/EUBCE_2019).

For citation purposes, cite each article independently as indicated on the article page online and as indicated below:

LastName, A.A.; LastName, B.B.; LastName, C.C. Article Title. *Journal Name* **Year**, *Volume Number*, Page Range.

ISBN 978-3-0365-0804-7 (Hbk)
ISBN 978-3-0365-0805-4 (PDF)

Cover image courtesy of Solange I. Mussatto.

© 2021 by the authors. Articles in this book are Open Access and distributed under the Creative Commons Attribution (CC BY) license, which allows users to download, copy and build upon published articles, as long as the author and publisher are properly credited, which ensures maximum dissemination and a wider impact of our publications.

The book as a whole is distributed by MDPI under the terms and conditions of the Creative Commons license CC BY-NC-ND.

Contents

About the Editors vii

Preface to "Selected Papers from 27th European Biomass Conference & Exhibition (EUBCE 2019)" ix

Moritz Von Cossel, Iris Lewandowski, Berien Elbersen, Igor Staritsky, Michiel Van Eupen, Yasir Iqbal, Stefan Mantel, Danilo Scordia, Giorgio Testa, Salvatore Luciano Cosentino, Oksana Maliarenko, Ioannis Eleftheriadis, Federica Zanetti, Andrea Monti, Dagnija Lazdina, Santa Neimane, Isabelle Lamy, Lisa Ciadamidaro, Marina Sanz, Juan Esteban Carrasco, Pilar Ciria, Ian McCallum, Luisa M. Trindade, Eibertus N. Van Loo, Wolter Elbersen, Ana Luisa Fernando, Eleni G. Papazoglou and Efthymia Alexopoulou
Marginal Agricultural Land Low-Input Systems for Biomass Production
Reprinted from: *Energies* **2019**, *12*, 3123, doi:10.3390/en12163123 1

Mariana Abreu, Alberto Reis, Patrícia Moura, Ana Luisa Fernando, António Luís, Lídia Quental, Pedro Patinha and Francisco Gírio
Evaluation of the Potential of Biomass to Energy in Portugal—Conclusions from the CONVERTE Project
Reprinted from: *Energies* **2020**, *13*, 937, doi:10.3390/en13040937 27

Cristina Moliner, Alberto Lagazzo, Barbara Bosio, Rodolfo Botter and Elisabetta Arato
Production, Characterization, and Evaluation of Pellets from Rice Harvest Residues †
Reprinted from: *Energies* **2020**, *13*, 479, doi:10.3390/en13020479 59

Dimitrios K. Sidiras, Antonios G. Nazos, Georgios E. Giakoumakis and Dorothea V. Politi
Simulating the Effect of Torrefaction on the Heating Value of Barley Straw
Reprinted from: *Energies* **2020**, *13*, 736, doi:10.3390/en13030736 71

Aidan Mark Smith, Ugochinyere Ekpo and Andrew Barry Ross
The Influence of pH on the Combustion Properties of Bio-Coal Following Hydrothermal Treatment of Swine Manure
Reprinted from: *Energies* **2020**, *13*, 331, doi:10.3390/en13020331 87

Mateusz Szul, Tomasz Iluk and Aleksander Sobolewski
High-Temperature, Dry Scrubbing of Syngas with Use of Mineral Sorbents and Ceramic Rigid Filters
Reprinted from: *Energies* **2020**, *13*, 1528, doi:10.3390/en13061528 107

Ekaterina Ovsyannikova, Andrea Kruse and Gero C. Becker
Feedstock-Dependent Phosphate Recovery in a Pilot-Scale Hydrothermal Liquefaction Bio-Crude Production
Reprinted from: *Energies* **2020**, *13*, 379, doi:10.3390/en13020379 129

Jéssica Siqueira Mancilha Nogueira, João Paulo Alves Silva, Solange I. Mussatto and Livia Melo Carneiro
Synthesis and Application of Heterogeneous Catalysts Based on Heteropolyacids for 5-Hydroxymethylfurfural Production from Glucose
Reprinted from: *Energies* **2020**, *13*, 655, doi:10.3390/en13030655 151

Abraham A. J. Kerssemakers, Pablo Doménech, Marco Cassano, Celina K. Yamakawa, Giuliano Dragone and Solange I. Mussatto
Production of Itaconic Acid from Cellulose Pulp: Feedstock Feasibility and Process Strategies for an Efficient Microbial Performance
Reprinted from: *Energies* **2020**, *13*, 1654, doi:10.3390/en13071654 169

Lelde Timma, Elina Dace and Marie Trydeman Knudsen
Temporal Aspects in Emission Accounting—Case Study of Agriculture Sector
Reprinted from: *Energies* **2020**, *13*, 800, doi:10.3390/en13040800 181

Sean O'Connor, Ehiaze Ehimen, Suresh C. Pillai, Gary Lyons and John Bartlett
Economic and Environmental Analysis of Small-Scale Anaerobic Digestion Plants on Irish Dairy Farms
Reprinted from: *Energies* **2020**, *13*, 637, doi:10.3390/en13030637 203

Lisa Mølgaard Lehmann, Magdalena Borzęcka, Katarzyna Żyłowska, Andrea Pisanelli, Giuseppe Russo and Bhim Bahadur Ghaley
Environmental Impact Assessments of Integrated Food and Non-Food Production Systems in Italy and Denmark
Reprinted from: *Energies* **2020**, *13*, 849, doi:10.3390/en13040849 223

Ricardo Luís Carvalho, Pooja Yadav, Natxo García-López, Robert Lindgren, Gert Nyberg, Rocio Diaz-Chavez, Venkata Krishna Kumar Upadhyayula, Christoffer Boman and Dimitris Athanassiadis
Environmental Sustainability of Bioenergy Strategies in Western Kenya to Address Household Air Pollution
Reprinted from: *Energies* **2020**, *13*, 719, doi:10.3390/en13030719 235

About the Editors

David Baxter is a Materials Engineer who joined the European Commission Joint Research Centre in 1991 after working in an industrial company supplying components for power generation and transport. He has been a member of the Sustainable Transport Unit in the Institute for Energy & Transport of the Joint Research Centre (Petten, The Netherlands) until retirement from the European Commission in August 2016. He was part of a team providing scientific and technical support to the development and maintenance of sustainability schemes for biomass and bioenergy, including biofuels, as well as member of the European Bioenergy Industrial Initiative (EIBI) team. He is a former chair of the scientific committee and one of the scientific organisers of the European Biomass Conference and Exhibition (EUBCE) over the last decade. He is also a former leader of the International Energy Agency Bioenergy Biogas Task 37, promoting economically and environmentally sustainable management of biogas production and utilization from agricultural residues, energy crops and municipal wastes.

Solange I. Mussatto is Professor and Head of the Research Group "Biomass Conversion and Bioprocess Technology" at the Technical University of Denmark. Her group conducts research in interdisciplinary areas including biotechnology, chemical engineering and sustainability, with the aim of creating innovative processes/solutions able to accelerate the development of a bioeconomy, with potential environmental, economic and social benefits (see more at: www.bcbtgroup.com). She is a reference person in the areas of bioeconomy and biorefinery, being the Danish representative in IEA Bioenergy Task 42—Biorefining in a Future Bioeconomy, and vice-chair of the Bioenergy and Bioeconomy section in the European Society of Biochemical Engineering Sciences. In 2020, she was ranked in the top 2% of scientists in the world (no. 1 in Denmark within the sub-field of biotechnology) in terms of career-long citation impact. She is/has been Editor of several scientific journals in her areas of specialization and also has active participation on the advisory boards of several leading international funding agencies.

Preface to "Selected Papers from 27th European Biomass Conference & Exhibition (EUBCE 2019)"

Biomass is all around us. All living things are biomass, although we might prefer to limit our definition of biomass to plant matter that grows in forests, fields and water and can eventually be used for essential day-to-day products and bioenergy after appropriate conversion. Before the Industrial Revolution and the age of increasing fossil fuel use, there was a well-established bioeconomy and at the same time a relatively constant level of carbon dioxide in the Earth's atmosphere. Global warming that we know today had not begun. The needs of plant life to absorb carbon dioxide from the atmosphere for growth while emitting oxygen provided the environmental balance essential for maintaining a stable ecosystem for all life on our planet Earth. Nature itself has been performing carbon capture and storage for millennia. The life we humans choose to live has put severe strain on the fine carbon balance, and we now realize the urgency of redressing that balance to avoid an environmental disaster. The European Biomass Conference and Exhibition (EUBCE) addresses the many key questions as to how biomass should be used in a modern post-fossil fuel age. The EUBCE has for about 40 years been involved in scientific research both into biomass utilization for bioenergy and increasingly into the role of biomass in an emerging global bioeconomy. Over the last couple of decades, the conference has also included a view of the many large-scale biomass utilization projects to have emerged from earlier research, the challenges of technology scale-up and the adaptations made to achieve commercial viability. Use of woody biomass directly as a fuel for heating is long-established but clean and efficient technologies are still developing, particularly with respect to emissions to air. Biofuels have long been used, and in recent times new processes have received close attention, with large volumes now being produced in many countries. While biorefineries are not new, there are many exciting and very promising innovative processes capable of yielding a very wide range of biochemicals and biomaterials that can not only be used to replace fossil-derived equivalents but can also create new opportunities in our fast-developing world. The potential for biomass harvests is increasingly better understood, whether in managed forests or on land used for farming. Biomass crops can be grown on low-grade or contaminated soils where food crops are not favored. The aquatic world is also a source of biomass, particularly for algae, and particularly in contaminated waters, for example in over-fertilized agricultural regions. Residues from many processes, for example food production, and wastes are a key source of biomass feedstock for conversion processes. There are so many options and possible ways in which all the available biomass can be used. The main questions are how biomass should actually be treated in the modern bioeconomy, and if it is used as a feedstock, what are the best options for utilization. It would be very nice if all the possible options for biomass use could be listed in order of priority according to some simple rules, but this is simply not possible. There are far too many variables and far too many possible pathways from raw biomass to final product. However, substantial progress in narrowing down options has been made and many good technologies have been developed, some of which do work at industrial scale. Nevertheless, there are still questions about not only economic and technical viability, but also the wider environmental impacts of these processes over the long term. Political policy makers are in real need of sound technical guidance to build appropriate policy frameworks to ensure the success of the transition away from the fossil fuel era. There are sound pathways for biomass exploitation in the emerging modern bioeconomy. The editors of this book are keen to show some of the areas of research that are contributing to the emergence of this bioeconomy. The book contains a small selection of reports from a range of projects designed to push forward the knowledge base supporting sustainable biomass use in the bioeconomy. The contents of all the papers presented here are derived from projects presented and discussed at the EUBCE conference that was held in Lisbon, Portugal, from May 27th to 30th 2019.

David Baxter, Solange I. Mussatto
Editors

Article

Marginal Agricultural Land Low-Input Systems for Biomass Production

Moritz Von Cossel [1,*], Iris Lewandowski [1], Berien Elbersen [2], Igor Staritsky [2], Michiel Van Eupen [2], Yasir Iqbal [3], Stefan Mantel [4], Danilo Scordia [5], Giorgio Testa [5], Salvatore Luciano Cosentino [5], Oksana Maliarenko [6], Ioannis Eleftheriadis [7], Federica Zanetti [8], Andrea Monti [8], Dagnija Lazdina [9], Santa Neimane [9], Isabelle Lamy [10], Lisa Ciadamidaro [10], Marina Sanz [11], Juan Esteban Carrasco [11], Pilar Ciria [11], Ian McCallum [12], Luisa M. Trindade [13], Eibertus N. Van Loo [14], Wolter Elbersen [15], Ana Luisa Fernando [16], Eleni G. Papazoglou [17] and Efthymia Alexopoulou [7]

1. Biobased Products and Energy Crops, Institute of Crop Science, University of Hohenheim, Fruwirthstr. 23, 70599 Stuttgart, Germany
2. Team Earth Informatics, Wageningen Environmental Research, P.O. Box 47, 6700 AA Wageningen, The Netherlands
3. College of Bioscience and Biotechnology, Hunan Agricultural University, Changsha 410128, Hunan, China
4. International Soil Reference and Information Centre, P.O. Box 353, 6700 AJ Wageningen, The Netherlands
5. Alimentazione e Ambiente, Di3A—Dipartimento di Agricoltura, University of Catania, 95123 Catania, Italy
6. Institute of Bioenergy Crops and Sugar Beet NAAS, 03141 Kyiv, Ukraine
7. Center for Renewable Energy Sources, Biomass Department, 19009 Pikermi Attikis, Greece
8. Dept. of Agricultural and Food Sciences (DISTAL), Alma Mater Studiorum—University of Bologna, 40126 Bologna, Italy
9. Latvian State Forest Research Institute SILAVA, 2169 Salaspils, Latvia
10. Environment and Agronomy Division, French National Institute for Agricultural Research, 78850 Thiverval-Grignon, France
11. Centro de Investigaciones Energeticas, Medioambientales y Tecnologicas-CIEMAT, 28040 Madrid, Spain
12. International Institute for Applied Systems Analysis, A-2361 Laxenburg, Austria
13. Laboratory of Plant Breeding, Wageningen University & Research, P.O. Box 9101, 6700 HB Wageningen, The Netherlands
14. Wageningen Plant Research, Plant Breeding, 6708 PB Wageningen, The Netherlands
15. Wageningen Food & Biobased Research, Wageningen University and Research Centre, P.O. Box 17, 6700 AA Wageningen, The Netherlands
16. Faculdade de Ciências e Tecnologia, Universidade Nova de Lisboa, 2829-516 Caparica, Portugal
17. Department of Crop Science, Agricultural University of Athens, 118 55 Athens, Greece
* Correspondence: moritz.cossel@uni-hohenheim.de; Tel.: +49-711-459-23557

Received: 9 July 2019; Accepted: 9 August 2019; Published: 14 August 2019

Abstract: This study deals with approaches for a social-ecological friendly European bioeconomy based on biomass from industrial crops cultivated on marginal agricultural land. The selected crops to be investigated are: Biomass sorghum, camelina, cardoon, castor, crambe, Ethiopian mustard, giant reed, hemp, lupin, miscanthus, pennycress, poplar, reed canary grass, safflower, Siberian elm, switchgrass, tall wheatgrass, wild sugarcane, and willow. The research question focused on the overall crop growth suitability under low-input management. The study assessed: (i) How the growth suitability of industrial crops can be defined under the given natural constraints of European marginal agricultural lands; and (ii) which agricultural practices are required for marginal agricultural land low-input systems (MALLIS). For the growth-suitability analysis, available thresholds and growth requirements of the selected industrial crops were defined. The marginal agricultural land was categorized according to the agro-ecological zone (AEZ) concept in combination with the marginality constraints, so-called 'marginal agro-ecological zones' (M-AEZ). It was found that both large marginal agricultural areas and numerous agricultural practices are available for industrial crop cultivation on European marginal agricultural lands. These results help to further describe the suitability of industrial crops for the development of social-ecologically friendly MALLIS in Europe.

Keywords: bioeconomy; bio-based industry; biomass; bioenergy; industrial crop; perennial crop; low-input agriculture; marginal land; MALLIS; sustainable agriculture

1. Introduction

In the targeted 'ideal' bioeconomy, the production of biomass will take social, ecological, and health aspects into account [1] to help achieve the sustainable development goals 2015–2030. From the bioeconomy's ambitions and definitions, conclusions can be drawn that the growth of the bioeconomy demands both a reduction of waste and losses and an adequate supply of sustainably grown biomass [2]. However, an increasing biomass production also carries a higher risk of social-ecological threats, such as increased use of fertilizers and pesticides, negative impacts from land-use changes and additional pressure on water resources [3–5]. The EU Horizon 2020 project MAGIC (Grant agreement ID: 727698) was established with the ambition of supporting the mitigation of these risks. This study deals with the basic findings of the 'Low-input agricultural practices for industrial crops on marginal land'.

Low-input agriculture (Figure 1) generally provides a number of promising practices that can help improve the social-ecological sustainability of biomass production while maintaining economic feasibility [6]. Here, a key parameter is the ratio between on- and off-farm inputs. According to Biala et al. (2007) [6], in low-input agriculture, the use of on-farm inputs should be maximized and off-farm inputs minimized. Currently, there are four concrete and real farming system types which follow these low-input agriculture principles (taken from Reference [7]): (i) Integrated farming, (ii) organic farming, (iii) precision farming, and (iv) conservation farming.

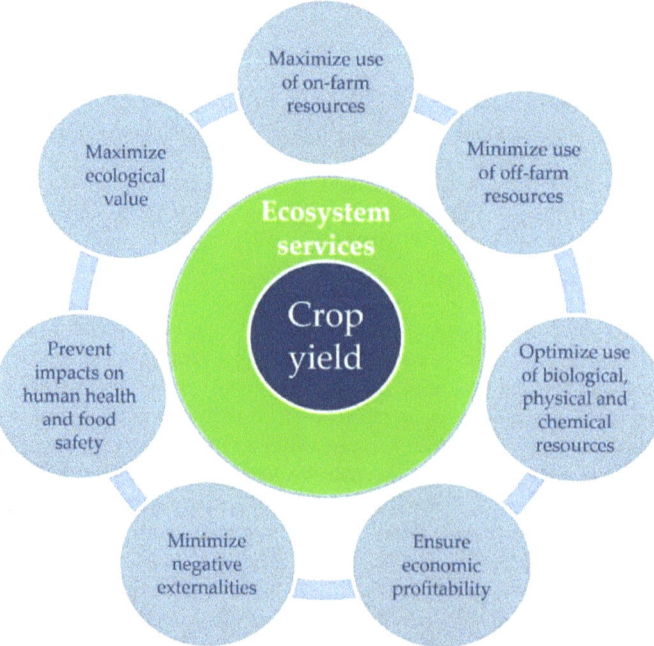

Figure 1. Principles of low-input agriculture (Source: This study).

For each of these farming systems, crop selection was found to be highly relevant for efficient use of resources during their cultivation [8–11]. The resource use efficiency becomes even more relevant for industrial crop cultivation on marginal agricultural lands (Figure 2). This is because both the yield potential and the resilience of the agro-ecosystems (their robustness against cropping failures) may be lower on marginal agricultural lands compared to fertile agricultural lands [9,12–17]. According to Elbersen et al. [12], marginal agricultural lands can be defined as *'lands having limitations which in aggregate are severe for sustained application of a given use and/or are sensitive to land degradation, as a result of inappropriate human intervention, and/or have lost already part or all of their productive capacity as a result of inappropriate human intervention and also include contaminated and potentially contaminated sites that form a potential risk to humans, water, ecosystems, or other receptors'*. The implementation of a low-input approach that can potentially reduce the risk to humans, water, ecosystems or other receptors is mainly dependent on the farming system and requires site-specific consideration [18,19].

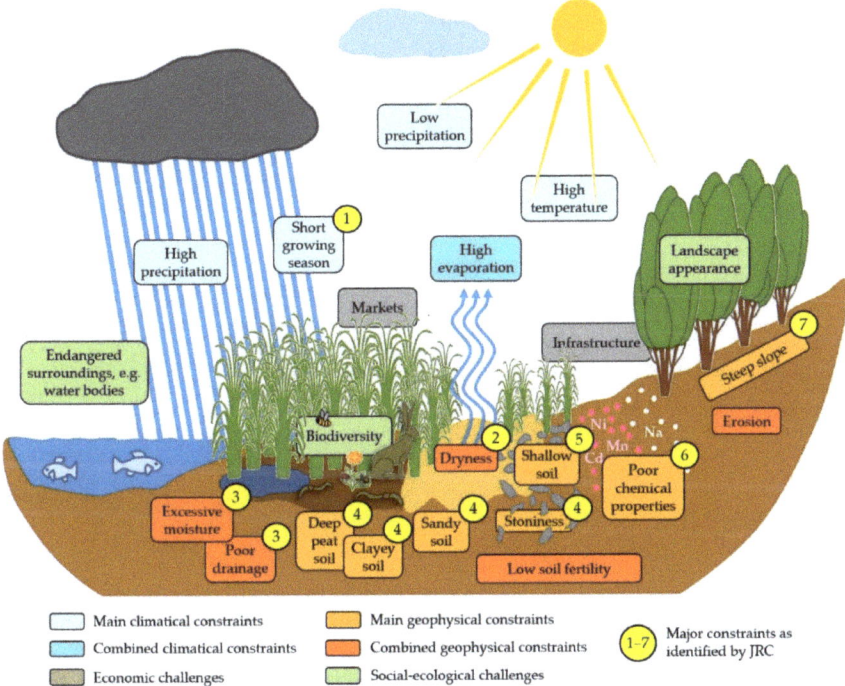

Figure 2. Illustration of relevant biophysical constraints and both economic and social-ecological challenges selected for marginal agricultural land low-input systems (Source: This study). Numbers 1–7 indicate the major biophysical constraints on marginal lands as defined by the Joint Research Centre (JRC) [20–22]. The other parameters either influence (main constraints) or follow on (combined constraints) from the major biophysical constraints, which limit the site-specific plant growth suitability (Table A1). The economic and social-ecological challenges have been added, due to their increasing relevance for modern agricultural systems [23–26]. These challenges can render a site marginal under both economic and social-ecological aspects, such as environmental protection, biodiversity conservation, infrastructure, markets and landscape appearance.

2. Material and Methods

For ethical reasons, low-input industrial crop cultivation on marginal agricultural lands is to be preferred in order to reduce competition for agricultural land use with both food crop cultivation and biodiversity conservation [9,27–30]. As favorable agricultural lands should primarily be used for food crop cultivation, this study focuses on the use of marginal agricultural lands for low-input industrial crop cultivation. Consequently, it aimed at:

1. Mapping the major climatic and biophysical constraints across European marginal agricultural lands;
2. Assessing the growth suitability of pre-selected industrial crops under the prevailing climatic and biophysical constraints; and
3. The development of social-ecologically friendly marginal agricultural land low-input systems (MALLIS) for industrial crop cultivation.

To address the above-mentioned research objectives, a thorough literature review was conducted using the search engines of SCOPUS (Elsevier, B.V.) and Google Scholar (Google LLC.). The pre-selection of the industrial crops (Table 1) which was based on a multi-criteria analysis (among others, the maturity of knowledge on industrial crops on marginal land and crops' productivity on marginal land) did not form part of this study. Instead, the study deals with the further evaluation of the growth suitability of 19 promising industrial crops (Table 1), and thus how they meet the requirements for successful development of MALLIS.

Table 1. Overview of physiological and technical characteristics of the industrial crops.

Crop		Physiology		
Common Name	Binomial Name	Life Cycle	Photo-Synzthetic Pathway	Purpose/Type of Use
Biomass sorghum	*Sorghum bicolor* L. Moench	Annual	C4	Multipurpose
Camelina	*Camelina sativa* L. Crantz	Annual	C3	Oil
Cardoon	*Cynara cardunculus* L.	Perennial	C3	Multipurpose
Castor bean	*Ricinus communis* L.	Annual	C3	Oil
Crambe	*Crambe abyssinica* Hochst Ex Re Fries	Annual	C3	Oil
Ethiopian mustard	*Brassica carinata* A. Braun	Annual	C3	Oil
Giant reed	*Arundo donax* L.	Perennial	C3	Lignocellulosic
Hemp	*Cannabis sativa* L.	Annual	C3	Multipurpose
Lupin	*Lupinus mutabilis* Sweet	Perennial	C3	Multipurpose
Miscanthus	*Miscanthus × giganteus* Greef et Deuter	Perennial	C4	Lignocellulosic
Pennycress	*Thlaspi arvense* L.	Annual	C3	Oil
Poplar	*Populus* spp.	Perennial	C3	Lignocellulosic
Reed canary grass	*Phalaris arundinacea* L.	Perennial	C3	Lignocellulosic
Safflower	*Carthamus tinctorius* L.	Annual	C3	Oil
Siberian elm	*Ulmus pumila* L.	Perennial	C3	Lignocellulosic
Switchgrass	*Panicum virgatum* L.	Perennial	C4	Lignocellulosic
Tall wheatgrass	*Thinopyrum ponticum* Podp. Z.-W. Liu and R.-C. Wang	Perennial	C3	Lignocellulosic
African fodder cane	*Saccharum spontaneum* L. ssp. *aegyptiacum* (Willd.) Hack.	Perennial	C4	Lignocellulosic
Willow	*Salix* spp.	Perennial	C3	Lignocellulosic

The following sub-sections present the concepts underlying the key elements of this study. These key elements are (i) the identification of marginal agro-ecological zones (M-AEZ), (ii) the determination of the growth suitability of the pre-selected industrial crops in the prevailing M-AEZ, and (iii) the development of MALLIS for industrial crop cultivation.

2.1. The Identification of Marginal Agro-Ecological Zones (M-AEZ)

To achieve the first two key elements, mapping was performed as follows: Marginal agricultural lands were mapped [31] according to the biophysical limitations defined and classified by JRC [20–22]. The mapping was limited to a so-called 'agricultural mask'. This mask includes all land that was classified in an agricultural land cover class in at least one of the four Corine Land Cover (CLC) versions (1990, 2000, 2006, and 2012). Further details of the methodological approaches are provided in the following sub-sections.

2.2. Determination of the Growth Suitability of the Pre-Selected Industrial Crops in the Prevailing M-AEZ

The approach to mapping the growth suitability of the 19 pre-selected crops involves the identification of the minimum and maximum climate and soil requirements per crop. The growth suitability requirements of the selected industrial crops were determined according to the literature [32,33]. They were used to map and calculate both the distribution and size of the crop-specific growth suitability areas across European marginal agricultural land. The thresholds for the suitability parameters were set as the starting point at which the crop can grow and survive. The suitable area is, thus, given as the area where all suitability factors are within the minimum and maximum range. In this mapping assessment, a distinction was made between suitable and unsuitable area per crop. However, no further classification of the suitable area was made, for example, into high to low suitability. For an easier interpretation of the results, the European land surface was divided into the three agro-ecological zones (AEZ): Mediterranean (AEZ1), Atlantic (AEZ2) and Continental and Boreal (AEZ3) (Figure 3, Table 2). Each combination of an AEZ with at least one biophysical constraint (Table A1) refers to as 'M-AEZ' (Table 2).

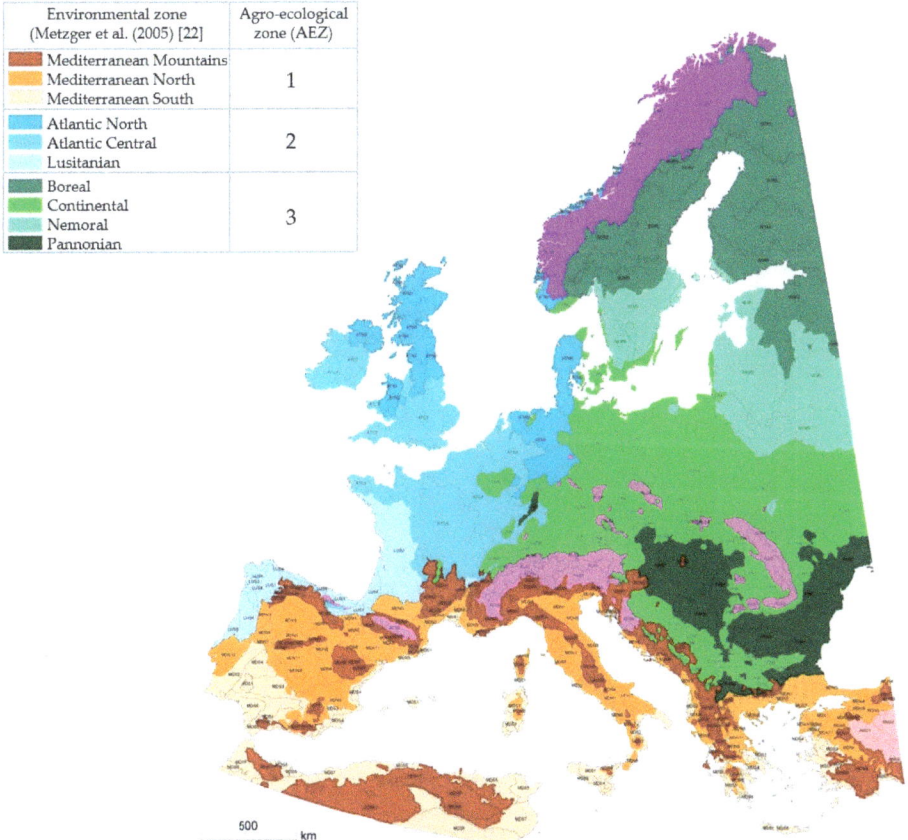

Figure 3. Distribution of agro-ecological zones (AEZ) taken into consideration for the development of marginal agricultural land low-input systems (MALLIS) for industrial crops across Europe (modified from Reference [34]).

The basic crop-specific biophysical growth requirements were compiled according to Ramirez-Almeyda et al. (2017) [32]. Each biophysical parameter was divided into a number of classes. For instance, the parameter "precipitation" was divided into eight classes (in mm a^{-1}): 0–100,

100–200, 200–300, 300–400, 400–500, 500–600, 600–800, and 800–1000 (Table A2). Afterwards, the growth suitability of each crop was ranked according to these classes based on available literature. Additionally, the basic climatic growth requirements of the crops were compiled (Table 3).

Table 2. Relevance of the constraints and constraint combinations expressed as agricultural area (km^2) per AEZ.

Constraint(-s) [a]	AEZ 1	AEZ 2	AEZ 3	AEZ 1-3
RT	62,247	51,823	41,449	155,519
CL	27,752	4564	79,780	112,096
WT	2526	65,322	40,233	108,081
TR	31,332	5710	11,362	48,404
RT-TR	15,636	14,656	2157	32,449
CL-RT	25,675	593	6064	32,332
CL-WT	701	13,141	16,263	30,105
FE	15,205	3087	5246	23,538
CH	6883	3642	11,987	22,512
CL-FE	14,527	291	3524	18,342
WT-RT	348	10,541	1745	12,634
CL-TR	2920	1577	4189	8686
CL-RT-TR	4240	1072	1150	6462
CL-WT-RT	95	1531	3472	5098
CL-WT-TR	12	4663	61	4736
CL-FE-RT	4272	47	97	4416
CL-FE-RT-TR	4272	47	97	4416
CL-WT-RT-TR	603	2361	1421	4385
CL-FE-TR	151	2361	1421	3933
WT-TR	51	1935	976	2962
FE-RT	1268	603	289	2160
CL-WT-FE	0	1344	594	1938
WT-RT-TR	4	1158	58	1220
WT-FE	11	986	198	1195
CL-CH	1173	0	0	1173
FE-CH	200	1	950	1151
CH-TR	273	46	654	973
CL-WT-FE-RT	0	185	697	882
CH-RT	280	107	195	582
WT-CH	37	239	154	430
CL-WT-FE-TR	0	417	1	418
CL-WT-FE-RT-TR	1	143	106	250
CL-FE-CH	244	0	0	244
FE-TR	117	49	51	217
WT-FE-RT	0	87	10	97
WT-FE-TR	0	77	1	78
CL-CH-RT	54	0	0	54
FE-RT-TR	7	32	6	45
CH-RT-TR	26	2	16	44
CL-CH-TR	18	0	0	18
FE-CH-TR	1	0	17	18
FE-CH-RT	4	0	7	11
CL-WT-CH	5	0	0	5
WT-FE-CH	0	0	1	1
Total marginal	218,962	192,302	235,569	646,833
Total not marginal	422,565	538,855	704,818	1,666,238
Total	641,527	731,157	940,387	2,313,071

[a] CH: Salinity or sodicity; CL: Low temperature, high temperature or dryness; FE: Acidity, alkalinity or soil organic matter; RT: Shallow rooting depth or unfavorable texture; TR: Steep slope; WT: Limited soil drainage or excess soil moisture.

Table 3. Main thermal growth requirements of the 19 pre-selected industrial crops.

Crop	Factors of Thermal Growth Requirements		
	Base Temperature (°C)	Minimum Length of Growing Season (d)	Minimum of Growing Degree Days [a] (Thermal Time, °C d)
Biomass sorghum	8	100	1500
Camelina	5	90	1000
Cardoon	7.5	120	1100
Castor bean	10	135	1500
Crambe	5	100	1200
Ethiopian mustard	5	120	2000
Giant reed	5	210	1843
Hemp	6	90	1400
Lupin	0	222	2260
Miscanthus	5	78	1700
Pennycress	4	90	1200
Poplar	0	180	2200
Reed canary grass	0	111	2000
Safflower	2	120	1800
Siberian elm	6	150	2000
Switchgrass	6	140	2060
Tall wheatgrass	4	90	1200
Wild sugarcane	10	210	2400
Willow	2	180	2000

[a] Accumulated mean daily temperature equal to or above than the crop-specific base temperature.

When mapping the crop-specific growth suitability areas, we only considered whether a crop could potentially grow. We did not take different yield levels into account. In the constraint-specific ranking, classes 0 and 1 were denoted as not suitable. Therefore, if any of the basic climatic growth requirements are not met or any of the constraint-specific rankings falls within class 0 or 1, the area is designated as 'not suitable'. The result was an overview of the potential growth suitability of the pre-selected industrial crops across European marginal agricultural land. This means that only agricultural areas were considered; woodlands and urban areas were excluded from the mapping of marginal agricultural land.

2.3. Definition and Methodology of Marginal Land Low-Input Systems (MALLIS) Development for Industrial Crops

In this sub-section, the definition of best-practice low-input management systems for the pre-selected industrial crops (Table 1) is elaborated. This ties in with current knowledge on best low-input agricultural practices for food crop production on good soils [6]. The concept of best-practice low-input agricultural cropping systems considers management approaches from many categories of agricultural production, including organic, integrated, conservation agriculture and mixed crop-livestock farming [35–37]. These all have one constant: Low-input agricultural practices seek to optimize the use of on-farm resources while minimizing off-farm resources [6,35,36]. This leads to a more 'closed' cycle of production (and less external inputs) [37]. Note, that this more closed production cycle requires both more advanced agronomic skills [38,39] and additional links within the value chain, such as application of biochar [40–49] or phosphate salt recovery from the digestates [50,51]. Therefore, practical guidelines for industrial crops are also under development within the MAGIC project.

Agronomic strategies for the successful application of low-input agricultural practices in a crop management system should be seen as a set of strategies that take into consideration both the interactions between plants, soil, the atmosphere and the efficient use of inputs to enable the highest output with minimal (on-farm and/or off-farm) input supply [6,52–54]. Agronomic strategies for low-input systems may also match good agricultural practices—cultivation practices that address economic, social and environmental sustainability [37] for high-quality food and non-food agricultural products [38,55]. Such practices include the implementation of appropriate crop rotations, pasture management, manure application, soil management that maintains or improves soil organic matter, and other land-use practices, as well as conservation tillage practices [8,37,48].

Diversity in crop rotations is a way to reduce reliance on synthetic chemicals, control weeds and pests, maintain soil fertility and reduce soil erosion, prevent soil-borne diseases, leading to the reduction of off-farm inputs [54]. Reduced soil tillage is a way to reduce soil erosion, improve water buffer capacity, and increase both soil fertility and organic matter [37]. Water management is a major challenge in the Common Agricultural Policy (CAP) and requires the monitoring of soil and crop water status to schedule irrigation efficiently. Fertilizers and agrochemicals should be applied following the good agricultural practices, e.g., to replace only the amount of nutrients that were extracted by harvest [37].

Crop protection should be done in a way that maximizes the biological prevention of pests and diseases, in particular by promoting integrated pest management (IPM) and though appropriate rates and timings of agrochemicals. Preventive crop protection can also be supported by the selection of resistant cultivars and varieties, crop sequences, crop associations (e.g., intercropping), and proper cultural practices [35].

The development of 'marginal agricultural land low-input systems', referred to as 'MALLIS', is based on the following definition: 'MALLIS is defined as a set of low-input practices which are relevant management components to form viable cropping systems on marginal (arable) lands under specific climatic conditions and are sustainable in both socio-economic and environmental terms'. The implementation of MALLIS should enable farmers to cultivate industrial crops on marginal agricultural lands, considering both economic and socio-environmental aspects. Consequently, MALLIS should not only allow for profitable net farm income under the challenging biophysical growth conditions of marginal lands. It also helps to (i) reduce off-farm inputs, such as synthetic fertilizer, pesticides and energy (e.g., for water pumps, fuel, crop harvest machinery, storage, processing, etc.) and (ii) mitigate negative macro-economic externalities (GHG emissions, biodiversity loss, ground- and surface water contamination, soil organic matter loss, erosion, degradation, land-use change), while (iii) ensuring feasible economic benefits at farm level. Therefore, the development of MALLIS considers not only the biophysical constraints, but also socio-economic and ecological demands of the respective areas.

The conceptualization of MALLIS development always begins with the selection of the most promising industrial crop, because all other agricultural practices (tillage, fertilization, weeding, irrigation, etc.) strongly depend on the type and site-specific performance of the crop. This MAEZ-specific growth-suitability ranking (and mapping) of the pre-selected industrial crops was based on the crop-suitability rankings. The basic climatical growth suitability thresholds are presented in Table 3. After the identification of suitable crops, the conceptualization of MALLIS for MAEZ was done on a general level (regional scale), since detailed best practice recommendations for the optimized management of agricultural practices very much depend on local conditions (field-to-farm scale) [56–60]. Therefore, the MALLIS for the new field trials to be conducted in the MAGIC project (field-to-farm scale) were developed considering three main MAEZ criteria:

- The crop's performance according to site-specific climatic and geographic conditions, especially under given biophysical constraints;
- The kind and quality of biomass required in the given infrastructure, processing industries and distribution channels (markets);
- The agricultural status of the farm(s), e.g., the techniques, knowledge and resources available to ensure successful cultivation of the crop.

3. Results and Discussion

3.1. Marginal Agro-Ecological Zones in Europe

As illustrated in Figure 2, there are various biophysical constraints and socio-economic challenges which need to be considered for MALLIS development. Table 2 shows the relevance of the numerous biophysical constraint combinations within each of the three AEZ. According to category 1 ('*natural constraints*'), the total marginal area across European land surface amounts to 646,833 km^2 (Table 2)—an area as large as France. However, this marginal agricultural land is widely scattered across Europe (Figure 4). Furthermore, there were 38 combinations of ≥ 2 constraints identified (Table 2). Across Europe, the most prevailing constraints are adverse rooting conditions, (155,519 km^2), adverse climatic conditions (112,096 km^2) and excessive soil wetness (108,081 km^2). The total marginal arable land characterized by soil constraints accounts for about 535,000 km^2. This is about 155,000 km^2 more than reported by Gerwin et al. (2018) (380,000 km^2) [56,61]. It is likely that this difference results from the use of different thresholds for determining what is marginal and what is not. However, both values are within the same range.

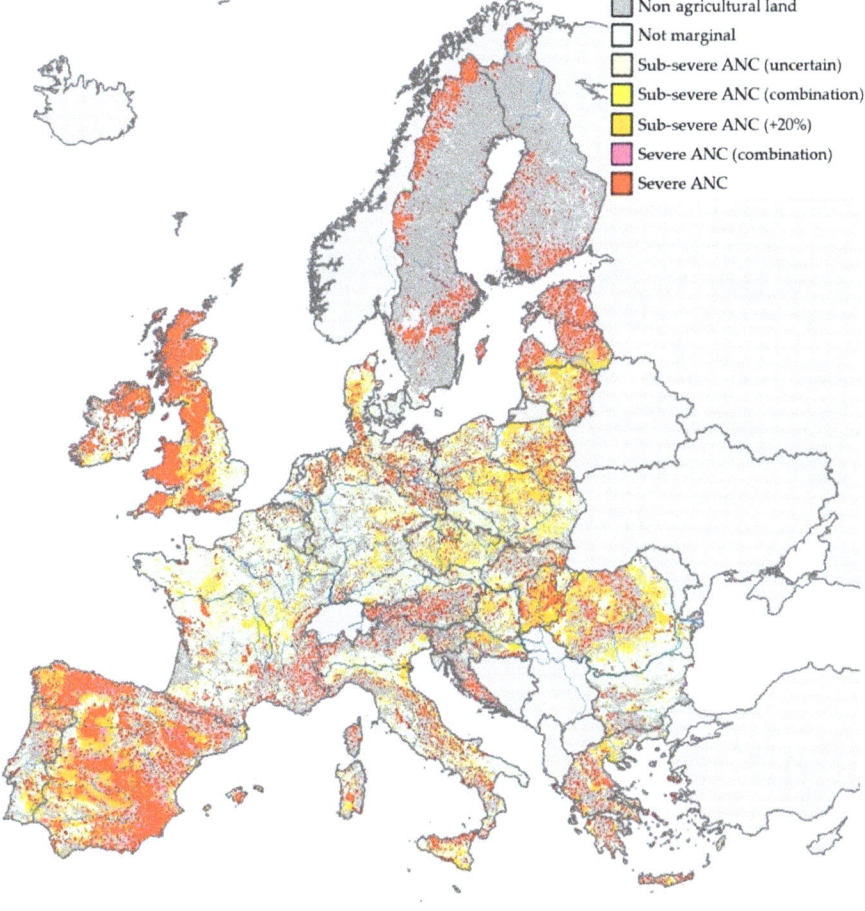

Figure 4. Marginal agricultural lands based on biophysical constraints in Europe (ANC = agricultural natural constraint) (Source: This study).

3.2. The Growth Suitability of the Pre-Selected Industrial Crops in the Prevailing M-AEZ

Potentially suitable industrial crops were identified for virtually all types of marginal agricultural land across Europe (Table A3). Each AEZ appears to have its own best-adapted industrial crops. A closer look at the type of biomass reveals that, for instance, oil crops are more suitable for Mediterranean regions than for the Atlantic region (Table A3). Among the woody lignocellulosic crops, Siberian elm outperforms poplar in the Mediterranean region (Table A3). The dominating lignocellulosic crops are tall wheatgrass, followed by reed canary grass and miscanthus (Table A3).

3.3. Marginal Agricultural Land Low-Input Systems (MALLIS) for Industrial Crop Cultivation

Sections 3.1 and 3.2 revealed both the major M-AEZ in Europe and the growth suitability of the pre-selected industrial crops. This section explains how MALLIS could be developed using the information on M-AEZ and the crops' growth suitabilities. Furthermore, it discusses which other aspects need to be taken into account for MALLIS development in order to improve both the economic and social-ecological sustainability of the MALLIS in the long term.

3.3.1. Agricultural Measures for MALLIS Development

The potential effects of structured and systematic agricultural measures on agriculture facing biophysical constraints are provided in Tables A4 and A5. Furthermore, the literature review revealed that there are several ways to overcome each of the biophysical constraints. Tables A4 and A5 provide an overview of the suitability of agricultural management options for dealing with the prevailing biophysical constraints on marginal agricultural lands. For example, the use of mulch helps to increase the soil thermal time, and thus increase the yield level in regions affected by water limitations and low temperatures [62].

3.3.2. Environmental Threats and Social Requirements

MALLIS implementations at a regional scale should also take both environmental threats and social requirements into consideration. Marginal agricultural lands could be characterized as fragile environments being highly susceptible to any types of external disturbance and input [6,12,63]. Key measures that can be highly recommended for the improvement of resilience include (i) the selection of low-demanding industrial crops (reduces the amount of fertilizers, and thus the risk of nutrient-leaching) [27], (ii) the development of heterogeneous landscape concepts (many small fields rather than only a few large fields) [64–67], and (iii) the implementation of agricultural diversification measures (intercropping, crop rotations, wildflower strips) [35,68,69]. Consequently, the assessment of the environmental performance of MALLIS should not be exclusively based on the global warming potential, but also on a number of other environmental impact categories, such as human toxicity threats, marine ecotoxicity, freshwater eutrophication and freshwater ecotoxicity, biodiversity and soil quality, pollution [70], and use of resources, e.g., water resources [71]. However, to enable a long-term sustainable implementation of MALLIS, besides the environmental impact categories, the social demands and the economic and market aspects must also be taken into account. The potential and viability of agricultural investments have to take into account land and labor costs, inputs, such as mechanical equipment costs, and income (which is linked with the market opportunities) [72]. The socioeconomic impacts can be measured via quantitative and qualitative parameters [73]. Moreover, aspects related to technological viability should also be taken into consideration. The yield loss associated with cultivation on marginal agricultural land may lead to higher contents of nutrients, such as nitrogen and potassium in the biomass, which may complicate further processing of the biomass [74]. Generally, this means that the prevailing structures of the existing agricultural systems [75], the farm typology [76], and the behavior patterns of the rural communities [24,77] require specific bottom-up research structures, such as the Integrated Renewable Energy Potential Assessment (IREPA) [78]. This would enable a better adaptation of MALLIS to the farm diversity [76,77] and the local community.

Finally, this could potentially have a positive influence on the overall public acceptance of the MALLIS [79].

3.3.3. Biodiversity Conservation

Another aspect worthy of discussion is the ecosystem functionality [80] of the pre-selected industrial crops in terms of biodiversity conservation. Concerning the soil ecological functions fulfilled by pedofauna, recent works on the following of bioenergy crops establishment on marginal contaminated soils showed that belowground fauna was stimulated [81]. Higher densities and diversity of soil invertebrates were found under miscanthus compared to annual cropping systems [82], as well as the positive effect on microbial diversity [83]. These crops were specifically selected as representative of those that deliver the most important crop-based biomass resources for current biomass industries. However, the recent (alarming) decrease in pollinator abundances across Central European landscapes [23,25] may induce changes in the priorities for crop selection, and thus the development of MALLIS in the future. For example, pollinator-supporting traits, such as nectar provision and high resistance to pests and diseases could become more important than economic traits, such as biomass yield and biomass quality if public awareness of this topic continues to increase [84]. There are a number of reports on alternative pollinator-supporting industrial crops, such as perennial wild plants [85–89], cup plant (*Silphium perfoliatum* L.) [90–92], sida (*Sida hermaphrodita* L.) [93–95], and amaranth (*Amaranthus hypochondriacus* L.) [96–98]. However, many of the pre-selected industrial crops are also expected to have positive effects on pollinators. These include camelina [99–101], crambe [100,102], safflower [103,104], lupin [105,106], cardoon [107,108] and willow [63,109,110]. In addition, the suitability of the MALLIS for habitat networking in combination with other highly diverse cropping systems, such as species-rich meadows [111] should be investigated to improve the overall efficiency of the MALLIS for biodiversity conservation. Also, marginal land can anchor rich biodiversity components (plants with high significance for locals, e.g., for medicinal or food purposes), and change of land use should take this element into account [112].

3.3.4. Explanatory Setup of a MALLIS on a Shallow Stony Soil

This section provides an example on how MALLIS could be implemented on a marginal agricultural site characterized by two biophysical constraints [21]: (i) Shallow soil (<35 cm topsoil depth); and (ii) stoniness (≥15% of topsoil volume is coarse material, rock outcrop or boulder). Due to these constraints, both the rooting conditions and the soil fertility are lower than in deep soils. It is economically not feasible to grow food crops under these conditions, and thus, the cultivation of certain industrial crops would not compete with food security on sites like this. However, not all industrial crops are able to grow well under these conditions either. Thus, the identification of a best-adapted industrial crop is the first step in developing a site-specifically suitable MALLIS. In this case, perennial crops, such as miscanthus and switchgrass are found to be suitable because (i) they do not require soil tillage and sowing each year compared to annual crops which helps both increasing soil fertility [113,114] and reducing erosion [115] in the long term, (ii) they can manage to root deep enough despite shallow soil, because their root systems are stronger and more developed than those of annual crops, and (iii) the climatic conditions meet the crop-specific growth requirements. In this case, the perennial C4-grass miscanthus (*Miscanthus* × *giganteus* Greef et Deuter) was chosen (Figure 5), due to its low demanding nature and high biomass yield potential under challenging conditions [116]. This is part of ongoing research on the cultivation of miscanthus on marginal agricultural lands in MAGIC [117]. In the EU-funded project 'GRACE' (Grant agreement ID: 745012), it is also investigated how the cultivation of miscanthus on marginal agricultural lands can be optimized [118].

Preliminary results of a field trial in southwest Germany indicate that miscanthus can establish well (Figure 5) under the given conditions [119]. The dry matter yield (DMY) averages 13 Mg ha^{-1} a^{-1} from the second year onwards [119]. This is a medium DMY level compared with miscanthus grown on good soil [116,120,121]. However, it should be mentioned that miscanthus requires very low nitrogen (N) fertilization [122], especially when harvested for combustion in winter [60,123]. This is because miscanthus has very efficient nutrient-recycling when harvested in winter [79,124]. The low demand for nitrogen fertilization renders a key low-input factor [6,32] of this MALLIS, due to an improvement of the on-farm/off-farm-ratio in favor of the on-farm inputs. Furthermore, low N fertilization levels help improve the ecosystem services of miscanthus cultivation, such as groundwater protection, environmental protection [26,80,120], while maintaining the soil nitrogen balance [125]. Overall, both the improved ecosystem services and low production costs justify the medium DMY level of miscanthus at comparable marginal agricultural sites (shallow soil, stoniness, etc.). Consequently, MALLIS must be developed under careful consideration of the given site-specific conditions [57]. Therefore, the major development steps are (i) the identification of the growth conditions and the biophysical constraints [20,20,22], (ii) the selection of best-adapted crops, and (iii) the conceptualization of best-adapted site-specific low-input agricultural practices.

Figure 5. Four-year old miscanthus (*Miscanthus × giganteus* Greef et Deuter) grown on a shallow stony soil in southwest Germany.

3.4. Recommendations and Outlook

The results of the suitability mapping are in line with the available literature [17,56,61,121,126–128]. Uncertainties were identified within the basic climatic requirements, because in some cases the distribution does not meet the expectations. This could be caused by the wide genetic variation within both perennial crop species, such as switchgrass and miscanthus, and annual crop species, such as camelina and safflower. To improve the representability, the basic climatic growth requirements should either include ranges (minimum–maximum) for each parameter per crop or different genotypes for each crop. For instance, there is a wide genetic variation among miscanthus genotypes with regard to their heat and cold tolerance [129–131]. For some annual industrial crops, such as camelina and safflower, winter-annual genotypes are also available [132–135]. It would very likely further increase the potential growth suitability of the pre-selected industrial crops to take these genetic variations into account. Nevertheless, this study provides valuable first insights into the potential distribution of growth suitability, contributing to an improved crop selection for the development of MALLIS across Europe.

The results of this study indicate that there are large areas potentially available for industrial crop cultivation. This is in line with available literature [17,56,61,126,127]. In many cases there are ≥2 crops suitable for the same area (Table A3). Thus, careful consideration should be given to the selection of crops or their most favorable combination according to the site conditions [136]. For an adequate crop selection, site-specific conditions other than the growth suitability should also be considered, such as the local social-ecological needs and the distance to the markets. For instance, if a site is prone to erosion, a perennial cropping system would be preferable to an annual (rotational) cropping system [115,137–139]. This could help ensure a more sustainable biomass production from both an environmental and economic point of view in the long term [140]. It would reduce the risk of further degradation through erosion, and thus help maintain or even improve the resilience of the given agroecosystem [14,141,142].

In this study, the growth suitability of the crops did not include yield and quality levels. This means that potential differences in yield or biomass quality between suitable industrial crops for the various types of marginal land across Europe remain unclear. Furthermore, the study did not cover macroeconomic aspects, such as infrastructure and market accessibility, which also play a vital role in the determination of the best site-specific crop selections across European marginal agricultural lands. In some cases, the suitability of an industrial crop also depends on the local conditions of the farms. For example, either the technical equipment or the know-how may impede an optimal MALLIS implementation. However, this study contributes to the ongoing research into how biomass for a growing bioeconomy can be provided in low-input systems, as the growth suitability of the crops forms the basis for the successful development of MALLIS. The site-specific growth suitability presented here are also available in the form of a decision support system [136]. This aims at enabling the selection of suitable case study regions for further optimization of site-specific MALLIS for industrial crop cultivation. In addition, the missing links mentioned above, including detailed information on the best crop- and site-specific harvesting technology and guidelines for farmers are also explored in the EU Horizon 2020 project MAGIC (GA 727698) [117]. As climate-change-forced shifts in the distribution of both marginal agricultural land and growth suitability of the industrial crops are to be expected [126,143–145], they are also under investigation [58,146]. This could help to better prepare European agriculture for the projected severe effects of climate change [143,144,147].

4. Conclusions

This study introduces the concept of marginal agricultural land low-input systems (MALLIS) for industrial crop cultivation. MALLIS are defined as sets of agricultural low-input practices to form viable cropping systems on marginal agricultural lands under specific climatic conditions. These sets of practices are intended to be holistically sustainable in both social-ecological and economic terms. The study identified the climatic and geophysical constraints on biomass production and the ability of 19 industrial crops to cope with these limitations. Overall, the industrial crops showed high suitability for low-input cultivation on marginal agricultural lands across Europe. However, further investigations of MALLIS are required to investigate their social-ecological sustainability and climate change effects on marginal agricultural lands.

Author Contributions: Conceptualization, M.V.C., I.L. (Iris Lewandowski), Y.I., D.S., S.L.C., O.M., I.E., F.Z., D.L., S.N., I.L. (Isabelle Lamy), L.C., M.S., J.E.C., P.C., I.M., L.M.T., E.N.V.L., W.E., A.L.F., E.G.P. and E.A.; Data curation, M.V.C., I.S., M.V.E., S.M., D.S., F.Z., I.M. and E.A.; Formal analysis, M.V.C., B.E., I.S. and M.V.E.; Funding acquisition, I.L. (Iris Lewandowski), S.L.C., A.M. and E.A.; Investigation, M.V.C., I.M., A.L.F. and E.A.; Methodology, M.V.C., I.L. (Iris Lewandowski), B.E., I.S., M.V.E., S.M., D.S., G.T., S.L.C., O.M., I.E., F.Z., D.L., I.L. (Isabelle Lamy), J.E.C., P.C., I.M., L.M.T., E.N.V.L., A.L.F., E.G.P. and E.A.; Project administration, E.A.; Resources, M.V.C., S.M., F.Z., I.M., E.N.V.L. and E.A.; Software, I.S., M.V.E. and I.M.; Supervision, I.L. (Iris Lewandowski), B.E., S.L.C., A.M., D.L., J.E.C., L.M.T., A.L.F. and E.A.; Validation, M.V.C., B.E., I.S., M.V.E., D.S., G.T., S.L.C., O.M., I.E., F.Z., A.M., D.L., I.L. (Isabelle Lamy), J.E.C., P.C., I.M., L.M.T., E.N.V.L., A.L.F., E.G.P. and E.A.; Visualization, M.V.C., B.E., I.S. and

M.V.E.; Writing—original draft, M.V.C., I.L. (Iris Lewandowski), B.E., Y.I., D.S., G.T., S.L.C., O.M., I.E., F.Z., D.L., I.L. (Isabelle Lamy), J.E.C., P.C., I.M., L.M.T., E.N.V.L., A.L.F., E.G.P. and E.A.; Writing—review and editing, M.V.C.

Funding: This research received funding from the European Union's Horizon 2020 research and innovation program under grant agreement No 727698. The article processing charge was funded by the European Union's Horizon 2020 research and innovation program under grant agreement No 727698.

Acknowledgments: The authors are thankful to Nicole Gaudet for proof-reading the manuscript.

Conflicts of Interest: The authors declare no conflict of interest. The funders had no role in the design of the study; in the collection, analyses, or interpretation of data; in the writing of the manuscript, or in the decision to publish the results.

Appendix A

Table A1. Overview of the three categories of marginality constraints as classified within this deliverable. Category 1 was adapted from Reference [20]. Categories 2 and 3 were developed based on the literature review.

Constraint Category	Factor Category	Thresholds/Specifications
Category 1: "Natural constraint based marginality"	Low temperature (insufficient thermal time)	Length of growing period ≤ 180 days Thermal time sum ≤ 1500 degree days
	Dryness—Too dry conditions	Precipitation/Potential Evapotranspiration (P/ET ≤ 0.5)
	Limited soil drainage and excess soil moisture	Wet 80 cm > 6 months Wet 40 cm > 11 months Poorly or very poorly drained Gleyic colour pattern within 40 cm Soil moisture above field capacity for >230 days (excessive soil moisture)
	Unfavorable soil texture and stoniness	Topsoil with stones (15% of topsoil volume is coarse material, rock outcrop, boulder) Texture class in half of the soil in a profile of 100 cm vertical depth is sand, loamy sand Organic soil, defined as having organic matter ≥ 30% of at least 40 cm Topsoil with 30% or more clay and presence of vertical properties within 100 cm
	Shallow rooting depth	The physical anchorage of the rooting system (rooting depth ≤ 30 cm) The provision/storage of nutrients and water The possibility of mechanized tillage
	Poor chemical properties (Soil salinity, soil sodicity, soil acidity)	The possibility of mechanized tillage Limitation to plant growth, due to toxic elements in soil Vulnerability to waterlogging Damage to soil structure (and consequently increase in risk of erosion) Limited availability of nutrients for plants Salinity ≥ 4 dS/m in topsoil Sodicity ≥ 6 ESP in half or more of the 100-cm surface layer Soil Acidity of topsoil with pH (H20) ≤ 5
	Steep slope	Slope ≥ 15%
Category 2: "Socio-economic-political constraints"		Lack of awareness (alternative strategies—lack of know-how, etc.) Social norms (adoption of same cropping patterns as done by elders) Economic viability, especially of set-aside, small land holdings Lack of infrastructure Lack of policies Lack of government programs, such as extension services
Category 3: "Endangered Sites"		Lands which are currently productive, but will be transformed into marginal lands in the long term if not managed properly (also, lack of know-how or lack of awareness from farmers/government).

Table A2. Crop-suitability ranking (from 0 = unsuitable to 4 = very suitable, whereas both 0 and 1 were defined as marginal) according to precipitation.

Crop	Precipitation Classes (mm year^{-1} or mm (Growth Season of Annuals)$^{-1}$)							
	0–100	100–200	200–300	300–400	400–500	500–600	600–800	800–1000
Biomass sorghum	0	1	2	3	4	4	4	4
Camelina	3	4	4	4	4	4	2	2
Cardoon	0	0	0	1	2	3	3	4
Castor bean	1	2	2	3	3	4	4	4
Crambe	3	4	4	4	4	4	2	2
Ethiopian mustard	2	3	3	3	3	4	4	4
Giant reed	0	0	1	1	2	3	4	4
Hemp	0	1	2	3	4	4	4	3
Lupin	0	1	2	2	3	4	4	4
Miscanthus	0	0	0	0	1	2	3	4
Pennycress	1	1	2	4	4	4	4	4
Poplar	0	0	0	0	0	0	2	3
Reed canary grass	0	0	0	0	0	0	2	3
Safflower	0	1	2	3	4	4	4	3
Siberian elm	0	0	1	2	3	4	4	4
Switchgrass	0	0	0	1	2	3	4	4
Tall wheatgrass	0	0	1	3	4	4	4	4
Wild sugarcane	0	1	1	2	3	4	4	4
Willow	0	0	0	0	0	0	2	3

Table A3. Total area (km^2) per selected industrial crop suitable for cultivation on marginal land across Europe (EU-28) and share (%) of marginal land suitable for cultivation of the crop.

Crop	AEZ 1		AEZ 2		AEZ 3		AEZ 1–3	
	km^2	%	km^2	%	km^2	%	Km^2	%
Biomass sorghum	193,118	88	31,322	16	6323	3	230,763	36
Camelina	209,761	96	186,018	97	183,667	78	579,446	90
Cardoon	172,804	79	71,822	37	83,249	35	327,875	51
Castor	160,990	74	10,658	6	3412	1	175,060	27
Crambe	216,577	99	175,244	91	130,959	56	522,780	86
Ethiopian mustard	184,988	84	43,177	22	10,111	4	238,276	37
Giant reed	129,501	59	2459	1	1173	0	133,133	21
Hemp	162,794	74	80,422	42	17,392	7	260,608	41
Lupin	201,888	92	36,790	19	37,162	16	275,840	43
Miscanthus	130,634	60	83,820	44	88,010	37	302,464	48
Pennycress	208,388	95	64,812	34	76,465	32	349,665	56
Poplar	48,166	22	159,938	83	150,428	64	358,532	60
Reed canary grass	45,863	21	124,828	65	147,470	63	318,161	53
Safflower	201,689	7	145,382	76	16,164	92	363,235	58
Siberian elm	179,148	82	20,611	11	28,261	12	228,020	36
Switchgrass	160,238	73	19,732	10	26,628	11	206,598	32
Tall wheatgrass	211,255	96	151,166	79	172,355	73	534,776	88
Wild sugarcane	46,516	21	252	0	0	0	46,768	7
Willow	56,880	26	164,191	85	119,536	51	340,607	56
Average	153,747	66	82,771	43	68,356	33	304,874	49

Table A4. Suitability ranking of selected agricultural management options for competing with the prevailing biophysical constraints on arable marginal lands (from -3 = strong negative effect to $+3$ = strong positive effect).

Agricultural Management Options	Biophysical Constraints												
	Climatical				Soil/Terrain								
	Low Temperature	High Temperature	Dryness	Excessive Moisture	Poor Soil Drainage	Unfav. Texture/Stoniness	Shallow Rooting Depth	Steep Slope	Low Soil Fertility	Alkalinity	Acidity	Salinity	Other Contamination
Structured measures													
Line irrigation	−1	0	1	−3	−3	−1	1	−3	1	0	0	1	0
Sprinkler irrigation	0	2	2	−3	0	2	1	1	1	1	1	1	0
Microirrigation (drip irrigation)	0	0	3	−3	1	3	3	2	2	0	0	2	0
Deficit irrigation technique	0	0	3	−3	1	2	2	3	3	1	1	−1	0
Terracing	0	0	0	0	0	0	0	3	0	0	0	0	0
Field shaping, planting density and geometry	0	0	1	1	0	2	2	1	0	0	0	0	0
Hedges	1	0	0	1	1	0	0	1	0	0	0	0	0
Water channels	−1	0	−3	2	3	0	0	−3	0	1	1	0	0
Systematic measures													
Catch crops	1	2	1	2	3	1	−1	0	1	0	0	1	1
Crop rotation	0	0	1	0	1	2	1	−1	2	0	0	1	2
Agroforestry	0	0	1	2	2	2	1	3	2	2	2	0	2
Intercropping	1	1	2	2	2	1	0	2	1	1	1	1	1
Mixed cropping	1	1	1	2	1	2	0	1	2	1	1	1	1
Crop selection													
Deep roots	1	2	3	2	3	−1	−3	2	1	1	1	1	0
Shallow roots	1	0	−3	−1	−2	2	3	1	0	0	0	0	0
C3-Metabolism	3	1	0	0	0	0	0	0	0	0	0	0	0
C4-Metabolism	0	3	2	0	0	0	0	0	0	0	0	0	1
Annual lifecycle	1	0	0	−2	−3	0	0	0	−1	0	0	1	1
Biennial lifecycle	1	1	1	1	1	0	1	1	2	0	0	1	1
Perennial lifecycle	2	1	3	2	2	1	2	2	3	1	1	2	1

Table A5. Suitability ranking of selected components of agricultural management systems for competing with the prevailing biophysical constraints on arable marginal lands (from −3 = strong negative effect to +3 = strong positive effect).

Components of Management System	Climatic				Biophysical Constraints			Soil/Terrain					
	Low Temperature	High Temperature	Dryness	Excessive Moisture	Poor Soil Drainage	Unfav. Texture/Stoniness	Shallow Rooting Depth	Steep Slope	Low Soil Fertility	Alkalinity	Acidity	Salinity	Other Contamination
Soil cultivation													
Full till	1	−2	−2	−3	−1	−1	−1	−3	−1	1	1	0	1
Reduced till	−1	1	1	1	1	1	1	−1	1	0	0	0	1
Precision tillage	3	2	2	0	2	2	1	1	2	0	0	0	1
No till	2	3	3	2	1	2	2	3	2	0	0	0	0
Living mulch	1	−2	−2	3	−2	1	−1	2	2	1	1	−1	−1
Cover soil with film	2	1	2	−1	1	2	1	1	1	1	0	1	−1
Harvest residuals	2	−1	−1	0	1	1	−1	−3	2	1	1	0	−1
Establishment													
Pesticides [a]	1	0	1	1	2	1	0	0	−1	1	1	1	0
Micronutrients [a]	1	0	1	1	1	2	1	0	3	0	1	1	−1
Bio-stimulators [a]	0	0	1	1	1	2	1	0	2	−1	1	1	0
Rhizomes	1	1	1	−1	0	0	1	0	1	0	−1	0	0
Plantlets	1	2	2	1	1	2	2	2	2	0	0	1	0
Collars	1	2	2	−1	−1	0	−1	1	1	−1	−1	−1	0
Unrooted cuttings	1	2	0	2	1	0	2	2	0	−1	1	1	0
Crop protection													
Pesticides	1	1	1	1	−1	0	0	2	1	1	1	1	0
Biological pest control	1	0	0	2	2	2	0	1	1	0	0	0	0
Crop rotation strategy	2	1	2	1	2	2	1	1	2	1	1	1	2
Mechanical weeding	1	1	−1	−1	0	−1	1	−3	1	0	−1	−1	0
Thermal weeding	3	1	1	2	2	2	0	1	2	0	0	0	0
Chemical weeding	1	1	−1	0	−1	1	−1	0	0	0	0	1	0
Biological weeding	2	1	−1	2	2	1	1	2	1	−1	1	1	−1
Cover soil with film	2	1	1	0	0	2	2	2	−2	0	0	0	0
Fertilization													
Broadcast application	−1	1	1	−1	1	1	0	1	1	0	0	0	0
Ground level application	0	1	1	0	0	1	0	−1	1	0	0	−1	0
Injection	1	2	3	0	2	2	−1	2	3	2	2	2	3
Organic fertilizer	2	3	3	1	1	2	0	0	2	−1	3	−2	0
Liming	1	0	0	−1	−1	1	0	0	3	−1	−1	−2	0
Chemical fertilizer	1	1	0	1	1	1	−1	0	3	0	0	1	0
Solid	2	2	2	1	1	2	−1	0	1	0	0	0	0
Liquid	2	−3	−3	1	1	1	0	0	−1	0	1	0	0
Spring application	−1	2	2	−1	1	1	0	0	1	0	0	0	0
Summer application	−2	1	2	1	1	0	0	0	1	0	0	0	0
Autumn application	0	0	0	−1	−1	0	0	0	−1	0	0	0	0
Winter application	0	0	0	−2	−1	−1	0	−1	0	0	0	0	0
One application	1	0	0	−1	−1	0	0	−1	1	0	0	0	0
>1 applications	1	1	1	1	1	1	−1	−1	2	0	0	0	0

[a] Priming of seeds and planting material.

References

1. Staffas, L.; Gustavsson, M.; McCormick, K. Strategies and policies for the bioeconomy and bio-based economy: An analysis of official national approaches. *Sustainability* **2013**, *5*, 2751–2769. [CrossRef]
2. Lewandowski, I. Securing a sustainable biomass supply in a growing bioeconomy. *Glob. Food Secur.* **2015**, *6*, 34–42. [CrossRef]
3. Scarlat, N.; Dallemand, J.F.; Monforti-Ferrario, F.; Nita, V. The role of biomass and bioenergy in a future bioeconomy: Policies and facts. *Environ. Dev.* **2015**, *15*, 3–34. [CrossRef]
4. Scarlat, N.; Dallemand, J.F.; Fahl, F. Biogas: Developments and perspectives in Europe. *Renew. Energy* **2018**, *129*, 457–472. [CrossRef]
5. Fernando, A.L.; Boléo, S.; Barbosa, B.; Costa, J.; Duarte, M.P.; Monti, A. Perennial Grass Production Opportunities on Marginal Mediterranean Land. *Bioenergy Res.* **2015**, *8*, 1523–1537. [CrossRef]
6. Biala, K.; Terres, J.M.; Pointereau, P.; Paracchini, M.L. Low Input Farming Systems: An opportunity to develop sustainable agriculture. *Proc. JRC Summer Univ. Ranco.* **2007**, 2–5. [CrossRef]
7. Lewandowski, I.; Lippe, M.; Castro-Montoya, J.; Dickhöfer, U.; Langenberger, G.; Pucher, J.; Schließmann, U.; Derwenskus, F.; Schmid-Staiger, U.; Lippert, C. Primary Production. In *Bioeconomy*; Springer: Cham, Switzerland, 2018; pp. 95–175.
8. Pulighe, G.; Bonati, G.; Fabiani, S.; Barsali, T.; Lupia, F.; Vanino, S.; Nino, P.; Arca, P.; Roggero, P.P. Assessment of the Agronomic Feasibility of Bioenergy Crop Cultivation on Marginal and Polluted Land: A GIS-Based Suitability Study from the Sulcis Area, Italy. *Energies* **2016**, *9*, 895. [CrossRef]
9. Dale, V.H.; Kline, K.L.; Wiens, J.; Fargione, J. *Biofuels: Implications for Land Use and Biodiversity*; Ecological Society of America: Washington, DC, USA, 2010.
10. Liu, T.T.; McConkey, B.G.; Ma, Z.Y.; Liu, Z.G.; Li, X.; Cheng, L.L. Strengths, Weaknessness, Opportunities and Threats Analysis of Bioenergy Production on Marginal Land. *Energy Procedia* **2011**, *5*, 2378–2386. [CrossRef]
11. Zhuang, D.; Jiang, D.; Liu, L.; Huang, Y. Assessment of bioenergy potential on marginal land in China. *Renew. Sustain. Energy Rev.* **2011**, *15*, 1050–1056. [CrossRef]
12. Elbersen, B.; Van Verzandvoort, M.; Boogaard, S.; Mucher, S.; Cicarelli, T.; Elbersen, W.; Mantel, S.; Bai, Z.; MCallum, I.; Iqbal, Y.; et al. *Definition and Classification of Marginal Lands Suitable for Industrial Crops in Europe (EU Deliverable)*; WUR: Wageningen, The Netherlands, 2018; p. 44.
13. Edrisi, S.A.; Abhilash, P.C. Exploring marginal and degraded lands for biomass and bioenergy production: An Indian scenario. *Renew. Sustain. Energy Rev.* **2016**, *54*, 1537–1551. [CrossRef]
14. Folke, C. Resilience: The emergence of a perspective for social–ecological systems analyses. *Glob. Environ. Chang.* **2006**, *16*, 253–267. [CrossRef]
15. Elmqvist, T.; Folke, C.; Nyström, M.; Peterson, G.; Bengtsson, J.; Walker, B.; Norberg, J. Response diversity, ecosystem change, and resilience. *Front. Ecol. Environ.* **2003**, *1*, 488–494. [CrossRef]
16. Ciria, C.S.; Sanz, M.; Carrasco, J.; Ciria, P. Identification of Arable Marginal Lands under Rainfed Conditions for Bioenergy Purposes in Spain. *Sustainability* **2019**, *11*, 1833. [CrossRef]
17. Krasuska, E.; Cadórniga, C.; Tenorio, J.L.; Testa, G.; Scordia, D. Potential land availability for energy crops production in Europe. *Biofuels Bioprod. Biorefin.* **2010**, *4*, 658–673. [CrossRef]
18. Elbersen, B.S.; Andersen, E. Low-input farming systems: Their general characteristics, identification and quantification. In *Low Input Farming Systems: An. Opportunity to Develop Sustainable Agriculture*; OPOCE: Brussels, Belgium, 2008; p. 12.
19. Fernando, A.L.; Costa, J.; Barbosa, B.; Monti, A.; Rettenmaier, N. Environmental impact assessment of perennial crops cultivation on marginal soils in the Mediterranean Region. *Biomass Bioenergy* **2018**, *111*, 174–186. [CrossRef]
20. Van Orshoven, J.; Terres, J.M.; Tóth, T. Updated common bio-physical criteria to define natural constraints for agriculture in Europe. In *JRC Scientific and Technical Reports*; Publications Office of the European Union: Brussels, Belgium, 2012.
21. Van Orshoven, J.; Terres, J.M.; Tóth, T. Updated common bio-physical criteria to define natural constraints for agriculture in Europe—Definition and scientific justification for the common biophysical criteria. In *JRC Scientific and Technical Reports*; Publications Office of the European Union: Brussels, Belgium, 2014. [CrossRef]

22. Terres, J.M.; Hagyo, A.; Wania, A. Scientific contribution on combining biophysical criteria underpinning the delineation of agricultural areas affected by specific constraints: Methodology and factsheets for plausible criteria combinations. In *JRC Scientific and Technical Reports*; Publications Office of the European Union: Brussels, Belgium, 2014.
23. Hallmann, C.A.; Sorg, M.; Jongejans, E.; Siepel, H.; Hofland, N.; Schwan, H.; Stenmans, W.; Müller, A.; Sumser, H.; Hörren, T. More than 75 percent decline over 27 years in total flying insect biomass in protected areas. *PLoS ONE* **2017**, *12*, e0185809. [CrossRef] [PubMed]
24. Huth, E.; Paltrinieri, S.; Thiele, J. Bioenergy and its effects on landscape aesthetics—A survey contrasting conventional and wild crop biomass production. *Biomass Bioenergy* **2019**, *122*, 313–321. [CrossRef]
25. Potts, S.G.; Imperatriz-Fonseca, V.L.; Ngo, H.T.; Biesmeijer, J.C.; Breeze, T.D.; Dicks, L.V.; Garibaldi, L.A.; Hill, R.; Settele, J.; Vanbergen, A.J. *Summary for Policymakers of the Assessment Report of the Intergovernmental Science-Policy Platform on Biodiversity and Ecosystem Services on Pollinators, Pollination and Food Production*; Bonn, Germany, 2016; ISBN 978-92-807-3568-0.
26. Svoboda, N.; Taube, F.; Kluß, C.; Wienforth, B.; Kage, H.; Ohl, S.; Hartung, E.; Herrmann, A. Crop production for biogas and water protection—A trade-off? *Agric. Ecosyst. Environ.* **2013**, *177*, 36–47. [CrossRef]
27. Lewandowski, I. The role of perennial biomass crops in a growing bioeconomy. In *Perennial Biomass Crops for a Resource-Constrained World*; Springer: Cham, Switzerland, 2016; pp. 3–13. [CrossRef]
28. Monti, A.; Alexopoulou, E. Non-food crops in marginal land: An illusion or a reality? *Biofuels Bioprod. Biorefin.* **2017**, *11*, 937–938. [CrossRef]
29. Tilman, D.; Socolow, R.; Foley, J.A.; Hill, J.; Larson, E.; Lynd, L.; Pacala, S.; Reilly, J.; Searchinger, T.; Somerville, C. Beneficial biofuels—The food, energy, and environment trilemma. *Science* **2009**, *325*, 270–271. [CrossRef]
30. Araújo, K.; Mahajan, D.; Kerr, R.; Silva, M.D. Global biofuels at the crossroads: An overview of technical, policy, and investment complexities in the sustainability of biofuel development. *Agriculture* **2017**, *7*, 32. [CrossRef]
31. Elbersen, B.; Van Eupen, M.; Verzandvoort, S.; Boogaard, H.; Mucher, S.; Cicarreli, T.; Elbersen, W.; Mantel, S.; Bai, Z.; Mcallum, I.; et al. *Methodological Approaches to Identify and Map Marginal Land Suitable for Industrial Crops in Europe*; WUR: Wageningen, The Netherlands, 2018; p. 142.
32. Ramirez-Almeyda, J.; Elbersen, B.; Monti, A.; Staritsky, I.; Panoutsou, C.; Alexopoulou, E.; Schrijver, R.; Elbersen, W. Assessing the Potentials for Nonfood Crops. In *Modeling and Optimization of Biomass Supply Chains*; Elsevier: Amsterdam, The Netherlands, 2017; pp. 219–251.
33. FAO Ecocrop. Food and Agriculture Organization of the UN 2007. Available online: http://ecocrop.fao.org/ecocrop/srv/en/cropSearchForm (accessed on 13 August 2019).
34. Metzger, M.J.; Bunce, R.G.H.; Jongman, R.H.; Mücher, C.A.; Watkins, J.W. A climatic stratification of the environment of Europe. *Glob. Ecol. Biogeogr.* **2005**, *14*, 549–563. [CrossRef]
35. Altieri, M.A.; Nicholls, C.I.; Montalba, R. Technological Approaches to Sustainable Agriculture at a Crossroads: An Agroecological Perspective. *Sustainability* **2017**, *9*, 349. [CrossRef]
36. Altieri, M.A.; Nicholls, C.I.; Henao, A.; Lana, M.A. Agroecology and the design of climate change-resilient farming systems. *Agron. Sustain. Dev.* **2015**, *35*, 869–890. [CrossRef]
37. Arthurson, V.; Jäderlund, L. Utilization of natural farm resources for promoting high energy efficiency in low-input organic farming. *Energies* **2011**, *4*, 804–817. [CrossRef]
38. Francis, C.; Lieblein, G.; Gliessman, S.; Breland, T.A.; Creamer, N.; Harwood, R.; Salomonsson, L.; Helenius, J.; Rickerl, D.; Salvador, R.; et al. Agroecology: The ecology of food systems. *J. Sustain. Agric.* **2003**, *22*, 99–118. [CrossRef]
39. Altieri, M.A.; Merrick, L. In situ conservation of crop genetic resources through maintenance of traditional farming systems. *Econ. Bot.* **1987**, *41*, 86–96. [CrossRef]
40. De Jesus Duarte, S.; Glaser, B.; Cerri, C.E.P. Effect of biochar particle size on physical, hydrological and chemical properties of loamy and sandy tropical soils. *Agronomy* **2019**, *9*, 165. [CrossRef]
41. Sánchez-Monedero, M.A.; Cayuela, M.L.; Sánchez-García, M.; Vandecasteele, B.; D'Hose, T.; López, G.; Martínez-Gaitán, C.; Kuikman, P.J.; Sinicco, T.; Mondini, C. Agronomic evaluation of biochar, compost and biochar-blended compost across different cropping systems: Perspective from the European project FERTIPLUS. *Agronomy* **2019**, *9*, 225. [CrossRef]

42. Horel, Á.; Tóth, E.; Gelybó, G.; Dencso, M.; Farkas, C. Biochar amendment affects soil water and CO_2 regime during Capsicum annuum plant growth. *Agronomy* **2019**, *9*, 58. [CrossRef]
43. Speratti, A.B.; Johnson, M.S.; Sousa, H.M.; Torres, G.N.; Couto, E.G. Impact of different agricultural waste biochars on maize biomass and soil water content in a Brazilian Cerrado Arenosol. *Agronomy* **2017**, *7*, 49. [CrossRef]
44. Zhang, Y.; Idowu, O.J.; Brewer, C.E. Using agricultural residue biochar to improve soil quality of desert soils. *Agriculture* **2016**, *6*, 10. [CrossRef]
45. O'toole, A.; Moni, C.; Weldon, S.; Schols, A.; Carnol, M.; Bosman, B.; Rasse, D.P. Miscanthus biochar had limited effects on soil physical properties, microbial biomass, and grain yield in a four-year field experiment in Norway. *Agriculture* **2018**, *8*, 171. [CrossRef]
46. Guizani, C.; Jeguirim, M.; Valin, S.; Limousy, L.; Salvador, S. Biomass chars: The effects of pyrolysis conditions on their morphology, structure, chemical properties and reactivity. *Energies* **2017**, *10*, 796. [CrossRef]
47. Qian, K.; Kumar, A.; Patil, K.; Bellmer, D.; Wang, D.; Yuan, W.; Huhnke, R.L. Effects of biomass feedstocks and gasification conditions on the physiochemical properties of char. *Energies* **2013**, *6*, 3972–3986. [CrossRef]
48. Lehmann, J.; Rillig, M.C.; Thies, J.; Masiello, C.A.; Hockaday, W.C.; Crowley, D. Biochar effects on soil biota—A review. *Soil Biol. Biochem.* **2011**, *43*, 1812–1836. [CrossRef]
49. Ahmad, M.; Rajapaksha, A.U.; Lim, J.E.; Zhang, M.; Bolan, N.; Mohan, D.; Vithanage, M.; Lee, S.S.; Ok, Y.S. Biochar as a sorbent for contaminant management in soil and water: A review. *Chemosphere* **2014**, *99*, 19–33. [CrossRef] [PubMed]
50. Ehmann, A.; Bach, I.M.; Laopeamthong, S.; Bilbao, J.; Lewandowski, I. Can phosphate salts recovered from manure replace conventional phosphate fertilizer? *Agriculture* **2017**, *7*, 1. [CrossRef]
51. Bergfeldt, B.; Morgano, M.T.; Leibold, H.; Richter, F.; Stapf, D. Recovery of phosphorus and other nutrients during pyrolysis of chicken manure. *Agriculture* **2018**, *8*, 187. [CrossRef]
52. Tilman, D.; Hill, J.; Lehman, C. Carbon-Negative Biofuels from Low-Input High-Diversity Grassland Biomass. *Science* **2006**, *314*, 1598–1600. [CrossRef]
53. Weigelt, A.; Weisser, W.W.; Buchmann, N.; Scherer-Lorenzen, M. Biodiversity for multifunctional grasslands: Equal productivity in high-diversity low-input and low-diversity high-input systems. *Biogeosciences* **2009**, *6*, 1695–1706. [CrossRef]
54. Altieri, M.A. The ecological role of biodiversity in agroecosystems. *Agric. Ecosyst. Environ.* **1999**, *74*, 19–31. [CrossRef]
55. Mockshell, J.; Kamanda, J. Beyond the Agroecological and Sustainable Agricultural Intensification Debate: Is Blended Sustainability the Way Forward? In *Discussion Paper*; Deutsches Institut für Entwicklungspolitik gGmbH: Bonn, Germany, 2017; pp. 1–42.
56. Galatsidas, S.; Gounaris, N.; Vlachaki, D.; Dimitriadis, E.; Kiourtsis, F.; Keramitzis, D.; Gerwin, W.; Repmann, F.; Rettenmaier, N.; Reinhardt, G. Revealing Bioenergy Potentials: Mapping Marginal Lands in Europe-The SEEMLA Approach. In Proceedings of the 26th European Biomass Conference and Exhibition, Copenhagen, Denmark, 14–18 May 2018; Available online: https://opus4.kobv.de/opus4-UBICO/frontdoor/index/index/docId/22081 (accessed on 9 July 2019).
57. Monti, A.; Zegada-Lizarazu, W.; Zanetti, F.; Casler, M. Chapter Two—Nitrogen Fertilization Management of Switchgrass, Miscanthus and Giant Reed: A Review. In *Advances in Agronomy*; Sparks, D.L., Ed.; Academic Press: Cambridge, MA, USA, 2019; Volume 153, pp. 87–119.
58. Von Cossel, M.; Winkler, B.; Wagner, M.; Lask, J.; Magenau, E.; Bauerle, A.; Von Cossel, V.; Warrach-Sagi, K.; Elbersen, B.; Staritsky, I.; et al. The future of bioenergy crops cultivation. *Agronomy*. unpublished.
59. Sun, Y.; Druecker, H.; Hartung, E.; Hueging, H.; Cheng, Q.; Zeng, Q.; Sheng, W.; Lin, J.; Roller, O.; Paetzold, S.; et al. Map-based investigation of soil physical conditions and crop yield using diverse sensor techniques. *Soil Tillage Res.* **2011**, *112*, 149–158. [CrossRef]
60. Kiesel, A.; Nunn, C.; Iqbal, Y.; Van der Weijde, T.; Wagner, M.; Özgüven, M.; Tarakanov, I.; Kalinina, O.; Trindade, L.M.; Clifton-Brown, J.; et al. Site-specific management of miscanthus genotypes for combustion and anaerobic digestion: A comparison of energy yields. *Front. Plant Sci.* **2017**, *8*, 927. [CrossRef] [PubMed]
61. Gerwin, W.; Repmann, F.; Galatsidas, S.; Vlachaki, D.; Gounaris, N.; Baumgarten, W.; Volkmann, C.; Keramitzis, D.; Kiourtsis, F.; Freese, D. Assessment and quantification of marginal lands for biomass production in Europe using soil-quality indicators. *Soil* **2018**, *4*, 267–290. [CrossRef]

62. Bu, L.; Liu, J.; Zhu, L.; Luo, S.; Chen, X.; Li, S.; Lee Hill, R.; Zhao, Y. The effects of mulching on maize growth, yield and water use in a semi-arid region. *Agric. Water Manag.* **2013**, *123*, 71–78. [CrossRef]
63. Lazdina, D.; Bardule, A.; Lazdins, A.; Stola, J. Use of waste water sludge and wood ash as fertiliser for Salix cultivation in acid peat soils. *Agron. Res.* **2011**, *9*, 305–314.
64. Holzschuh, A.; Dainese, M.; González-Varo, J.P.; Mudri-Stojnić, S.; Riedinger, V.; Rundlöf, M.; Scheper, J.; Wickens, J.B.; Wickens, V.J.; Bommarco, R.; et al. Mass-flowering crops dilute pollinator abundance in agricultural landscapes across Europe. *Ecol. Lett.* **2016**, *19*, 1228–1236. [CrossRef]
65. Allan, J.D. LANDSCAPES AND RIVERSCAPES: The Influence of Land Use on Stream Ecosystems. *Annu. Rev. Ecol. Evol. Syst.* **2004**, *35*, 257–284. [CrossRef]
66. Fahrig, L. How much habitat is enough? *Biol. Conserv.* **2001**, *100*, 65–74. [CrossRef]
67. Fischer, J.; Brosi, B.; Daily, G.C.; Ehrlich, P.R.; Goldman, R.; Goldstein, J.; Lindenmayer, D.B.; Manning, A.D.; Mooney, H.A.; Pejchar, L.; et al. Should agricultural policies encourage land sparing or wildlife-friendly farming? *Front. Ecol. Environ.* **2008**, *6*, 380–385. [CrossRef]
68. Von Cossel, M. *Agricultural Diversification of Biogas Crop Cultivation*; University of Hohenheim: Stuttgart, Germany, 2019; Available online: http://opus.uni-hohenheim.de/volltexte/2019/1600/ (accessed on 13 August 2019).
69. Von Cossel, M.; Mangold, A.; Iqbal, Y.; Hartung, J.; Lewandowski, I.; Kiesel, A. How to Generate Yield in the First Year—A Three-Year Experiment on Miscanthus (Miscanthus × giganteus (Greef et Deuter)) Establishment under Maize (Zea mays L.). *Agronomy* **2019**, *9*, 237. [CrossRef]
70. Wagner, M. Methodological Approaches for Assessing the Environmental Performance of Perennial Crop-Based Value Chains. Dissertation, University of Hohenheim, Hohenheim, Germany, 2018. Available online: http://opus.uni-hohenheim.de/volltexte/2018/1433/ (accessed on 13 August 2019).
71. Barbosa, B.; Costa, J.; Fernando, A.L.; Papazoglou, E.G. Wastewater reuse for fiber crops cultivation as a strategy to mitigate desertification. *Ind. Crop. Prod.* **2015**, *68*, 17–23. [CrossRef]
72. Soldatos, P. Economic aspects of bioenergy production from perennial grasses in marginal lands of South Europe. *Bioenergy Res.* **2015**, *8*, 1562–1573. [CrossRef]
73. Fernando, A.L.; Rettenmaier, N.; Soldatos, P.; Panoutsou, C. Sustainability of Perennial Crops Production for Bioenergy and Bioproducts. In *Perennial Grasses for Bioenergy and Bioproducts*; Alexopoulou, E., Ed.; Academic Press: Cambridge, MA, USA, 2018; pp. 245–283.
74. Barbosa, B.; Costa, J.; Fernando, A.L. Production of Energy Crops in Heavy Metals Contaminated Land: Opportunities and Risks. In *Land Allocation for Biomass Crops: Challenges and Opportunities with Changing Land Use*; Li, R., Monti, A., Eds.; Springer International Publishing: Cham, Switzerland, 2018; pp. 83–102.
75. Leopold, A. *A Sand County Almanac*; Oxford University Press: Oxford, UK, 1949.
76. Alvarez, S.; Timler, C.J.; Michalscheck, M.; Paas, W.; Descheemaeker, K.; Tittonell, P.; Andersson, J.A.; Groot, J.C.J. Capturing farm diversity with hypothesis-based typologies: An innovative methodological framework for farming system typology development. *PLoS ONE* **2018**, *13*, e0194757. [CrossRef] [PubMed]
77. Michalscheck, M. *On Smallholder Farm and Farmer Diversity*; Wageningen University & Research: Wageningen, The Netherlands, 2019.
78. Winkler, B.; Lemke, S.; Ritter, J.; Lewandowski, I. Integrated assessment of renewable energy potential: Approach and application in rural South Africa. *Environ. Innov. Soc. Transit.* **2017**, *24*, 17–31. [CrossRef]
79. Kiesel, A.; Wagner, M.; Lewandowski, I. Environmental performance of miscanthus, switchgrass and maize: Can C4 perennials increase the sustainability of biogas production? *Sustainability* **2017**, *9*, 5. [CrossRef]
80. De Groot, R.S.; Wilson, M.A.; Boumans, R.M. A typology for the classification, description and valuation of ecosystem functions, goods and services. *Ecol. Econ.* **2002**, *41*, 393–408. [CrossRef]
81. Chauvat, M.; Perez, G.; Hedde, M.; Lamy, I. Establishment of bioenergy crops on metal contaminated soils stimulates belowground fauna. *Biomass Bioenergy* **2014**, *62*, 207–211. [CrossRef]
82. Hedde, M.; van Oort, F.; Renouf, E.; Thénard, J.; Lamy, I. Dynamics of soil fauna after plantation of perennial energy crops on polluted soils. *Appl. Soil Ecol.* **2013**, *66*, 29–39. [CrossRef]
83. Bourgeois, E.; Dequiedt, S.; Lelièvre, M.; van Oort, F.; Lamy, I.; Maron, P.A.; Ranjard, L. Positive effect of the Miscanthus bioenergy crop on microbial diversity in wastewater-contaminated soil. *Environ. Chem. Lett.* **2015**, *13*, 495–501. [CrossRef]
84. TEEB. *Guidance Manual for TEEB Country Studies-Version 1.0*; Institute for European Environmental Policy: Geneva, Switzerland, 2013.

85. Von Cossel, M.; Lewandowski, I. Perennial wild plant mixtures for biomass production: Impact of species composition dynamics on yield performance over a five-year cultivation period in southwest Germany. *Eur. J. Agron.* **2016**, *79*, 74–89. [CrossRef]
86. Vollrath, B.; Werner, A.; Degenbeck, M.; Illies, I.; Zeller, J.; Marzini, K. *Energetische Verwertung von Kräuterreichen Ansaaten in der Agrarlandschaft und im Siedlungsbereich-Eine Ökologische und Wirtschaftliche Alternative bei der Biogasproduktion*; Energie aus Wildpflanzen; Bayerische Landesanstalt für Weinbau und Gartenbau: Veitshöchheim, Germany, 2012; p. 207.
87. Von Cossel, M.; Steberl, K.; Hartung, J.; Agra Pereira, L.; Kiesel, A.; Lewandowski, I. Methane yield and species diversity dynamics of perennial wild plant mixtures established alone, under cover crop maize (Zea mays L.) and after spring barley (Hordeum vulgare L.). *GCB Bioenergy* **2019**. [CrossRef]
88. Weißhuhn, P.; Reckling, M.; Stachow, U.; Wiggering, H. Supporting Agricultural Ecosystem Services through the Integration of Perennial Polycultures into Crop Rotations. *Sustainability* **2017**, *9*, 2267. [CrossRef]
89. Emmerling, C.; Schmidt, A.; Ruf, T.; von Francken-Welz, H.; Thielen, S. Impact of newly introduced perennial bioenergy crops on soil quality parameters at three different locations in W-Germany. *J. Plant Nutr. Soil Sci.* **2017**, *180*, 759–767. [CrossRef]
90. Gansberger, M.; Montgomery, L.F.R.; Liebhard, P. Botanical characteristics, crop management and potential of Silphium perfoliatum L. as a renewable resource for biogas production: A review. *Ind. Crop. Prod.* **2015**, *63*, 362–372. [CrossRef]
91. Mast, B.; Lemmer, A.; Oechsner, H.; Reinhardt-Hanisch, A.; Claupein, W.; Graeff-Hönninger, S. Methane yield potential of novel perennial biogas crops influenced by harvest date. *Ind. Crop. Prod.* **2014**, *58*, 194–203. [CrossRef]
92. Bufe, C.; Korevaar, H. *Evaluation of Additional Crops for Dutch List of Ecological Focus Area: Evaluation of Miscanthus, Silphium Perfoliatum, Fallow Sown in with Melliferous Plants and Sunflowers in Seed Mixtures for Catch Crops*; Wageningen Research Foundation (WR) business unit Agrosystems Research: Lelystad, The Netherlands, 2018.
93. Nabel, M.; Barbosa, D.B.P.; Horsch, D.; Jablonowski, N.D. Energy Crop (Sida Hermaphrodita) Fertilization Using Digestate under Marginal Soil Conditions: A Dose-response Experiment. *Energy Procedia* **2014**, *59*, 127–133. [CrossRef]
94. Nabel, M.; Temperton, V.M.; Poorter, H.; Lücke, A.; Jablonowski, N.D. Energizing marginal soils—The establishment of the energy crop Sida hermaphrodita as dependent on digestate fertilization, NPK, and legume intercropping. *Biomass Bioenergy* **2016**, *87*, 9–16. [CrossRef]
95. Jablonowski, N.D.; Kollmann, T.; Nabel, M.; Damm, T.; Klose, H.; Müller, M.; Bläsing, M.; Seebold, S.; Krafft, S.; Kuperjans, I.; et al. Valorization of Sida (Sida hermaphrodita) biomass for multiple energy purposes. *GCB Bioenergy* **2017**, *9*, 202–214. [CrossRef]
96. Von Cossel, M.; Möhring, J.; Kiesel, A.; Lewandowski, I. Methane yield performance of amaranth (Amaranthus hypochondriacus L.) and its suitability for legume intercropping in comparison to maize (Zea mays L.). *Ind. Crops Prod.* **2017**, *103*, 107–121. [CrossRef]
97. Eberl, V.; Fahlbusch, W.; Fritz, M.; Sauer, B. *Screening und Selektion von Amarantsorten und Linien als Spurenelementreiches Biogassubstrat*; Berichte aus dem TFZ; Technologie-und Förderzentrum im Kompetenzzentrum für Nachwachsende Rohstoffe: Straubing, Germany, 2014; p. 120.
98. Eberl, V.; Fritz, M. *Amarant als Spurenelementreiches Biogassubstrat*; Biogas Forum Bayern; Technologie-und Förderzentrum (TFZ) im Kompetenzzentrum für Nachwachsende Rohstoffe: Straubing, Germany, 2018.
99. Righini, D.; Zanetti, F.; Martínez-Force, E.; Mandrioli, M.; Toschi, T.G.; Monti, A. Shifting sowing of camelina from spring to autumn enhances the oil quality for bio-based applications in response to temperature and seed carbon stock. *Ind. Crop. Prod.* **2019**, *137*, 66–73. [CrossRef]
100. Stolarski, M.J.; Krzyżaniak, M.; Kwiatkowski, J.; Tworkowski, J.; Szczukowski, S. Energy and economic efficiency of camelina and crambe biomass production on a large-scale farm in north-eastern Poland. *Energy* **2018**, *150*, 770–780. [CrossRef]
101. Stolarski, M.J.; Krzyżaniak, M.; Tworkowski, J.; Załuski, D.; Kwiatkowski, J.; Szczukowski, S. Camelina and crambe production – Energy efficiency indices depending on nitrogen fertilizer application. *Ind. Crop. Prod.* **2019**, *137*, 386–395. [CrossRef]
102. Righini, D.; Zanetti, F.; Monti, A. The bio-based economy can serve as the springboard for camelina and crambe to quit the limbo. *OCL* **2016**, *23*, 23. [CrossRef]

103. Dordas, C.A.; Sioulas, C. Dry matter and nitrogen accumulation, partitioning, and retranslocation in safflower (Carthamus tinctorius L.) as affected by nitrogen fertilization. *Field Crop. Res.* **2009**, *110*, 35–43. [CrossRef]
104. Bassil, E.S.; Kaffka, S.R. Response of safflower (Carthamus tinctorius L.) to saline soils and irrigation II. Crop response to salinity. *Agric. Water Manag.* **2002**, *54*, 81–92. [CrossRef]
105. Rodrigues, M.L.; Pacheco, C.M.A.; Chaves, M.M. Soil-plant water relations, root distribution and biomass partitioning in Lupinus albus L. under drought conditions. *J. Exp. Bot.* **1995**, *46*, 947–956. [CrossRef]
106. Huyghe, C. White lupin (Lupinus albus L.). *Field Crop. Res.* **1997**, *53*, 147–160. [CrossRef]
107. Mauromicale, G.; Sortino, O.; Pesce, G.R.; Agnello, M.; Mauro, R.P. Suitability of cultivated and wild cardoon as a sustainable bioenergy crop for low input cultivation in low quality Mediterranean soils. *Ind. Crop. Prod.* **2014**, *57*, 82–89. [CrossRef]
108. Francaviglia, R.; Bruno, A.; Falcucci, M.; Farina, R.; Renzi, G.; Russo, D.E.; Sepe, L.; Neri, U. Yields and quality of Cynara cardunculus L. wild and cultivated cardoon genotypes. A case study from a marginal land in Central Italy. *Eur. J. Agron.* **2016**, *72*, 10–19. [CrossRef]
109. Pučka, I.; Lazdiņa, D. Review about investigations of Salix spp. in Europe. In Proceedings of the Annual 19th International Scientific Conference Proceedings, "Research for Rural Development", Jelgava, Latvia, 15–17 May 2013; Latvia University of Agriculture: Jelgava, Latvia, 2013; Volume 2, pp. 13–19.
110. Stolarski, M.J.; Niksa, D.; Krzyżaniak, M.; Tworkowski, J.; Szczukowski, S. Willow productivity from small-and large-scale experimental plantations in Poland from 2000 to 2017. *Renew. Sustain. Energy Rev.* **2019**, *101*, 461–475. [CrossRef]
111. Boob, M.; Truckses, B.; Seither, M.; Elsäs ser, M.; Thumm, U.; Lewandowski, I. Management effects on botanical composition of species-rich meadows within the Natura 2000 network. *Biodivers. Conserv.* **2019**, *28*, 729–750. [CrossRef]
112. Dauber, J.; Brown, C.; Fernando, A.L.; Finnan, J.; Krasuska, E.; Ponitka, J.; Styles, D.; Thrän, D.; Van Groenigen, K.J.; Weih, M. Bioenergy from" surplus" land: Environmental and socio-economic implications. *BioRisk* **2012**, *7*, 5–50. [CrossRef]
113. Felten, D.; Emmerling, C. Effects of bioenergy crop cultivation on earthworm communities—A comparative study of perennial (Miscanthus) and annual crops with consideration of graded land-use intensity. *Appl. Soil Ecol.* **2011**, *49*, 167–177. [CrossRef]
114. Emmerling, C.; Pude, R. Introducing Miscanthus to the greening measures of the EU Common Agricultural Policy. *Gcb Bioenergy* **2017**, *9*, 274–279. [CrossRef]
115. Cosentino, S.L.; Copani, V.; Scalici, G.; Scordia, D.; Testa, G. Soil erosion mitigation by perennial species under Mediterranean environment. *BioEnergy Res.* **2015**, *8*, 1538–1547. [CrossRef]
116. Anderson, E.; Arundale, R.; Maughan, M.; Oladeinde, A.; Wycislo, A.; Voigt, T. Growth and agronomy of Miscanthus x giganteus for biomass production. *Biofuels* **2011**, *2*, 71–87. [CrossRef]
117. MAGIC. Marginal Lands for Growing Industrial Crops: Turning a Burden into an Opportunity. Available online: http://magic-h2020.eu/ (accessed on 14 June 2019).
118. GRACE. GRowing Advanced Industrial Crops on Marginal Lands for Biorefineries. Available online: https://www.grace-bbi.eu/project/ (accessed on 14 June 2019).
119. Von Cossel, M.; Lewandowski, I. Miscanthus (Miscanthus x giganteus Greef et Deuter) cultivation on a shallow stony soil in southwest Germany. Manuscript unpublished.
120. Mangold, A.; Winkler, B.; Von Cossel, M.; Iqbal, Y.; Kiesel, A.; Lewandowski, I. Implementing miscanthus into sustainable farming systems: A review on agronomic practices, capital and labor demand. Review article, under review, unpublished.
121. Fajardy, M.; Chiquier, S.; Mac Dowell, N. Investigating the BECCS resource nexus: Delivering sustainable negative emissions. *Energy Environ. Sci.* **2018**, *11*, 3408–3430. [CrossRef]
122. Heaton, E.; Voigt, T.; Long, S.P. A quantitative review comparing the yields of two candidate C4 perennial biomass crops in relation to nitrogen, temperature and water. *Biomass Bioenergy* **2004**, *27*, 21–30. [CrossRef]
123. Iqbal, Y.; Kiesel, A.; Wagner, M.; Nunn, C.; Kalinina, O.; Hastings, A.F.S.J.; Clifton-Brown, J.C.; Lewandowski, I. Harvest Time Optimization for Combustion Quality of Different Miscanthus Genotypes across Europe. *Front. Plant Sci.* **2017**, *8*. [CrossRef] [PubMed]
124. Lewandowski, I.; Schmidt, U. Nitrogen, energy and land use efficiencies of miscanthus, reed canary grass and triticale as determined by the boundary line approach. *Agric. Ecosyst. Environ.* **2006**, *112*, 335–346. [CrossRef]

125. Sastre, C.M.; Carrasco, J.; Barro, R.; González-Arechavala, Y.; Maletta, E.; Santos, A.M.; Ciria, P. Improving bioenergy sustainability evaluations by using soil nitrogen balance coupled with life cycle assessment: A case study for electricity generated from rye biomass. *Appl. Energy* **2016**, *179*, 847–863. [CrossRef]
126. Tuck, G.; Glendining, M.J.; Smith, P.; House, J.I.; Wattenbach, M. The potential distribution of bioenergy crops in Europe under present and future climate. *Biomass Bioenergy* **2006**, *30*, 183–197. [CrossRef]
127. Cosentino, S.L.; Testa, G.; Scordia, D.; Alexopoulou, E. Future yields assessment of bioenergy crops in relation to climate change and technological development in Europe. *Ital. J. Agron.* **2012**, *7*, 22. [CrossRef]
128. Cai, X.; Zhang, X.; Wang, D. Land availability for biofuel production. *Environ. Sci. Technol.* **2011**, *45*, 334–339. [CrossRef] [PubMed]
129. Iqbal, Y.; Lewandowski, I. Inter-annual variation in biomass combustion quality traits over five years in fifteen Miscanthus genotypes in south Germany. *Fuel Process. Technol.* **2014**, *121*, 47–55. [CrossRef]
130. Kalinina, O.; Nunn, C.; Sanderson, R.; Hastings, A.F.S.; van der Weijde, T.; Özgüven, M.; Tarakanov, I.; Schüle, H.; Trindade, L.M.; Dolstra, O.; et al. Extending Miscanthus Cultivation with Novel Germplasm at Six Contrasting Sites. *Front. Plant Sci.* **2017**, *8*, 185. [CrossRef]
131. Clifton-Brown, J.; Hastings, A.; Mos, M.; McCalmont, J.P.; Ashman, C.; Awty-Carroll, D.; Cerazy, J.; Chiang, Y.-C.; Cosentino, S.; Cracroft-Eley, W.; et al. Progress in upscaling Miscanthus biomass production for the European bio-economy with seed-based hybrids. *GCB Bioenergy* **2017**, *9*, 6–17. [CrossRef]
132. Johnson, R.C.; Petrie, S.E.; Franchini, M.C.; Evans, M. Yield and yield components of winter-type safflower. *Crop. Sci.* **2012**, *52*, 2358–2364. [CrossRef]
133. Jamshidmoghaddam, M.; Pourdad, S.S. Genotype$\times$ environment interactions for seed yield in rainfed winter safflower (Carthamus tinctorius L.) multi-environment trials in Iran. *Euphytica* **2013**, *190*, 357–369. [CrossRef]
134. Gesch, R.W.; Matthees, H.L.; Alvarez, A.L.; Gardner, R.D. Winter camelina: Crop growth, seed yield, and quality response to cultivar and seeding rate. *Crop. Sci.* **2018**, *58*, 2089. [CrossRef]
135. Walia, M.K.; Wells, M.S.; Cubins, J.; Wyse, D.; Gardner, R.D.; Forcella, F.; Gesch, R. Winter camelina seed yield and quality responses to harvest time. *Ind. Crop. Prod.* **2018**, *124*, 765–775. [CrossRef]
136. MAGIC DSS MAGIC Decision Support System Marginal Lands and Industrial Crops. Available online: https://iiasa-spatial.maps.arcgis.com/apps/webappviewer/index.html?id=a813940c9ac14c298238c1742dd9dd3c (accessed on 28 April 2019).
137. Kort, J.; Collins, M.; Ditsch, D. A review of soil erosion potential associated with biomass crops. *Biomass Bioenergy* **1998**, *14*, 351–359. [CrossRef]
138. Vaughan, D.H.; Cundiff, J.S.; Parrish, D.J. Herbaceous crops on marginal sites Erosion and economics. *Biomass* **1989**, *20*, 199–208. [CrossRef]
139. Fagnano, M.; Impagliazzo, A.; Mori, M.; Fiorentino, N. Agronomic and environmental impacts of giant reed (Arundo donax L.): Results from a long-term field experiment in hilly areas subject to soil erosion. *Bioenergy Res.* **2015**, *8*, 415–422. [CrossRef]
140. Dauber, J.; Jones, M.B.; Stout, J.C. The impact of biomass crop cultivation on temperate biodiversity. *Gcb Bioenergy* **2010**, *2*, 289–309. [CrossRef]
141. Folke, C.; Carpenter, S.; Walker, B.; Scheffer, M.; Elmqvist, T.; Gunderson, L.; Holling, C.S. Regime shifts, resilience, and biodiversity in ecosystem management. *Annu. Rev. Ecol. Evol. Syst.* **2004**, *35*, 557–581. [CrossRef]
142. Deutsch, L.; Folke, C.; Skånberg, K. The critical natural capital of ecosystem performance as insurance for human well-being. *Ecol. Econ.* **2003**, *44*, 205–217. [CrossRef]
143. Teuling, A.J. A hot future for European droughts. *Nat. Clim. Chang.* **2018**, *8*, 364. [CrossRef]
144. Samaniego, L.; Thober, S.; Kumar, R.; Wanders, N.; Rakovec, O.; Pan, M.; Zink, M.; Sheffield, J.; Wood, E.F.; Marx, A. Anthropogenic warming exacerbates European soil moisture droughts. *Nat. Clim. Chang.* **2018**, *8*, 421. [CrossRef]
145. Garbolino, E.; Daniel, W.; Hinojos Mendoza, G. Expected Global Warming Impacts on the Spatial Distribution and Productivity for 2050 of Five Species of Trees Used in the Wood Energy Supply Chain in France. *Energies* **2018**, *11*, 3372. [CrossRef]

146. Von Cossel, M.; Mohr, V.; Elbersen, B.; Staritsky, I.; Van Eupen, M.; Mantel, S.; Iqbal, I.; Happe, S.; Scordia, D.; Cosentino, S.L.; et al. How to feed the European bioeconomy in the future? Climate change-forced shifts in growth suitability of industrial crops until 2100. unpublished.
147. Pachauri, R.K.; Allen, M.R.; Barros, V.R.; Broome, J.; Cramer, W.; Christ, R.; Church, J.A.; Clarke, L.; Dahe, Q.; Dasgupta, P. *Climate Change 2014: Synthesis Report, Contribution of Working Groups I, II and III to the Fifth Assessment Report of the Intergovernmental Panel on Climate Change*; IPCC: Geneva, Switzerland, 2014.

© 2019 by the authors. Licensee MDPI, Basel, Switzerland. This article is an open access article distributed under the terms and conditions of the Creative Commons Attribution (CC BY) license (http://creativecommons.org/licenses/by/4.0/).

Article

Evaluation of the Potential of Biomass to Energy in Portugal—Conclusions from the CONVERTE Project

Mariana Abreu [1,*], Alberto Reis [1], Patrícia Moura [1], Ana Luisa Fernando [2], António Luís [3], Lídia Quental [3], Pedro Patinha [3] and Francisco Gírio [1]

1. Unidade de Bioenergia, Laboratório Nacional de Energia e Geologia-LNEG, I.P., 1649-038 Lisboa, Portugal; alberto.reis@lneg.pt (A.R.); patricia.moura@lneg.pt (P.M.); francisco.girio@lneg.pt (F.G.)
2. MEtRICs, Departamento de Ciências e Tecnologia da Biomassa, Faculdade de Ciências e Tecnologia, Universidade Nova de Lisboa, 2829-516 Caparica, Portugal; ala@fct.unl.pt
3. Unidade de Informação Geocientífica, Laboratório Nacional de Energia e Geologia-LNEG, I.P., 2610-999 Amadora, Portugal; gabriel.luis@lneg.pt (A.L.); lidia.quental@lneg.pt (L.Q.); pedro.patinha@lneg.pt (P.P.)
* Correspondence: mariana.abreu@lneg.pt

Received: 31 December 2019; Accepted: 13 February 2020; Published: 21 February 2020

Abstract: The main objective of the Portuguese project "CONVERTE-Biomass Potential for Energy" is to support the transition to a low-carbon economy, identifying biomass typologies in mainland Portugal, namely agri-forest waste, energy crops and microalgae. Therefore, the aim was to design and construct a georeferenced (mapping) database for mainland Portugal, to identify land availability for the implementation of energy crops and microalgae cultures, and to locate agricultural and forestry production areas (including their residues) with potential for sustainable exploitation for energy. The ArcGIS software was used as a Geographic Information System (GIS) tool, introducing the data corresponding to the type of soil, water needs and edaphoclimatic conditions in shapefile and raster data type, to assess the areas for the implantation of the biomass of interest. After analysing the data of interest in each map in ArcGIS, the intersection of all maps is presented, suggesting adequate areas and predicting biomass productions for the implementation of each culture in mainland Portugal. Under the conditions of the study, cardoon (72 kha, 1085 kt), paulownia (81 kha, 26 kt) and microalgae (29 kha, 1616 kt) presented the greater viability to be exploited as biomass to energy in degraded and marginal soils.

Keywords: biomass; energy crops; miscanthus; cardoon; *Paulownia tomentosa*; microalgae; marginal land; contaminated soils; geographic information systems (GIS); ArcGIS

1. Introduction

In the past few years, a significant increase in the demand for agricultural species for biofuels production that compete with the food and feed sectors have been reported, such as, starch-rich crops (corn, wheat, barley, oats as well as tubers and roots such as sweet potatoes, yams, cassava and potatoes), sugar-rich crops (sorghum, sugar beet and sugar cane) and oil-rich crops (sunflower, soybean, coconut, palm, sesame and olive), increasing the pressure on suitable soils for agriculture [1]. To avoid the risk of conflicts on land use due to competition for food and feed, it is necessary to limit and even prohibit the use of land presenting high carbon stock for the implementation of non-food crops or directed to the production of energy. The greater relevance is to utilize uncultivated land (or wasteland) and degraded soils that are not implemented in conventional agriculture [2].

Portuguese Decree-Law n. 152-C/2017 (created from European Directive 2015/1513), highlights the need to reduce the use of conventional biofuels obtained from food raw materials and from species grown on agricultural land or land with a high organic load. Another principle is to encourage the promotion, production and use of advanced biofuels for energy production obtained from waste, wood-pulp materials (forest biomass including their waste), non-food cellulosic material such as waste derived from agricultural food species (straw, stover, husks and shells) or grassy species (miscanthus, ryegrass, arundinaria gigantean, panicum), waste from human and animal food sector and, finally, algae [3]. These types of feedstocks can be used in thermochemical conversion technologies such as gasification, combustion, pyrolysis and hydrothermal liquefaction and in biochemical conversion technologies such as fermentation for the production of bioethanol, biogas, biohydrogen or biodiesel.

Portugal is a country characterized by 39% of forests, 26.3% of agriculture-based land, 12.4% of bush and 8% of agroforestry systems, with the remainder corresponding to pastures (6.5%), artificial territories (5.1%) and other (2.7%) (data obtained from the Portuguese Carta de Ocupação do Solo or Land Use Mapping (COS 2015) for the continental territory, developed by the Direção-Geral do Território, DGT) [4]. Considering those values, it can be said that Portugal is a biomass producer because most of its territory (more than 85%) is covered by vegetation, utilized in several economic sectors, including the production of biofuels and others forms of energy such as electricity and heat [4]. In Portugal, the installed power derived from biomass in July 2015, with and without cogeneration, was 474 MW, including those obtained from the use of agricultural waste, forest waste, and pulp and paper industry waste [5]. According to Ferreira and colaborators (2017) [6], the total biomass resources potential estimated for the country in 2017 was of 42,489.7 GW h/year, and Portugal intends to have 60% of its generated electricity coming from renewable resources by 2020, in order to satisfy 31% of its final energy consumption by the same year. However, the current biomass status is not enough to reach this target. Energy crops and microalgae are considered a good option to cover the existing deficit. However, in Portuguese territory the production of dedicated crops for energy is negligible and more studies and investment in R&D are needed, and the same applies for microalgae production for energy.

In order to identify suitable areas for the implementation of energy crops and microalgae, it is necessary to take advantage of geographic information science through the ArcGIS software of the Geographic Information System (GIS), developed by ESRI (Environmental Systems Research Institute). It is characterized by a multiplicity of functions such as the capture, collection, measurement, storage, organization, modeling, editing, analysis, treatment, mapping, sharing and publication of data with relevant information of potential zones for the planting of energy crops and microalgae production according to certain parameters, such as the type of area, soil, water needs and edaphoclimatic conditions. For these reasons, geo-referencing is fundamental and must be integrated in studies that promote the development of biomass for energy, so that productivities can be accurately estimated to help model the potential of bioenergy production.

The main objective of the Portuguese Project "CONVERTE-Biomass Potential for Energy" was to support the transition to a low-carbon economy, identifying the existing and still to be explored biomass typologies in mainland Portugal, namely urban waste, industrial waste (such as agro-food waste including sludge from wastewater treatment plants) and energy crops and microalgae. Therefore, the aim of this work is to present the design and construction of georeferenced databases (mapping) in mainland Portugal to evaluate areas/soils for the implementation of energy crops, areas/soils/waters for microalgae production and areas of cultivated agricultural/silvicultural species (including their residues) with energy potential. To our knowledge, no such studies have merged these three types of biomass in the same work. Moreover, in the construction of those maps, the focus will be on the cultivation of energy crops and microalgae production with low indirect land-use change-risk, taking into account also its sustainable use (environmental, social and economical).

2. Materials and Methods

The database for georeferenced mapping of the mainland territory, to evaluate areas of potential interest for the cultivation of energy crops, microalgae, as well as to map the cultivated agricultural/silvicultural species (including their residues), was created with ArcGIS software, a tool for GIS.

The applied methodology was:

- To select the energy crops to implement in the mainland Portugal;
- To search which types of soils are of interest and present a low ILUC (indirect land-use change) risk;
- To search and download all colletected maps finded in shapefile or raster format from official websites of Portuguese Institutions like Agência Portuguesa do Ambiente (APA), Instituto Superior de Agronomia da Universidade de Lisboa (ISA-UL), Instituto da Conservação da Natureza e das Florestas (ICNF), Empresa de Desenvolvimento Mineiro (EDM) and European Institutions too as the European Environment Agency (EEA);
- To create the georeferenced databases on ArcGIS, an ArcMAP document (tool of ArcGIS software) has to be created for each chosen culture, introducing only the selected maps for specif properties and/or attributes of interest such as temperature, precipitation, frost, land steepness, soil texture, soil pH, soil thickness, presence of physical obstacles, ecological soil value, current permeability, natural and semi-natural vegetation with conservation value, soil-morphological aptitude to irrigated agriculture and silviculture, soil susceptibility to desertification, protected areas, land use and land cover (COS 2010 and COS 2015), corine land cover (CLC 2012), contaminated soils, wastewater treatment plant capacity, CO_2 production in the energy and industrial sectors in mainland Portugal, among others. Bearing in mind the characteristics of growth and adaptation of each culture combined with the intersection of all maps, output data have been obtained suggesting available and appropriate areas for the cultivation of each culture. The productivity forecasting and predicted bioenergy generation are presented and critically discussed;
- Lastly, the publication on Laboratório Nacional de Energia e Geologia–LNEG´s spatial data infrastructure, i.e., institutional geoportal of energy and geology, to access all the created maps and related information.

Figure 1 represents a summary of the applied methodology.

Figure 1. Applied methodology to determine suitable areas for the implementation of energy crops.

Each of one the phases specified in Figure 1 are described in more detail in the following sub-chapters.

2.1. Selected Energy Crops and Microalgae Culture

Energy crops are species intended for biomass production for subsequent generation of energy in the form of biofuels, electricity or heat. These are species that should not compete with those used for food and feed (maize, cereals and tubers, among others) and, therefore, should not be cultivated on high carbon or agricultural land. These crops should be mainly non-food and lignocellulosic species for the production of 2nd-generation fuels [7], which can be divided in two groups:

- Herbaceous crops: perennial crops which can last for 15 years or longer, being usually harvested annualy. Within this category are species such as switchgrass (*Panicum virgatum*), reed canary grass (*Phalaris arundinacea*), giant reed (*Arundo donax*) [7], perennial ryegrass (*Lolium perenne*) as well as cardoon (*Cynara cardunculus*) and miscanthus (*Miscanthus* x *giganteus*) [8];
- Short-rotation coppice: fast-growing woody species that have a short cycle, being cut and regenerated every three to five years to a total of 25 years, with the idea of obtaining high yields in a short time for energy production. Among the species classified within this criterion are eucaliptus (*Eucaliptus* spp.), willow (*Salix* spp.), poplar (*Populus* spp.) [7], among others. paulownia (*Paulownia tomentosa*) may also be included as a short-cycle species.

The energy crops selected (and also microalgae cultures) and evaluated in the frame of the CONVERTE Project are listed below, explaining the rationale for this selection according to their characteristics, advantages and benefits for the bioenergy production.

2.1.1. Cardoon (Cynara cardunculus)

Cardoon is a perennial and herbaceous plant with a productive life of 10 years (in some cases attaining 15 years) and an annual growing cycle. In the Mediterranean region, the crop presents a productivity of 10 t/ha dry matter (DM) in the first year and between 12 to 15 t/ha DM from the second year, being easily adapted to a wide range of climatic variations [9]. The cardoon is a species native of the Mediterranean basin, that support the drought stress, it can be grown in drylands, being defined as a multifunctional crop due to its characteristics that allow its use for several options, e.g., energy production in the form of biofuels, heat and power, cellulose and pulp and paper, phytochemicals and pharmacological products, among others [10,11]. The cardoon seeds composition presents nearly 24% (dry mass basis) of oil and 5.6% of water. Seeds have been traditionally considered a feedstock for biodiesel production [10]. Moreover, cardoon has shown a low impact on the environment, in the marginal soils of the Mediterranean region, particularly in certain parameters such as landscape diversity, cause of the flowering season, and use of water resources, due to the low water requirements [12]. In addition, cardoon can be irrigated with wastewater to avoid yield drop due to water stress, as it has been tested in a plantation located in Alcázar de San Juan in Spain [13]. In this study it was concluded that no effects were perceived on the energy outputs when cardoon was irrigated with wastewater.

Therefore, this crop was chosen for the study, not only because of its multiple uses but also due to its capacity to support the dry characteristics of the Portuguese summers.

2.1.2. Miscanthus (*Miscanthus* x *giganteus*)

Miscanthus is an herbaceous perennial C4 plant, native from East Asia, and long-lasting non-food crop. It is able to reach 3 m of height and produces between 20 to 30 t/ha of dry matter, in Portugal, performing best with a precipitation between 500 and 600 mm per year [9]. The following characteristics have promoted this crop as a sustainable energy crop:

- It can be cropped with the existing machinery;
- It requires low levels of fertilizers and presents high levels of carbon sequestration rates when compared with other species;

- The nitrogen and other nutrients are translocated to the roots and rhizomes, when the crop starts to lignify, thus presenting a high nutrient-use efficiency;
- It is a species that has a low incidence of plant disease and attack of pests, being considered a non-invasive plant, factor that allows plantation and utilization for energy;
- It is a species that adapt easily to various types of soils including marginal land [14].

In the last few years, miscanthus has shown a high potential concerning its implementation in marginal land and degraded and contaminated soils. A study was reported in which various miscanthus genotypes (*M. x giganteus*, *M. sinensis* and *M. floridulus*) were evaluated in a soil contaminated with 450 and 900 mg de zinc (Zn)/kg, over two years. In the contaminated soil, the *M. sinensis* and *M. floridulus* did not present changes in yields but in the case of *M. x giganteus*, the production was 20% lower. However, this last genotype presented, even in the contaminated soils, higher yields than the other two species [15]. Moreover, the deep and extensive rooting system of the plant allows this crop to be irrigated with wastewater with success, since the growth and productivity of the species were not affected and the polluting elements were removed from the wastewater by the plants, indicating that this might represent a solution for its cultivation in semi-arid regions with a high scarcity of water [16], such as those of the Mediterranean Region.

Several technologies can be applied to the miscanthus for energy production: heat, electricity or both (combined heat and power, CHP) through combustion in cogeneration systems (its most widespread application in Europe); biogas and bioethanol production [17].

Consequently, miscanthus can be considered a promising crop in mainland Portugal, not only due to the high yields but also due to the lower fertilizer and pesticide requirements. Moreover, the crop presents a good tolerance to a variety of soils, including marginal, contaminated and degraded areas. The major drawback of this crop is the annual water demand which is above 500 mm per year [18,19].

2.1.3. Paulownia (*Paulownia tomentosa*)

Paulownia tomentosa is a large deciduous hardwood and fast growing tree that is native from China [20]. Paulownia species are found naturally growing and under cultivated conditions at several regions of the world. It quickly spreaded to other parts of Asia, being cultivated in particular in Japan and Korea. Presently, it can also be found mainly in central Europe, north and central America, and Australia. Its main uses are industrial applications of the wood, due to its high ignition point, as well as to its dimensional stability and life time maintenance its characteristics [21].

Paulownia tomentosa is widely studied and utilized for the rehabilitation of contaminated soils and abandoned agricultural soils with low water needs, adapting itself to a great variety of climatic conditions and diversity of soils. It helps, too, in soil recuperation and stabilization including erosion control [22]. In Portugal, the Government recommends its afforestation in "Annex II-Non-indigenous species with interest for afforestation" of Decree-Law n. ° 565/99, being a non-indigenous species, with non-invasive character [23]. Concerning the growth, it was found that in a period of only 5–7 years after planting 2000 trees/ha under favorable conditions a significant annual production of 150–300 tons of wood was achieved [24]. Due to its high cellulose content, paulownia has shown its feasibility for use in the pulp industry and in lignin applications, combining both delignification and auto hydrolysis processes [25,26]. A wood analysis showed a composition of 50.55% cellulose, 13.6% hemicellulose, 21.36% lignin, 14% extractable and only 0.49% ash [27]. The energetic valorization of paulownia may be through its direct use as solid biomass for the production of heat and electricity or as a raw material to second generation advanced biofuels, such as bioethanol [28,29]. For those reasons, paulownia was included in this list of options for bioenergy in Portugal.

2.1.4. Microalgae Culture

Microalgae are single-cell photosynthetic micro-organisms mainly found (but not only) in aquatic environments. These organisms are able to synthetise important amounts of lipids, proteins, carbohydrates as well as other compounds with biological activity in a very short timeframe from three basic ingredients: solar radiation, carbon dioxide (CO_2) and fertilizers/nutrient-rich water. Microalgae have been reported over a long time to be very important in the biofixation and storage of CO_2, as a way to neutralize the huge production and release of greenhouse gases (GHG) worldwide, derived from the enormous use of fossil fuels at various scales, including the industrial sector. The several benefits of microalgae biomass as a feedstock for biofuel production compared to traditional biomass resources are listed as follows:

- Microalgae have a fast growth rate (short biomass duplication time) compared to land-based crops and could be harvested the whole year, even daily. They are able to double their biomass in less than 24 h. Remarkably, some species can even double their biomass in periods shorter than 3.5 h [30];
- The photosynthetic machinery in microalgae is analogous to higher plants, but present a higher photosynthetic efficiency (about 4–7.5%) compared to 0.5% for land-based cultures [31];
- Microalgae cultivation requires less water and land resources than terrestrial crops. Water quality and salinity do not create any problems, it being possible to use brackish, sea or freshwater and non-arable land (even rocky and/or sandy areas) minimizing the environmental impacts, while not compromising the production of food crops [32], even very steep land is not a problem as photobioreactors can be tilted at any angle or even placed vertically;
- Microalgae can obtain nutrients (such as nitrogen and phosphorus) from wastewater, while providing treatment to domestic, food and agro-industrial effluents among others [32];
- Microalgae are able of biofixing CO_2 from the atmosphere, but may also utilize anthropogenic CO_2 emissions from power stations and other industries. Typically, 1 kg of microalgae is produced from the biofixation of, at least, 1.83 kg of CO_2 [31];
- Microalgae biomass can be converted into a wide range of valued products such as food, feed, nutraceuticals, cosmetics, and fuels such as biodiesel, gasoline, jet fuel, hydrogen, aviation gas, and bioethanol, among others. The leftover biomass may be recycled to be further used as feed or fertilizer [31];
- The microalgal biochemical composition can be changed simply through the variation (manipulation) of culture conditions [32].

Different microalgae were previously studied concerning their biofuel production potential. The conversion route and technology depend on different parameters such as the biomass composition, selected biofuel product, operating conditions, process time and production cost, in order to assure either economically viability or environmental sustainability.

Tables 1 and 2 resumes the main characteristics of each type of selected and studied biomass (including microalgae).

Table 1. Cultivation parameters of the selected and studied biomasses (including microalgae).

Biomass	Binomial Name	Species and Culture Type	Productivity/Yield	Growth Rate	Harvest	Number of Harvests/Year
Cardoon	*Cynara cardunculus*	Herbaceous (lignocellulosic and oleaginous type)	10 to 25 t/year, with 500 mm/year rainfall in the Mediterranean Region [33]. In the 1st year, it is possible to obtain 10 t (dry matter)/ha and in the 2nd year, 12–15 t (dry matter)/ha with 400 and 550 mm/year of rainfall during the vegetative period [9]	Life span from 10 to 15 years [33] and crop height up to 2 m [9]	Harvest occurs between July and September, with moisture in between 10% and 15% and before seed dissemination [33]	1 [33]
Miscanthus	*Miscanthus x giganteus*	Herbaceous (lignocellulosic type)	From 15 [34] to 30 t/ha of dry matter, with precipitation higher than 500 mm/year [18]	Life span between 10 and 15 years (in some cases can reach up to 20 years [9]) achieving maturity after 2 or 3 years. The stems in the 1st year, with favorable summer temperatures, reach a height of 1.5 to 2 m (sometimes up to 3 m [9]), with higher yields in Southern Europe (water cannot be a limiting factor) [18]	Rhizomes are planted between march and may (in Europe and depending on the weather). Harvesting is performed when the dry matter content is high (usually in autumn [34] or end of winter) [18]	1 [18]
Paulownia	*Paulownia tomentosa*	Woody (lignocellulosic type)	35 to 45 t/ha/year (30% moisture) with minimum precipitation of 500 mm/year. Planting density is on average 1600 trees/ha [22]	It is possible to reach a height of 20 m in rotation cycles of 4 and 5 years [22]	-	-
Microalgae	-	Aqueous medium	55 t/ha year (conservative)	Achievable biomass doubling time of 7h for mass culture	Daily	365

Table 2. Optimal conditions for the cultivation of the selected and studied biomasses (including microalgae).

Biomass	Insolation	Precipitation	Temperature	Land Steepness	Soil Type	Weather Limitations or other Aspects
Cardoon	-	The highest demand of water (precipitation higher than 450 mm) is from autumn until early spring [35], therefore, it can growth in dry areas [34]. However, with lower rainfall, a decrease in productivity is observed [35]	It supports temperatures down to −5 °C, as long as it has developed four leaves [33]	The maximum value between 8 and 15% [36]	It supports pH above 6 in the soil. Preference for well drained deep soils, with water retention in the subsoil between 1 and 3 m [35]. It is a little sensitive to stony conditions [35]	Winter frost causes leaf death, however, it survives and recovers as soon this period ends. Requires water (in the Mediterranean) in the late spring [33]. It isn't very exigent in relation to the type of soil and water. It requires high temperature climate [34]
Miscanthus	-	It presents moderate needs, therefore, rainfall has to exceed 500 mm [18]	It can grow with soil temperature above 8 °C. Negative temperatures can cause rhizomes death [18]	The maximum value between 8 and 15% [36]	Optimum pH between 5.5 and 7.5. Sandy soils: maximum productivity attained in 3 years. Loamy or clay soils: higher yields obtained after 5 years. Preference for loamy–sandy soils (up to 10% clay, well-aerated and with high water retention capacity) with high organic matter content [18]	-
Paulownia	-	It requires minimum average precipitation of 500 mm [22]	It supports cold very well (down to −17 °C) and heat (up to 45 °C) [22]	It stands slope up to 25% [22]	It requires a soil pH of 5.5 to 8. Prefers well-drained soils, with water table higher than 2 or 2.5 m and not very clayey. It adapts to a great variety of climates and soils and resists moderately periods of dryness between the 1st and 2nd year [22]	It resists fires very well, being a species free from pests and diseases [22]
Microalgae	Elevated	Not relevant [a]	From 10 to 40 °C	-	Not relevant [a]	Not relevant [a]

[a] The microalgae are able to be mass produced at confined photobioreactors placed over land with any water regardless the quality.

The use of energy crops and microalgae for bioenergy presents some opportunities for rural areas, such as employment generation, land recuperation from an abandonment state, rural development especially of isolated regions, increased income for farmers and economic benefits for companies or entities with an interest in developing or using these species for the production of bioenergy. Yet, implementation of these biomasses can also lead to some constraints. For this reason, some steps have to be taken before implementation of such projects:

- To study the areas where the species will be cultivated, so as not to generate negative impacts;
- To analyze specific areas considered sensitive within the zones of interest as a form of protection;

To study the possibility of the implementation of crops in an intercalated form of species with different ages or of several other typologies, given the rotation of these, and avoid extensive use and massive soil damage.

2.2. Low Indirect Land-Use Change (ILUC) Risk Soils

Low indirect land-use change (ILUC) risk soils refers to areas that can be considered for the implementation of dedicated energy crops [7], once the soils present a low quality for food and feed species, caused by a multitude of natural or human factors. There is a wide diversity of soils considered to be at low ILUC risk such as contaminated soils, wasted land, devastated land, moderate and highly degraded soils, abandoned land such as pasture or arable soils, marginal and fallow lands [37], all of which are suitable for the implementation of energy crops or microalgae for bioenergy production. Wastes from farmland species and from food and feed species can also be harnessed for energy production. Fertility and productivity are different for each soil type, as shown in Figure 2.

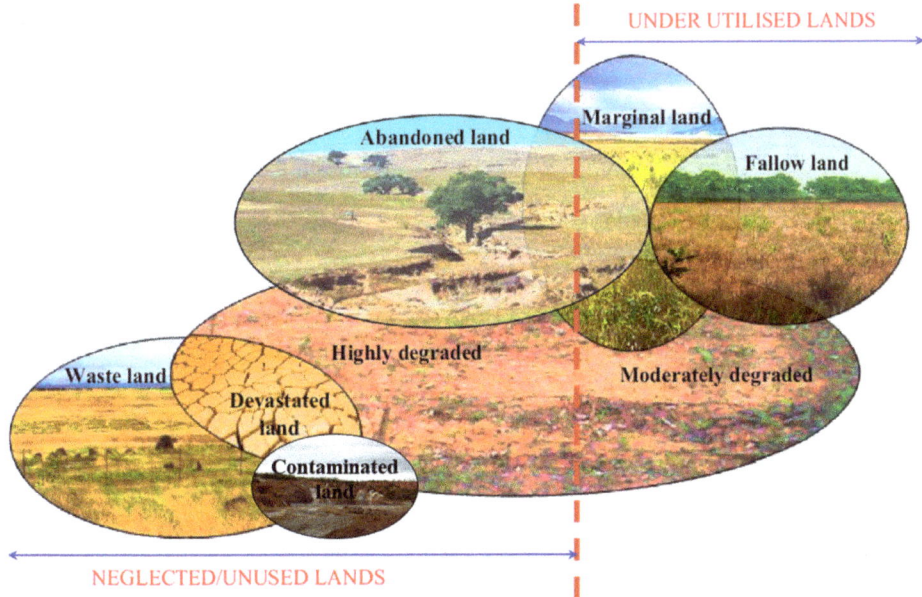

Figure 2. Low indirect land-use change (ILUC) risk soils for biomass and biofuels production. X axis refers to increasing biomass (bioenergy) productivity potential. The red vertical dash line separates bad (low) and fairly good (moderately high) quality lands (Adapted from [37]).

Based on the literature, the more studied soils that represent low ILUC risk are the degraded, marginal and contaminated soils, which according to their properties meet the sustainability requirements. With the objective of implementing the energy crops previously mentioned in this type of soils for to reduce the risk of land use conflicts due to competition for food and feed, that they can bring additional revenue to land owners, thus contributing positively to economic growth. These three types of soils were then evaluated and selected in this study, being described below:

- Degraded land: areas that suffer from a continuous deterioration process that can be caused naturally where lands with high carbon-laden are converted to dry land, causing changes in the physical, chemical and biological characteristics of the soil, reducing soil quality and causing severe erosion, namely, nutrient loss, soil infiltration problems [7] and wind erosion. Degradation can also occur by human action that generates progressive and continuous soil depletion causing biological and economic loss by decreasing the value of the land [38].

Considering the aforementioned definition, it can be confirmed that desertification is a state of the soil that is associated with degraded land and is characterized by very dry areas (dry sub-humid, arid and semi-arid) that reach this state for environmental or human reasons [39], being a factor that depending on the zone, is gradually increasing. Desertification causes alteration and destruction of ecosystems and increases the presence of invasive species, with loss of suitable areas for agriculture and decreased groundwater [40].

In this study, soil affected by natural factors that cause desertification [41] of these areas will be considered degraded;

- Contaminated soil: land with high concentrations of pollutants such as asbestos, gold, tin and tungsten, polymetallic, coal, base metals, iron and manganese, radioactive, among others [42], caused by human action, namely, in industrialized areas, intensively applied agriculture [7] and in areas where mining has occurred for a certain period of time.

Currently, in Portugal, two types of situations are identified in contaminated areas. In the first case the area remains contaminated without any possible use, becoming an abandoned area and in the second case, the recovery and valorization of this area is being accomplished, by companies, such as EDM, that have to monitor and control the recovery process along and after the application of gradual remediation methods [8]. The recovery process may take years to complete land reclamation, either for agriculture, recreational areas or residential areas.

Contaminated soils are also considered degraded soils [7]. However, in this study, degraded (desertified) and contaminated areas will be assessed separately. The areas massively exploited by human activity as mining areas characterized by the presence of heavy metals [42] were considered as contaminated soils;

- Marginal areas: although there is no clear and accurate concept of this type of soil [8], the APEC (Asia-Pacific Economic) Energy Working Group report presents a very broad definition of these areas which are characterized by poor weather conditions (low rainfall and high temperatures) and very poor soil physical-chemical characteristics (low quality and with physical constraints such as mountainous, extremely dry areas, saline, drenched, glacial and rocky areas) [43].

Based on data found in the literature, the saline soils are considered marginal, therefore, areas with moderate and high concentrations of saline elements are inadequate for the implementation of food crops [7]. The term salinization refers to areas with low precipitation and high evapotranspiration that causes salt accumulation making it impossible to wash on the soil surface. These areas can be found in the coastal part of the territory [8]. Much of the marginal land could be used for agriculture due to the quality and type of soil, however, many of them, are found in high zones, with high slopes, hard-to-reach areas or abandoned land, that are no longer used for this purpose [8] and now are considered suitable for other purposes such as the implementation of energy crops. For these reasons, in this study, we consider as marginal lands the pasture areas such as natural herbaceous vegetation, areas with dense, light dense undergrowth, dense and dense sclerophyte vegetation, other woody formations and, lastly, areas related to uncovered spaces or with sparse vegetation [44].

Based on a report by the 2014 Joint Research Center that presents an analysis of the extent of degraded areas in European Union (EU) countries, it can be stated that in the case of Portugal, the contaminated land has the largest area of the three types of soils presented above. The total extent of these areas is 2318 kha of which 12.3% represents highly saline soils (marginal land); 33.8% are areas with severe erosion (degraded areas) and the remaining, 53.9%, are contaminated soils. This document also identifies an area of 93 kha that will be available in Portugal by 2050 for the implementation of energy crops, but this implies the conversion of 19% arable and 44% forested areas in these lands [7]. According to the European Court of Auditors of 2018, 8% of EU territory which includes countries such as Bulgaria, Cyprus, Greece, Italy, Spain and Portugal, have areas with very high values highly sensitive to desertification (degraded soils), causing a decrease in land use for the agricultural sector [45].

2.3. Colleted Maps

In order to obtain the appropriate areas for the crops implementation, it was necessary to compile as many data or factors as possible, according to information available from various sources, mainly on official websites of Portuguese and European institutions. The administrative maps of the territory and those related to land use and occupation provided by DGT have been considered; environmental aspects such as temperature, precipitation, sunshine and frost provided by APA, including the map created with CO_2 production in each municipality; various ecological factors of soil and subsoil and the edapho-morphological aptitude for agriculture and forestry of the Ecological Planning, Investigation and Cartography - EPIC WebGIS platform (ISA-UL); protected areas and soils susceptible to desertification from ICNF; contaminated soils based on the mining areas managed by EDM and the capacity and treatment applied in the wastewater treatment plants (WWTPs) in mainland Portugal, according to the EEA platform, each factor being considered a spatial thematic layer.

In the Supplementary Materials, each considered map will be described as well as the selected data for every species, considering the listed characteristics in Tables 1 and 2.

3. Results and Discussions

Figure 3 shows a scheme summarizing the results that are of interest in this study.

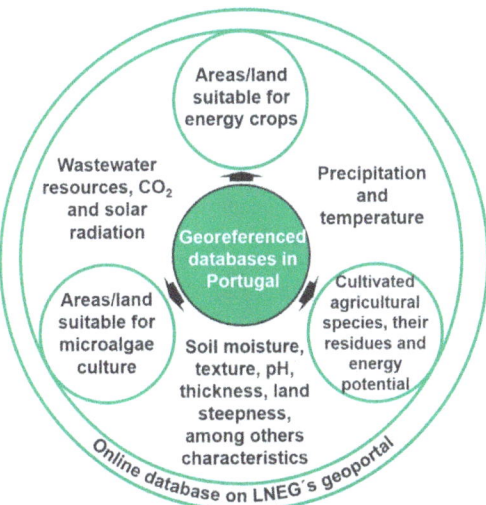

Figure 3. Summary of the results that are of interest in this study.

3.1. Georreferenced Databases on ArcGIS

An infrastructure of spatial data was developed in order to facilitate the exchange and the use of the information among all of the beneficiary agents, with visualization and consultation through the web.

Several maps were created corresponding to scenarios from the most restrictive to the most comprehensive for each species, being these related to the suitable areas for the implementation of the respectives crops, detailed below.

3.1.1. Suitable Areas/Soils to the Cardoon, Miscanthus and Paulownia Cultivation

- 1st Scenario

The map presents the smallest suitable area for cardoon, miscanthus and paulownia in mainland Portugal, since it corresponds to the more restrictive data selection, namely, conditions and characteristics of the less appropriate soil or more penalizing in the maps of "Ecological Soil Value", "Current Permeability", "Natural and semi-natural vegetation with conservation value" and "Soil susceptibility to desertification". The map of "Soil-morphological aptitude to irrigated agriculture" was applied only to cardoon and miscanthus and in the case of paulowinia, it utilized the data of "Soil-morphological aptitude to silviculture". All these maps represents in consequence, inadequate areas for the cultivation of agricultural species.

For the cardoon, miscanthus and paulownia, the parameters related to "Temperature", "Precipitation (total amount)", "Frost (number of days of the year)", "Land steepness" (to the miscanthus were not considered due to the low area that was obtained), "Soil texture", "Soil pH" and "Soil thickness" were obtained from several sources.

Concerning the data of the "COS 2010", "CLC 2012" and "COS 2015", the areas that have not been identified as artificialized territories, cultivated areas for agricultural and forestry species, wetlands and water bodies were selected, with the areas considered as marginal soils, due to their suitability for non-food crops, remaining of interest.

- 2nd Scenario

Cardoon: the same maps identified in the 1st Scenario were considered, being added only one more parameter in the following maps (in brackets the value added is specified): "Ecological Soil Value" [Variable (E.V 3)], "Current Permeability" (Moderate-Class 3), "Natural and semi-natural vegetation with conservation value" (Moderate), "Soil-morphological aptitude to irrigated agriculture" (Moderate-4.5%) and "Soil susceptibility to desertification" (2-Moderate), in order to guarantee a wider area compared to the area previously obtained in the 1st Scenario.

Miscanthus: the same maps identified in the 1st Scenario, but incorporating a few parameters in the following maps (in brackets the added valueis specified): "Ecological Soil Value" [Variable (E.V 3)], "Current Permeability" (Low to Moderate-Class 2 and Moderate-Class 3), "Natural and semi-natural vegetation with conservation value" (Moderate), "Soil-morphological aptitude to irrigated agriculture" (Moderate-4.5%) and "Soil susceptibility to desertification" (2-Moderate), to obtain a wider area than the 1st Scenario.

Paulownia: the same maps identified for the 1st Scenario were considered, adding only one more parameter in the following maps (in brackets is specified the added value): "Ecological Soil Value" [Variable (E.V 3)], "Current Permeability" (Moderate-Class 3), "Natural and semi-natural vegetation with conservation value" (Moderate), "Soil-morphological aptitude to silviculture" (Undifferentiated silviculture-20.3%)] and "Soil susceptibility to desertification" (2-Moderate), in order to ensure a wider area when compared to that obtained for the 1st Scenario.

- 3rd Scenario

Cardoon: the same maps of the 2nd Scenario were considered, adding only a few values in the "Temperature" map (equal or superior to 0 °C), "Precipitation" (equal or superior to 400 mm/year), "Soil texture" [Coarse (more than 35% clay, or less 35% clay and less than 15% sand)] and "Soil pH" (4.5 to 5.0; 5.0 to 5.5; 5.5 to 6.0; 4.5 to 6.0).

This methodology was applied in order to compare the edapho-climatic characteristics previously consulted and initially selected as adequate for the cardoon growth, with other data presented on the website www.cabi.org that corresponds to a "Directory of Invasive Species" with detailed information about the most important characteristics of the cardoon [46].

Miscanthus: the same maps of the 2nd Scenario were considered, adding only one value in the map of "Soil pH" (4.5 to 5.0) and "Soil thickness" (25–50 cm) to evaluate how they can affect the areas suggested for miscanthus cultivation.

Paulownia: the same maps of the 2nd Scenario were considered, changing only the "Soil pH" map, being added the next values 4.5 to 5.0; 5.0 to 5.5 and removed the following 7.5 to 8.0; ≥7.5, therefore, the pH parameters on the map are between 4.5 and 7.5.

This scenario was established to compare initially selected and consulted edapho-climatic characteristics for the paulownia planting, with other data presented on the Centre for Agriculture and Bioscience International—CABI website [47].

- 4th Scenario

For the cardoon and miscanthus the same maps of the 2nd Scenario were considered, with the exception of the "Natural and semi-natural vegetation with conservation value" map and those related to the "COS 2010", "CLC 2012" and "COS 2015" not considered at the intersection as it presents a smaller number of data, limiting the area obtained as suitable for the cultivation of this crops. In the case of paulownia, the same as the two crops occurs but with the difference that the map of "Natural and semi-natural vegetation with conservation value" was considered.

In Table 3 all the data considered to obtain the adequate areas for the implementation of cardoon, miscanthus and paulownia in mainland Portugal are presented.

In Figure 4 the 4 scenarios created for the cardoon are shown, being visible the evolution and difference of the area obtained in each case.

The 4 scenarios created for miscanthus are identified in Figure 5, showing the evolution and difference of the area obtained in each case.

In Figure 6, the 4 scenarios created from paulownia are identified, showing the evolution and difference of the area obtained in each case.

Table 3. Data considered in each scenario of the cardoon, miscanthus and paulownia.

Maps	Crops	1st Scenario	2nd Scenario	3rd Scenario	4th Scenario
Districts	3 species		Aveiro; Beja; Braga; Bragança; Castelo Branco; Coimbra; Évora; Faro; Guarda; Leiria; Lisboa; Portalegre; Porto; Santarém; Setúbal; Viana do Castelo; Vila Real; Viseu		
Land Use and Land Cover - COS 2010	3 species		3.2.1.01.1 Natural herbaceous vegetation; 3.2.2.01.1 Dense bushes; 3.2.2.02.1 Low dense bushes; 3.2.3.01.1 Dense sclerophyte vegetation; 3.2.3.02.1 Sclerophyte vegetation not very dense; 3.2.4.07.1 Other woody formations; 3.3.3.01.1 Vegetation sparse		×
Corine Land Cover - CLC 2012	3 species		Bushes; Sclerophyllic vegetation; Sparse vegetation; Natural herbaceous vegetation		×
COS 2015	3 species		Natural herbaceous vegetation; Bushes; Spaces discovered or with little vegetation		×
Temperature	Cardoon		Equal or superior than 7.5 °C	Equal or superior than 0 °C	Equal or superior than 7.5 °C
	Miscanthus		Equal or superior than 10 °C		
	Paulownia		Equal or superior than 0 °C		
Precipitation	Cardoon		Equal or superior than 500 mm/year	Equal or superior than 400 mm/year	Equal or superior than 500 mm/year
	Miscanthus		Equal or superior than 500 mm/year		
	Paulownia				
Frost	Cardoon		Up to 60 days		
	Miscanthus				
	Paulownia		Up to 80 days		
Land steepness	Cardoon		0–3%; 3–5%; 5–8%; 8–12%; 12–16%		
	Miscanthus	×		0–3%; 3–5%; 5–8%; 8–12%; 12–16%	
	Paulownia		0–3%; 3–5%; 5–8%; 8–12%; 12–16%; 16–25%		

Table 3. Cont.

Maps	Crops	1st Scenario	2nd Scenario	3rd Scenario	4th Scenario
Soil texture	Cardoon	Fine; Median	Fine; Median	Fine; Median; Coarse	Fine; Median
	Miscanthus		Coarse		
	Paulownia			Fine; Median	
Soil pH	Cardoon	6.0–6.5; 6.5–7.0; 7.0–7.5; 6.0–7.5; 7.5–8.0; 7.5	6.0–6.5; 6.5–7.0; 7.0–7.5; 6.0–7.5; 7.5–8.0; 8.0–8.5; ≥ 7.5	4.5–5.0; 5.0–5.5; 5.5–6.0; 4.5–6.0; 6.0–6.5; 6.5–7.0; 7.0–7.5; 6.0–7.5; 7.5–8.0; 8.0–8.5; ≥7.5	6.0–6.5; 6.5–7.0; 7.0–7.5; 6.0–7.5; 7.5–8.0; 8.0–8.5; ≥7.5
	Miscanthus	5.5–6.0; 6.0–6.5; 6.5–7.0; 7.0–7.5; 6.0–7.5		5.5–6.0; 4.5–6.0; 6.0–6.5; 6.5–7.0; 7.0–7.5; 6.0–7.5	5.5–6.0; 6.0–6.5; 6.5–7.0; 7.0–7.5; 6.0–7.5
	Paulownia	5.5–6.0; 6.0–6.5; 6.5–7.0; 7.0–7.5; 6.0–7.5; 7.5–8.0; ≥7.5		4.5–5.0; 5.0–5.5; 5.5–6.0; 4.5–6.0; 6.0–6.5; 6.5–7.0; 7.0–7.5; 6.0–7.5	5.5–6.0; 6.0–6.5; 6.5–7.0; 7.0–7.5; 6.0–7.5; 7.5–8.0; ≥7.5
Soil thickness (cm)	Cardoon		0–10; 0–25; 0–30; 10–25; 25–50; 30–50; 50–100; >100		
	Miscanthus	0–10; 0–25; 0–30; 10–25		0–10; 0–25; 0–30; 10–25; 25–50	0–10; 0–25; 0–30; 10–25
	Paulownia		0–10; 0–25; 0–30; 10–25; 25–50; 30–50; 50–100; >100		
Ecological Soil Value	3 species	Very reduced (E.V 1); Reduced (E.V 2)	Very reduced (E.V 1); Reduced (E.V 2)	Very reduced (E.V 2); Variable (E.V 3)	
Current Permeability	Cardoon	Moderate to High-Class 4; High-Class 5		Moderate-Class 3; Moderate to High-Class 4; High-Class 5	
	Miscanthus	Low-Class 1		Low-Class 1; Low to Moderate-Class 2; Moderate-Class 3	
	Paulownia	Moderate to High-Class 4; High-Class 5			
Natural and semi-natural vegetation with conservation value	Cardoon	Very low; Low	Very low; Low; Moderate	×
	Miscanthus				
	Paulownia				Very low; Low
Soil-morphological aptitude to irrigated agriculture	Cardoon	Without aptitude-25.3%; Low-34.7%	Without aptitude-25.3%; Low-34.7%; Moderate-4.5%		
	Miscanthus				

Table 3. *Cont.*

Maps	Crops	1st Scenario	2nd Scenario	3rd Scenario	4th Scenario
Soil-morphological aptitude to silviculture	Paulownia	Forestry not recommended-15.7%	Forestry not recommended-15.7%; Undifferentiated silviculture-20.3%		
Unprotected areas	3 species	Zones that not included the protected areas, namely, Important Community Sites (SIC); Protected Special Zones (ZPE); Protected Areas National Network (RNAP); Ramsar Sites (wetlands) and Biosphere Reserve			
Soil susceptibility to desertification	Cardoon	3-High; 4-Very High		2-Moderate; 3-High; 4-Very High	
	Miscanthus				
	Paulownia			3-High; 4-Very High	
Classification of soils according to each mine group	3 species	As (Others); Asbestos (Asbestos); Au (Gold); Au, Ag (Gold); Ba, Pb, W, Sn (Tin and Wolfram); Barium (Polymetallic); Coal (Coal); Cu (Basic Metals); Cu, Pb, Ag (Basic Metals); Cu, Pb, Zn (Polymetallic); Fe (Iron and Manganese); Fe, Cu, Pb, Zn (Iron and Manganese); Mn (Iron and Manganese); Pb (Basic Metals); Pb, Au (Basic Metals); Pb, Zn (Basic Metals); Pb, Zn, Ag (Basic Metals); Pb, Zn, Ag (Polymetallic); Qz, Felds (Others); Ra (Radioactive); Ra, U (Radioactive); Sb (Others); Sb, Au (Gold); Sn (Tin and Wolfram); Sn, Nb, Ta, W (Tin and Wolfram); Sn, W (Tin and Wolfram); Sn, W, Ti (Tin and Wolfram); Sn, W, Ti (Tin and Volphamium); W (Tin and Wolfram); W, As (Tin and Wolfram); W, Mo (Tin and Wolfram); W, Sb, Au (Gold); W, Sn (Tin and Wolfram); W, Sn, Li (Tin and Wolfram); W, Sn, Qz (Tin and Wolfram); Zn, Pb (Basic Metals)			
Wastewater Treatment Plant - WWTP capacity (p.e—population equivalent)	3 species	400–31,500; 31,500–97,200; 97,200–216,000; 216,000–400,000; 400,000–920,000			
Treatments applied to wastewater facilities	3 species	Primary; Secondary; Tertiary			

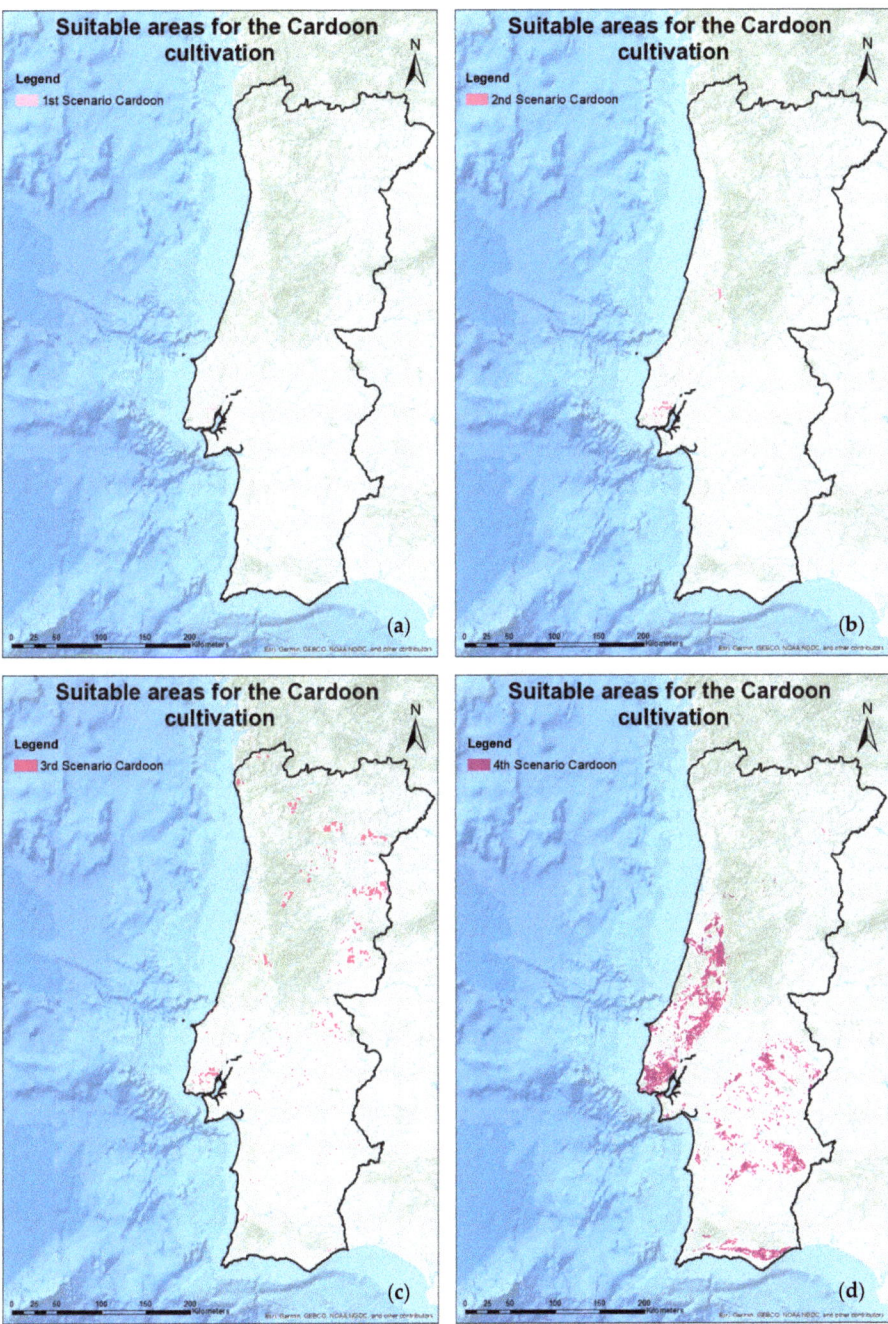

Figure 4. Scenarios displaying the adequate areas for cardoon cultivation: (**a**) 1st scenario; (**b**) 2nd scenario; (**c**) 3rd scenario; (**d**) 4th scenario.

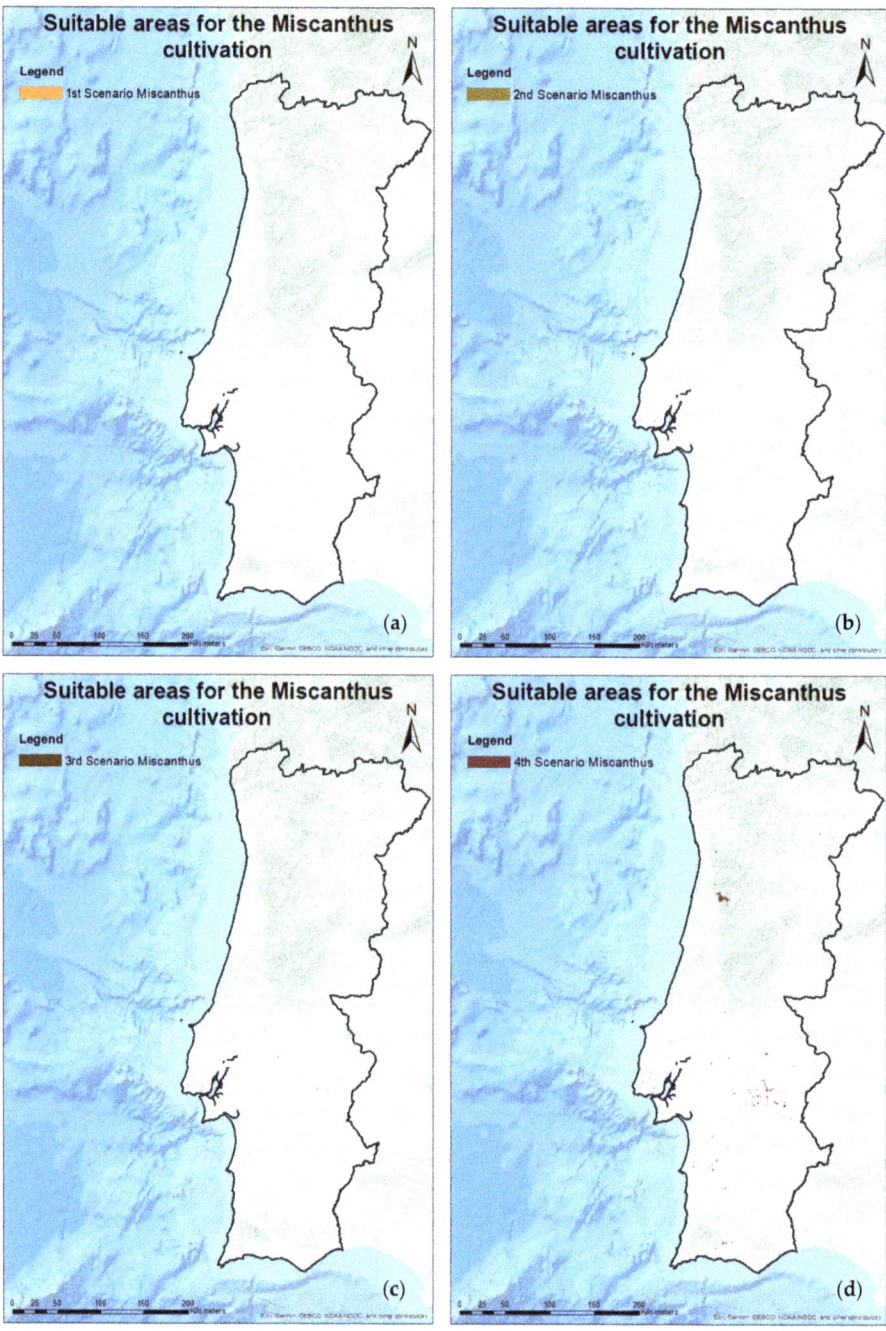

Figure 5. Scenarios presenting the suggested areas for miscanthus cultivation: (**a**) 1st scenario; (**b**) 2nd scenario; (**c**) 3rd scenario; (**d**) 4th scenario.

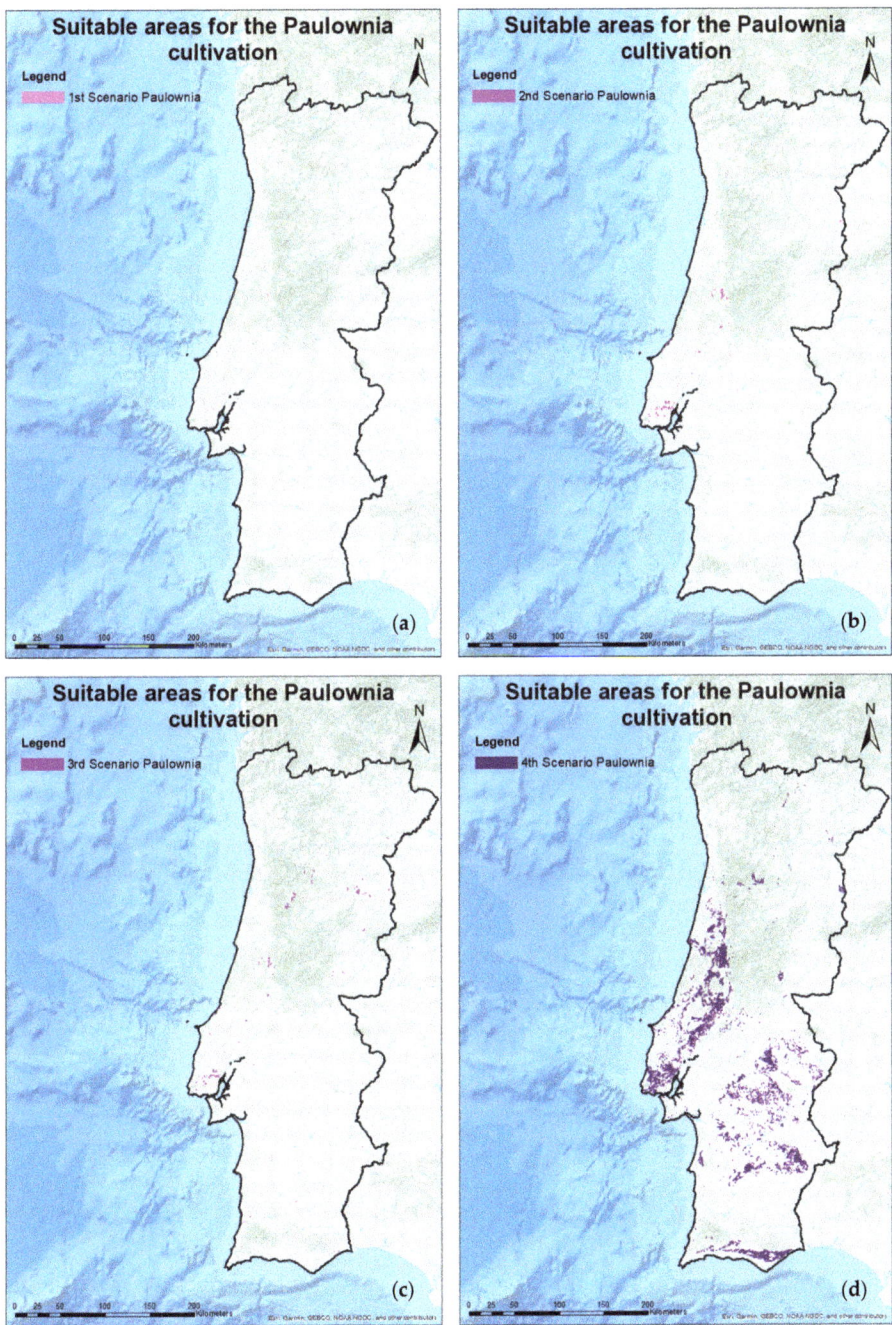

Figure 6. Scenarios with the suitable areas for paulownia cultivation: (**a**) 1st scenario; (**b**) 2nd scenario; (**c**) 3rd scenario; (**d**) 4th scenario.

3.1.2. Suitable Areas/Soils for Microalgae Cultivation

- 1st Scenario

Only for the microalgae case the map designated as "Presence of physical obstacles" was considered, with mostly stony areas, which means unsuitable areas for cultures of other species besides microalgae. This map, as well as those relating to "Insolation", "Temperature" and "Land steepness" (obtained from the same source) and those of "Municipalities of mainland Portugal with CO_2 production in the energy and industrial sector" were considered for all scenarios.

Regarding the "COS 2010" and "CLC 2012", only those areas identified as bare rock were selected. The "COS 2015" was not included because this criterion was not considered on the map.

The protected areas were considered, since it is unknown if it is possible or not to use these zones for the development of microalgae.

- 2nd Scenario

The same maps identified in the 1st Scenario were considered, not including the protected areas, to verify in what extent this factor can affect the appropriate area for the development of microalgae.

- 3rd Scenario

The same maps identified in the 2nd Scenario were considered, without the parameter's integration of the "COS 2010" and "CLC 2012".

- 4th Scenario

The same maps identified in the 3rd Scenario were considered, without including the protected areas, to verify how the suggested area for the development of microalgae could be affected.

All considered maps and data to obtain the appropriate areas for the implementation of microalgae in mainland Portugal are shown in Table 4.

The 4 created scenarios for the microalgae culture are identified in Figure 7, showing the similarity of the area obtained in all cases.

Table 4. Data considered in each scenario for the microalgae culture.

Maps	1st Scenario	2nd Scenario	3rd Scenario	4th Scenario
Districts		Aveiro; Beja; Braga; Bragança; Castelo Branco; Coimbra; Évora; Faro; Guarda; Leiria; Lisboa; Portalegre; Porto; Santarém; Setúbal; Viana do Castelo; Vila Real; Viseu		
Land Use and Land Cover - COS 2010	3.3.2.01.1 bare rock		×	
Corine Land Cover - CLC 2012	Bare rock		×	
Insolation			Superior than 2500 h	
Temperature			Equal or superior than 10 °C	
Municipalities with CO_2 production in the energy sector	28.184080–224.886097; 224.886098–345.873456; 345.873457–974.625606; 974.625607–3602.772540; 3602.772541–8665.540705			
Municipalities with CO_2 production in the industrial sector	26.595721–82.594012; 82.594013–226.494754; 226.494755–551.636051; 551.636052–1481.167615; 1481.167616–2788.265732			
Pollutant Release and Transfer Register (PRTR)	Steel; Biomass; Hydraulic lime; Non-hydraulic lime; Coal; Iron; Fuel oil; Natural gas; Kraft with bleaching; Unbleached Kraft; No specific (are just of interest the identified as Paste production, Manufacture of refined petroleum products; Manufacture of industrial gases; Manufacture of cement and iron and steel industry and manufacture of ferro-alloys); Other [it is only of interest Paste production, Manufacture of paper and paperboard (except corrugated paper); Manufacture of paper and paper products for domestic and sanitary purposes; Manufacture of refined petroleum products; Manufacture of petroleum products from waste; Manufacture of cement and production of thermal origin electricity)]; Other-Biofuels; Other-Biodiesel; Other-Co-incineration; Other-Incineration of hazardous waste; Other-Fuel Treatment; Printing and writing paper; Paper, cardboard and packaging; Bleached sulphite; Tissue (paper mills); Glass packaging; Domestic glass and Oils (Biodiesel Manufacturing Only)			
Land steepness			Superior than 25%	
Presence of physical obstacles		P-Stony phase; R2-Rock outcrops exceeding 25–40% of the area; R3-Rock outcrops exceeding 50–70% of the area		
Unprotected areas		×	Zones that not included the protected areas, namely, Important Community Sites (SIC); Protected Special Zones (ZPE); Protected Areas National Network (RNAP); Ramsar Sites (wetlands) and Biosphere Reserves	×
Wastewater Treatment Plant - WWTP capacity (p.e—population equivalent)		400–31,500; 31,500–97,200; 97,200–216,000; 216,000–400,000; 400,000–920,000		
Treatments applied to Wastewater Facilities			Primary; Secondary; Tertiary	

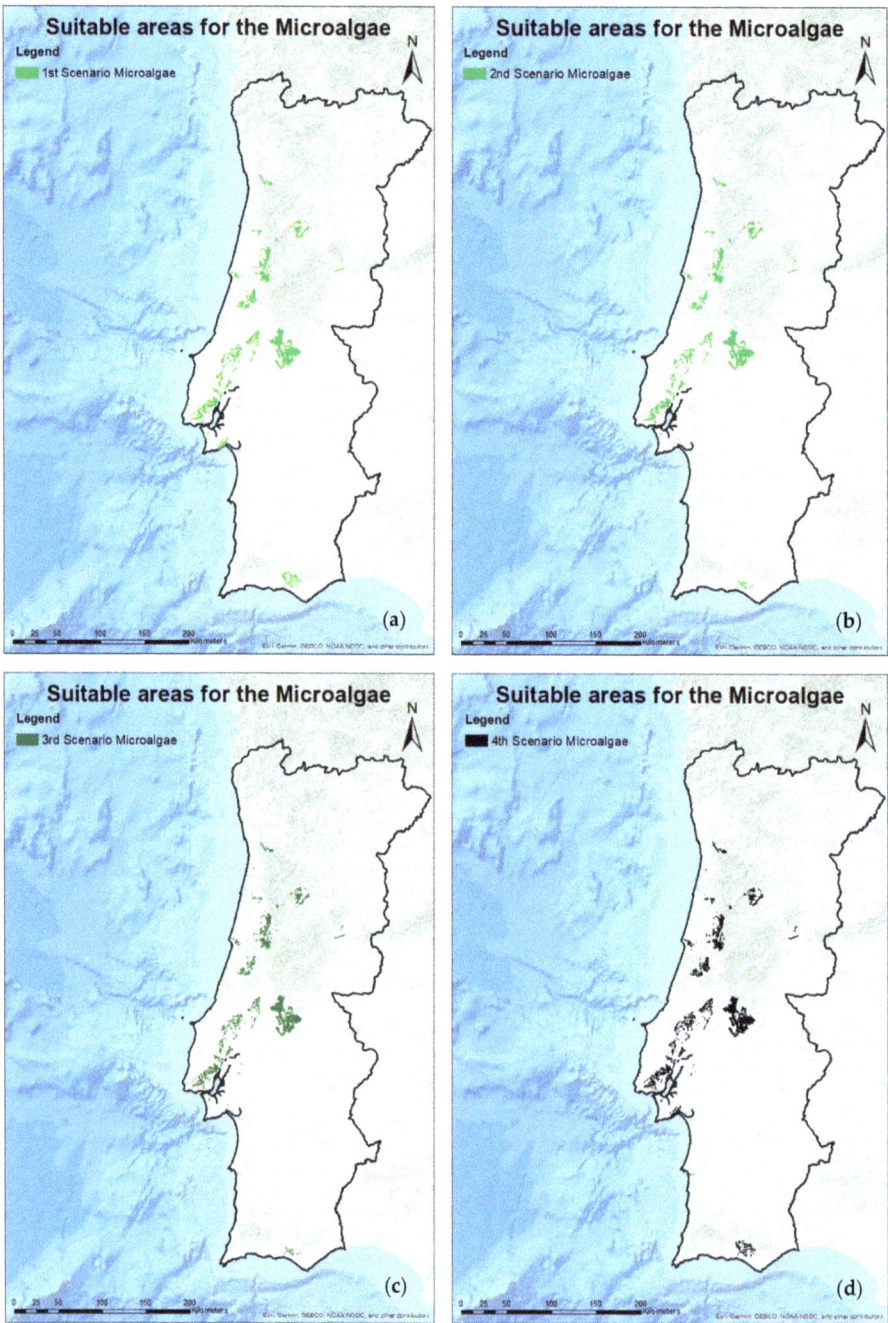

Figure 7. Scenarios with the appropriate areas for the microalgae: (**a**) 1st scenario; (**b**) 2nd scenario; (**c**) 3rd scenario; (**d**) 4th scenario.

3.1.3. Cultivated Crops (Agricultural/Silviculture) and their Residues with Potential Interest for Portuguese Inland Territory

- 1st Scenario

The map presents the areas constituted by the cultivated species identified in the "COS 2010", the "CLC 2012" and the "COS 2015", which do not belong to the categories of artificialized territories, pastures, open spaces or sparse vegetation, wetlands and water bodies. Also, the protected areas were not considered, with the entire obtained area being presented in a single map.

- 2nd Scenario

The same maps identified in the 1st Scenario were considered, including protected areas, to evaluate the effect of alteration of the area occupied by cultivated species, with all parameters being represented in one map.

- 3rd Scenario

Three maps with the same data identified in the 2nd Scenario have been created: a map with all the cultivated areas of the "COS 2010", another map with the represented parameters in "CLC 2012" and a last map, with the identified data in the "COS 2015". These maps are only for know of cultivated area with agricultural and silvicultural species in each map of mainland Portugal.

Table 5 presents all the maps and data considered to obtain the areas with agricultural and forestry species cultivated in mainland Portugal.

In Figure 8, the first and second scenario created for agricultural and forestry species grown in mainland Portugal are identified.

Figure 9 represents the third scenario with the 3 created maps, namely, with the agricultural and forestry species grown only for "COS 2010", "CLC 2012" and "COS 2015".

Table 5. Data considered in each scenario for cultivated agricultural-silvicultural species.

Maps	1st Scenario	2nd Scenario	3rd Scenario		
			COS 2010	CLC 2012	COS 2015
Districts	Aveiro; Beja; Braga; Bragança; Castelo Branco; Coimbra; Évora; Faro; Guarda; Leiria; Lisboa; Portalegre; Porto; Santarém; Setúbal; Viana do Castelo; Vila Real; Viseu				
Land Use and Land Cover - COS 2010	All the parameters that correspond to the Agriculture (except 2.1.1.02 Greenhouses and nurseries), the areas classified as Forest (without the criteria that include aspects related to burned areas) and all the lands that correspond the agroforestry systems (SAF)		x		x
Corine Land Cover - CLC 2012		Agriculture with natural and semi-natural spaces; Rice paddies; Temporary irrigated crops; Temporary rainfed crops; Temporary crops and/or pastures associated with permanent crops; Open forests, cuts and new plantations; Hardwood forests; Softwood forests; Mixed forests; Olive groves; Orchards; Agro-forestry systems; Cultural systems and complex parcel and Vineyards.	x	Agriculture with natural and semi-natural spaces; Rice paddies; Temporary irrigated crops; Temporary rainfed crops; Temporary crops and/or pastures associated with permanent crops; Open forests, cuts and new plantations; Hardwood forests; Softwood forests; Mixed forests; Olive groves; Orchards; Agro-forestry systems; Cultural systems and complex parcel and Vineyards.	
COS 2015		Criteria relating to agriculture as rainfed and irrigated temporary crops were considered; Rice paddies; Vineyards; Orchards; Olive groves; Temporary crops and/or pastures associated with permanent crops; Cultural systems and complex parcel and lastly, Agriculture with natural and semi-natural spaces. The SAFs of Cork oak; Holm oak; Other oaks; Stone pine; Other species; Cork oak with holm oak and Other mixed. The forest areas of Cork oak; Holm oak; Other oaks; Chestnut; Eucalyptus; Invasive species; Other hardwood; Maritime pine; Stone pine and Other	x		Criteria relating to agriculture as rainfed and irrigated temporary crops were considered; Rice paddies; Vineyards; Orchards; Olive groves; Temporary crops and/or pastures associated with permanent crops; Cultural systems and complex parcel and lastly, Agriculture with natural and semi-natural spaces. The SAFs of Cork oak; Holm oak; Other oaks; Stone pine; Other species; Cork oak with holm oak and Other mixed. The forest areas of Cork oak; Holm oak; Other oaks; Chestnut; Eucalyptus; Invasive species; Other hardwood; Maritime pine; Stone pine and Other
Unprotected areas	Zones that not included the protected areas, namely, Important Community Sites (SIC); Protected Special Zones (ZPE); Protected Areas National Network (RNAP); Ramsar Sites (wetlands) and Biosphere Reserves	Zones that included the protected areas, namely, Important Community Sites (SIC); Protected Special Zones (ZPE); Protected Areas National Network (RNAP); Ramsar Sites (wetlands) and Biosphere Reserves			

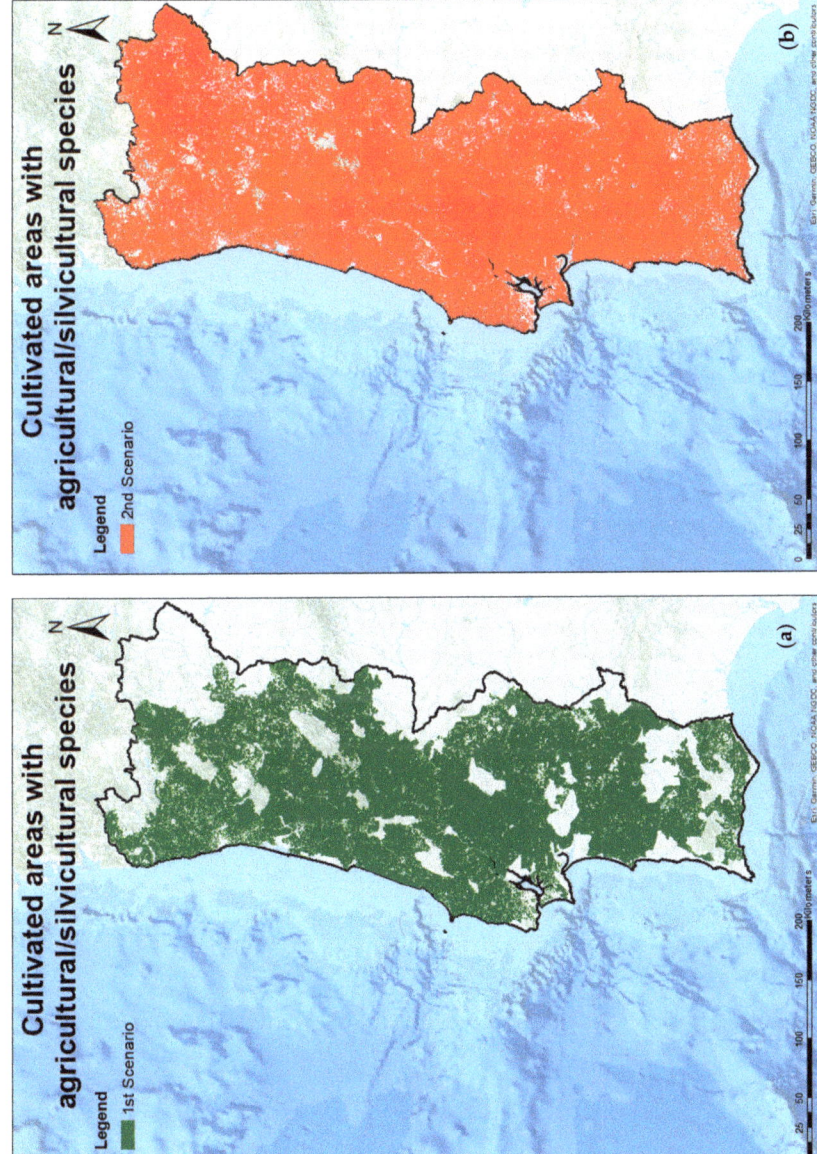

Figure 8. Scenarios displaying suggested areas for cultivated agricultural and forestry species: (**a**) 1st scenario; (**b**) 2nd scenario.

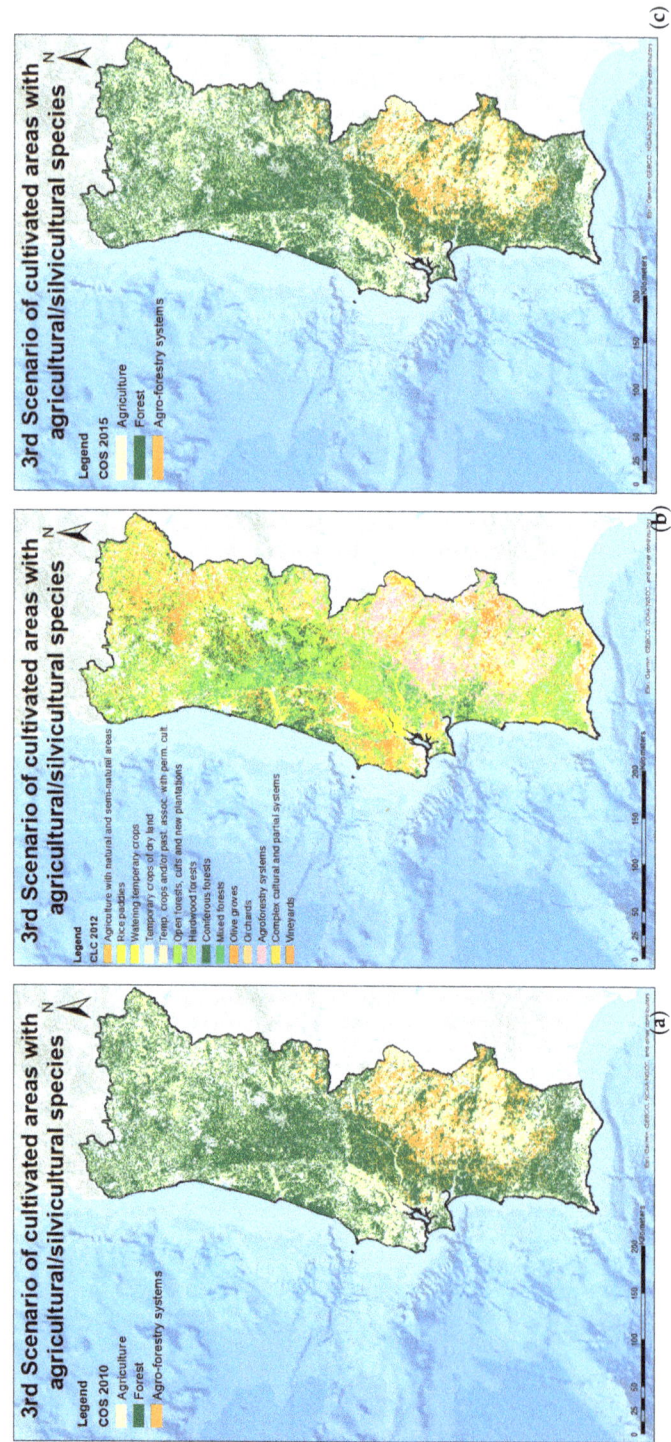

Figure 9. 3rd scenario exhibiting areas from cultivated agricultural and forestry species according to: (a) COS 2010; (b) CLC 2012; (c) COS 2015.

3.1.4. Estimated Production of Energy Crops

Considering the areas obtained for each created scenario with the suggested zones for the cultivation of energy crops and the implementation of microalgae crops, and bearing in mind the realistic yield data found in the literature, it is possible to estimate the production values theoretically for each species and scenario. From this starting point, Table 6 presents the value of the areas obtained (according to ArcGIS software), as well as the percentage of these areas out of the whole Portugal mainland area of 89,015 km^2, the minimum and maximum productivity (according to Table 1) and the estimated minimum and maximum production for each crop. It is important to specify that the calculated productivity values are overestimated, considering a productivity of 100%.

The following text is a guideline for the calculation of either minimum or maximum production, including the percentage of the land area based on the mainland territory, all of this related to the cardoon 4th scenario, for example.

$$Minimum\ yield\ (t) = Area(ha) \times Minimum\ productivity\ (t/ha)$$

$$Minimum\ production = 72,312.73 ha \times 10\ t/ha$$

$$Minimum\ production = 723,127\ t \approx 723\ kt$$

$$Maximum\ yield\ (t) = Area(ha) \times Maximum\ productivity\ (t/ha)$$

$$Maximum\ production = 72,312.73 ha \times 15\ t/ha$$

$$Maximum\ production = 1,084,691\ t \approx 1085\ kt$$

$$89,015\ km^2 \rightarrow 100\%\ da\ area$$

$$723.13\ km^2 \rightarrow X$$

$$X = 0.81\%\ of\ the\ area\ obtained\ for\ the\ Cardoon\ 4th\ scenario$$

Based on the results presented in Table 6, the scenarios with the largest proposed area for each species are the 4th scenario for the cardoon, miscanthus and paulownia crops, with it being very limited by the selected data in "COS 2010", "CLC 2012", "COS 2015" and by the map of "Natural and semi-natural vegetation with conservation value".

In the case of microalgae cultures, there is not a significant difference among the 4 obtained scenarios, so it can be guaranteed that the areas would not diverge too much if new scenarios were created. However, the largest area identified was obtained in the 1st and 4th scenarios.

For the last case, the use of waste for bioenergy, approximately 73% of the area corresponds to land with already cultivated species, thus ensuring the existence of residual material that can be used in biomass for further energy conversion systems.

It should be pointed out that the yielded area from the 4th scenario for either cardoon or paulownia is around half the area of the previously reported unproductive soils (2% of mainland Portugal).

Table 6. Estimated production of each species and scenario.

Energy Crops/Species		Scenario	Area (km²)	Area (kha)	% of the Obtained Area in Mainland Portugal	Minimum Productivity (t/ha)	Maximum Productivity (t/ha)	Minimum Production (kt)	Maximum Production (kt)
Cardoon		1st	5.21	0.521	0.01	10	15	5	8
		2nd	13.82	1	0.02			14	21
		3rd	128.84	13	0.14			129	193
		4th	723.13	72	0.81			723	1085
Miscanthus		1st	0.00	0	0.00	15	30	0.001	0.002
		2nd	0.04	0.004	0.00			0.054	0.109
		3rd	0.17	0.017	0.00			0.262	0.523
		4th	8.74	0.874	0.01			13	26
Paulownia		1st	0.03	0.003	0.00	35	45	0.105	0.134
		2nd	7.64	0.764	0.01			27	34
		3rd	9.71	0.971	0.01			34	44
		4th	80.75	81	0.91			2834	3644
Microalgae		1st	293.95	29	0.33		55	1617	
		2nd	234.77	23	0.26			1291	
		3rd	234.69	23	0.26			1291	
		4th	293.87	29	0.33			1616	
Cultivated agricultural/ silvicultural species		1st	40,280.43	4028	45.25				
		2nd	55,734.56	5573	62.61				
	3rd	COS 2010	64,761.63	6476	72.75				
		CLC 2012	73,664.68	7366	82.76				
		COS 2015	65,316.13	6532	73.38				

54

3.2. Publicação on Laboratório Nacional de Energia e Geologia – LNEG's geoportal

After obtaining all maps of interest, all the created maps were compiled and published in the LNEG's geoportal [48], on a dedicated theme for the CONVERTE project.

4. Conclusions and Recommendations

The aforementioned 4th scenario represents the largest proposed area for cardoon, miscanthus and paulownia cultures: 72,313 ha (0.81% of the total area of mainland Portugal), identified in the Regions of Estremadura and Ribatejo, Lisbon, Algarve and part of Beira Litoral and Alentejo, 874 ha (0.01% of the total area of the continent) being located in the Beira Litoral and Alentejo Regions and 80,975 ha (0.91% of the mainland Portugal total area), identified in the Beira Litoral, Estremadura and Ribatejo, Lisbon, Alentejo, Algarve and part of Trás-os-Montes and Alto Douro and Beira Interior, respectively.

Concerning microalgae cultures, the 1st and 4th scenario represent the largest proposed area for its implementation with approximately 29,395 ha in both cases (0.33% of the total area of mainland Portugal) identified in the Beira Litoral, Estremadura and Ribatejo Regions, Lisbon and Setúbal, Algarve and part of Beira Interior.

Concerning cultivated species, the most significant value of the agricultural-forestry area is 6,531,613 ha (73.38% of the total continent) from COS 2015, identified in all regions, anticipating a huge potential for waste generation and recovery.

The previously identified areas for energy crops production correspond only to marginal and degraded soils. The scenarios created yielded very restricted areas which fulfill all predefined parameters for each species.

The GIS is a powerful tool for predicting areas for biomass production to feed energy-based biorefineries and geographical availability of the feedstock. It is an instrument for technicians, beneficiaries and decision-makers regarding the optimal location of future biomass power plants.

The implementation of energy crops in degraded and contaminated soils presents also a dual purpose: it allows the sustainable production of energy and soils can also be recovered for agriculture or forestry. This study combines GIS and a multiplicity of data in order to predict the availability of biomass for bioenergy, acting as support guidelines for further implementation elsewhere, as the methodology can be implemented in other countries or regions.

Supplementary Materials: The following are available online at http://www.mdpi.com/1996-1073/13/4/937/s1: S.1 Maps obtained from Direção Geral do Território (DGT); S.1.1 Districts of the Official Administrative Charter of Portugal-CAOP; S.1.2 Land Use and Land Cover; S.1.3 Corine Land Cover; Figure S1. Maps obtained from the DGT website: (a) Districts of mainland Portugal from CAOP; (b) COS 2010; (c) CLC 2012; (d) COS 2015; S.2 Maps consulted from Agência Portuguesa do Ambiente (APA); S.2.1 Insolation; S.2.2 Temperature; S.2.3 Precipitation; S.2.4 Frost; Figure S2. Maps presenting APA data (annual basis): (a) Insolation level measured in number of hours; (b) Average daily air temperature in °C; (c) Precipitation measured as the total amount (average values) in mm; (d) Map of frost in number of days; S.2.5 Municipalities of mainland Portugal with CO2 production in the energy and industrial sectors; S.2.6 Pollutant Release and Transfer Register—PRTR; Figure S3. Consulted maps from APA (annual basis): (a) Municipalities vs the quantity produced CO2 (emitted and released) in the energy sector; (b) Produced (emitted) CO2 (kt) in the industrial sector vs Municipalities; (c) Location of the companies by subsector identified in the Pollutant Release and Transfer Register—PRTR, that represents the GHG emission sources in mainland Portugal; S.3 Maps obtained from EPIC WebGIS platform (Instituto Superior de Agronomia, Universidade de Lisboa, ISA-UL); S.3.1 Land steepness; S.3.2 Soil texture; S.3.3 Soil pH; S.3.4 Soil thickness; Figure S4. Maps obtained from EPIC WebGIS platform (ISA-UL): (a) Land steepness map measured in %; (b) Soil texture represented as the surface layer up to 30 cm; (c) Soil pH map from soils considered high acid up to very alkaline; (d) Soil thickness in cm; S.3.5 Presence of physical obstacles; S.3.6 Ecological Soil Value; S.3.7 Current Permeability; Figure S5. Maps according to EPIC WebGIS platform (ISA-UL): (a) Map of the presence of physical obstacles; (b) Ecological soil value map; (c) Current permeability; S.3.8 Natural and semi-natural vegetation with conservation value; S.3.9 Soil-morphological aptitude to irrigated agriculture and silviculture; Figure S6. Consulted maps from EPIC WebGIS platform (ISA-UL): (a) Natural and semi-natural vegetation with conservation value map; (b) Soil-morphological aptitude to irrigated agriculture; (c) Soil-morphological aptitude to silviculture; S.4 Consulted maps from Instituto da Conservação da Natureza e das Florestas (ICNF); S.4.1 Protected areas; S.4.2 Unprotected areas; S.4.3 Soil susceptibility to desertification; Figure S7. Maps obtained from ICNF: (a) Protected areas in mainland Portugal corresponding to areas not permitted for cultivation; (b) Unprotected areas, corresponding to zones where cultivation is allowed; (c) Soil susceptibility to desertification;

S.5 Maps from Empresa de Desenvolvimento Mineiro (EDM)-Classification of soils according to each mine group (Contaminated soils); S.6 Maps obtained from European Environment Agency (EEA); S.6.1 Wastewater Treatment Plant (WWTP) capacity; S.6.2 Applied treatments to Wastewater Facilities; Figure S8. Maps according to EDM and EEA data: (a) Mines that are being recovered by EDM (contaminated soils); (b) WWTP capacity in mainland Portugal; (c) Treatment applied in WWTPs.

Author Contributions: Conceptualization, A.R., P.M. and L.Q.; Methodology, A.R., L.Q., P.P. and P.M.; Validation, A.R., P.M., and L.Q.; Formal analysis, M.A., A.R. and P.P.; Investigation, M.A. and A.R.; Resources, all authors; Data curation, M.A. and A.R.; Writing—Original draft preparation, M.A., A.L.F. and A.R.; Writing—Review and editing, M.A., A.L.F. and A.R.; Supervision, P.M.; Project administration, F.G. and P.M.; Funding acquisition, P.M. All authors have read and agreed to the published version of the manuscript.

Funding: This work was integrated in the project CONVERTE, supported by POSEUR (POSEUR-01-1001-FC-000001) under the PORTUGAL 2020 Partnership Agreement. This research has been carried out at the Biomass and Bioenergy Research Infrastructure (BBRI)- LISBOA-01-0145-FEDER-022059, supported by Operational Programme for Competitiveness and Internationalization (PORTUGAL2020), by Lisbon Portugal Regional Operational Programme (Lisboa 2020) and by North Portugal Regional Operational Programme (Norte 2020) under the Portugal 2020 Partnership Agreement, through the European Regional Development Fund (ERDF). This work has also been supported by FCT – Fundação para a Ciência e Tecnologia within the R&D Unit Project Scope: UIDP/04077/2020 and UIDB/04077/2020.

Conflicts of Interest: The authors declare no conflict of interest. The funders had no role in the design of the study; in the collection, treatment analyses, or interpretation of data; in the writing of the manuscript; or in the decision to publish the results.

References and Notes

1. The European Parliament and the Council of the European Union. *Directive (EU) 2015/1513 of the European Parliament and of the Council of 9 September 2015*; Official Journal of the European Union: Brussels, Belgium, 2015; pp. L239/1–L239/29.
2. Barbosa, B.J.; Fernando, A.L. Production of Energy Crops in Heavy Metals Contaminated Land: Opportunities and Risks. In *Land Allocation for Biomass Crops*; Li, R., Monti, A., Eds.; Springer: Cham, Switzerland, 2018; pp. 83–102. ISBN 978-3-319-74536-7.
3. Decreto-Lei n.° 152-C/2017; Diário da República n.° 236/2017, 2° Suplemento, Série I de 2017-12-11. pp. 6584-(73)–6584-(88). Available online: https://dre.pt/application/file/a/114336874 (accessed on 26 August 2019).
4. Caetano, M.; Marcelino, F.; Igreja, C.; Girão, I. *Estatísticas e Dinâmicas Territoriais em Portugal Continental 1995–2007–2010–2015 com Base na Carta de Uso e Ocupação do Solo (COS)*; Direção-Geral do Território (DGT) Lisboa: Lisboa, Portugal, 2018.
5. Nunes, L.J.R.; Matias, J.C.O.; Catalão, J.P.S. Biomass in the generation of electricity in Portugal: A review. *Renew. Sustain. Energy Rev.* **2017**, *71*, 373–378. [CrossRef]
6. Ferreira, S.; Monteiro, E.; Brito, P.; Vilarinho, C. Biomass resources in Portugal: Current status and prospects. *Renew. Sustain. Energy Rev.* **2017**, *78*, 1221–1235. [CrossRef]
7. Castillo, C.; Baranzelli, C.; Maes, J.; Zulian, G.; Barbosa, A.; Vandecasteele, I.; Mari-Rivero, I.; Vallecillo, S.; Batista e Silva, F.; Jacobs-Crisioni, C.; et al. *An Assessment of Dedicated Energy Crops in Europe under the EU Energy Reference Scenario 2013. Application of the LUISA Modelling Platform-Updated Configuration 2014*; EUR 27644; European Union: Brussels, Belgium, 2016. [CrossRef]
8. Allen, B.; Kretschmer, B.; Baldock, D.; Menadue, H.; Nanni, S.; Tucker, G. *Space for Energy Crops–Assessing the Potential Contribution to Europe's Energy Future*; IEEP: London, UK, 2014.
9. SilvaPlus Espécies Energéticas Herbáceas e Arbustivas. Available online: http://www.silvaplus.com/pt/culturas-energeticas-florestais/especies-energeticas-herbaceas-e-arbustivas/ (accessed on 16 November 2018).
10. Gominho, J.; Dolores, M.; Lourenço, A.; Fernández, J.; Pereira, H. Cynara cardunculus L. as a biomass and multi-purpose crop: A review of 30 years of research. *Biomass Bioenergy* **2018**, *109*, 257–275. [CrossRef]
11. Mancini, M.; Lanza Volpe, M.; Gatti, B.; Malik, Y.; Moreno, A.C.; Leskovar, D.; Cravero, V. Characterization of cardoon accessions as feedstock for biodiesel production. *Fuel* **2019**, *235*, 1287–1293. [CrossRef]
12. Fernando, A.L.; Costa, J.; Barbosa, B.; Monti, A.; Rettenmaier, N. Environmental impact assessment of perennial crops cultivation on marginal soils in the Mediterranean Region. *Biomass Bioenergy* **2018**, *111*, 174–186. [CrossRef]

13. Mañas, P.; Castro, E.; de las Heras, J. Application of treated wastewater and digested sewage sludge to obtain biomass from Cynara cardunculus L. *J. Clean. Prod.* **2014**, *67*, 72–78. [CrossRef]
14. Chung, J.; Kim, D. Miscanthus as a Potential Bioenergy Crop in East Asia. *J. Crop Sci. Biotechnol.* **2012**, *15*, 65–77. [CrossRef]
15. Barbosa, B.; Boléo, S.; Sidella, S.; Costa, J.; Duarte, M.P.; Mendes, B.; Cosentino, S.L.; Fernando, A.L.; Fernando, A.L. Phytoremediation of Heavy Metal-Contaminated Soils Using the Perennial Energy Crops Miscanthus spp. and Arundo donax L. *Bioenergy Res.* **2015**, *8*, 1500–1511. [CrossRef]
16. Barbosa, B.; Costa, J.; Fernando, A.L.; Papazoglou, E.G. Wastewater reuse for fiber crops cultivation as a strategy to mitigate desertification. *Ind. Crop. Prod.* **2015**, *68*, 17–23. [CrossRef]
17. Alexopoulou, E.; Casler, M.D.; Christou, M.; Clifton-Brown, J.; Copani, V.; Cosentino, S.L.; Montaño, C.M.D.; Donnison, I.S.; Elbersen, H.W.; Farrar, K.; et al. *Perennial Grasses for Bioenergy and Bioproducts*; Academic Press: London, UK, 2018; ISBN 9780128129005.
18. Fernando, A.L.; Godovikova, V.; Oliveira, J.F.S. Miscanthus x giganteus: Contribution to a Sustainable Agriculture of a Future/Present-Oriented Biomaterial. *Mater. Sci. Forum Adv. Mater. Forum II* **2004**, *455*, 437–441. [CrossRef]
19. Oliveira, J.; Duarte, M.; Christian, D.; Eppel-Hotz, A.; Fernando, A.L. Environmental aspects of Miscanthus production. In *Miscanthus: For Energy and Fibre*; Jones, M.B., Walsh, M., Eds.; James & James (Science Publishers): London, UK, 2001; pp. 172–178. ISBN 9781849710978.
20. Zhu, Z.H.; Chao, C.J.; Lu, X.Y.; Xiong, Y.G. *Paulownia in China: Cultivation and Utilization*; Chinese Academy of Forestry Staff, Ed.; Asian Network for Biological Sciences and International Development Research Centre: Singapore, 1986; ISBN 9971845466.
21. El-Showk, N.; El-Showk, S. *The Paulownia Tree an Alternative for Sustainable Forestry*; The Farm—Crop Development.org: Rabat, Morocco, 2003.
22. Direção Nacional das Fileiras Florestais Culturas Energéticas Florestais-Primeira Abordagem do Levantamento da Situação Actual. Available online: http://www2.icnf.pt/portal/florestas/fileiras/resource/doc/biom/biomass-gtce-jun10 (accessed on 16 November 2018).
23. Decreto-Lei n. 565/99; Diário da República n.º 295/1999, Série I-A de 1999-12-21; pp. 9100–9115.
24. Jiménez, L.; Rodriguez, A.; Ferrer, J.L.; Pérez, A.; Angulo, V. Paulownia, a fast-growing plant, as a raw material for paper manufacturing. *Afinidad Barcelona* **2005**, *62*, 100–105.
25. López, F.; Pérez, A.; Zamudio, M.A.M.; De Alva, H.E.; García, J.C. Paulownia as raw material for solid biofuel and cellulose pulp. *Biomass Bioenergy* **2012**, *45*, 77–86. [CrossRef]
26. Zamudio, M.A.M.; Alfaro, A.; de Alva, H.E.; García, J.C.; García-Morales, M.; López, F. Biorefinery of paulownia by autohydrolysis and soda-anthraquinone delignification process. Characterization and application of lignin. *J. Chem. Technol. Biotechnol.* **2015**, *90*, 534–542. [CrossRef]
27. Yadav, N.K.; Vaidya, B.N.; Henderson, K.; Lee, J.F.; Stewart, W.M.; Dhekney, S.A.; Joshee, N. A Review of Paulownia Biotechnology: A Short Rotation, Fast Growing Multipurpose Bioenergy Tree. *Am. J. Plant Sci.* **2013**, *4*, 2070–2082. [CrossRef]
28. Domínguez, E.; Romaní, A.; Domingues, L.; Garrote, G. Evaluation of strategies for second generation bioethanol production from fast growing biomass Paulownia within a biorefinery scheme. *Appl. Energy* **2017**, *187*, 777–789. [CrossRef]
29. Vega, D.J.; Dopazo, R.; Ortiz, L. *Manual de Cultivos Energéticos*; Vigo University: Vigo, Spain, 2010.
30. Rawat, I.; Kumar, R.R.; Mutanda, T.; Bux, F. Biodiesel from microalgae: A critical evaluation from laboratory to large scale production. *Appl. Energy* **2013**, *103*, 444–467. [CrossRef]
31. Raheem, A.; Azlina, W.A.K.G.W.; Tau, Y.H.; Danquah, M.K. Thermochemical conversion of microalgal biomass for biofuel production. *Renew. Sustain. Energy Rev.* **2015**, *49*, 990–999. [CrossRef]
32. Brennan, L.; Owende, P. Biofuels from microalgae-A review of technologies for production, processing, and extractions of biofuels and co-products. *Renew. Sustain. Energy Rev.* **2010**, *14*, 557–577. [CrossRef]
33. El Bassam, N. *Handbook of Bioenergy Crops: A Complete Reference to Species, Development and Applications*, 1st ed.; Earthcan Ltd.: London, UK, 2010; ISBN 9781138975712.
34. Gonçalves, M. Aproveitamento da Biomassa Florestal para Fins Energéticos (o início de um novo ciclo para a floresta portuguesa)—Potencial das Culturas Energéticas. *Feira Nac. Agric.* **2007**.
35. AFG-Asociación Forestal de Galicia ECAS-Cultivos Energéticos en el Espacio Atlántico. Available online: http://enersilva.navegantes.info/areasubir/libros/FolletoECAS.pdf (accessed on 17 October 2017).

36. Almeida, A.R. Avaliação do potencial ecológico para a influência do perímetro de rega do Alqueva. Master's Thesis, Engenharia Agronómica—Agro-pecuária, Universidade Técnica de Lisboa, Lisboa, Portugal, 2009.
37. Edrisi, S.A.; Abhilash, P.C. Exploring marginal and degraded lands for biomass and bioenergy production: An Indian scenario. *Renew. Sustain. Energy Rev.* **2016**, *54*, 1537–1551. [CrossRef]
38. Olsson, L.; Barbosa, H.; Bhadwal, S.; Cowie, A.; Delusca, K.; Flores-Renteria, D.; Hermans, K.; Jobbagy, E.; Kurz, W.; Li, D.; et al. *Chapter 4: Land Degradation*; Special Report on Climate Change and Land of Intergovernmental Panel on Climate Change (IPCC); IPCC: Geneva, Switzerland, 2019.
39. United Nations Convention to Combat Desertification (UNCCD). *United Nations Convention to Combat Desertification in Those Countries Experiencing Serious Drought and/or Desertification, Particularly in Africa*; UNCCD: Paris, France, 1994.
40. Mirzabaev, A.; Wu, J.; Evans, J.; Garcia-Oliva, F.; Hussein, I.A.G.; Iqbal, M.M.; Kimutai, J.; Knowles, T.; Meza, F.; Nedjraoui, D.; et al. *Chapter 3: Desertification*; Special Report on Climate Change and Land of Intergovernmental Panel on Climate Change (IPCC); IPCC: Geneva, Switzerland, 2019; Volume 21.
41. ICNF Cartografia de Apoio ao PDR 2020-Suscetibilidade dos Solos à Desertificação. Available online: http://www2.icnf.pt/portal/pn/biodiversidade/ei/unccd-PT/pancd/o-pancd-2014-2020/cartografia-apoio-pdr2020 (accessed on 7 October 2018).
42. Empresa de Desenvolvimento Mineiro (EDM) Shapefiles sent by email with contaminated areas from EDM 2018.
43. Milbrandt, A.; Overend, R.P. *Assessment of Biomass Resources from Marginal Lands in APEC Economies*; Asia-Pacific Economic Cooperation (APEC): Singapore, 2009.
44. Direção-Geral do Território Carta de Uso e Ocupação do Solo de Portugal Continental Para 2010–(COS2010v1.0). Available online: mapas.dgterritorio.pt/inspire/atom/CDG_COS2015v1_Continente_Atom.xml (accessed on 7 January 2019).
45. ECA. *Desertification in the EU. Background Paper*; European Court of Auditors (ECA)—Guardians of the EU Finances: Luxembourg, 2018.
46. CABI Invasive Species Compendium-Cynara Cardunculus (Cardoon). Available online: https://www.cabi.org/isc/datasheet/17584 (accessed on 7 March 2019).
47. CABI Invasive Species Compendium-Paulownia Tomentosa (Paulownia). Available online: https://www.cabi.org/isc/datasheet/39100 (accessed on 7 March 2019).
48. Laboratório Nacional de Energia e Geologia—LNEG Projeto Converte. Available online: http://geoportal.lneg.pt/geoportal/mapas/index.html?mapa=converte (accessed on 25 September 2019).

© 2020 by the authors. Licensee MDPI, Basel, Switzerland. This article is an open access article distributed under the terms and conditions of the Creative Commons Attribution (CC BY) license (http://creativecommons.org/licenses/by/4.0/).

Article

Production, Characterization, and Evaluation of Pellets from Rice Harvest Residues †

Cristina Moliner *, Alberto Lagazzo, Barbara Bosio, Rodolfo Botter and Elisabetta Arato

Dipartimento di Ingegneria Civile, Chimica e Ambientale (DICCA), Università degli Studi di Genova, Via Opera Pia 15A, 16145 Genova, Italy; alberto.lagazzo@unige.it (A.L.); Barbara.bosio@unige.it (B.B.); Rodolfo.botter@unige.it (R.B.); elisabetta.arato@unige.it (E.A.)
* Correspondence: cristina.moliner@edu.unige.it
† This paper is an extended version of our paper published in 27th European Biomass Conference & Exhibition (EUBCE 2019), Lisbon, Portugal, 27–30 May 2019; pp. 1023–1028.

Received: 20 December 2019; Accepted: 14 January 2020; Published: 18 January 2020

Abstract: Pellets from residues from rice harvest (i.e., straw and husk) were produced and their main properties were evaluated. Firstly, rice straw pellets were produced at lab scale at varying operational conditions (i.e., load compression and wt % of feeding moisture content) to evaluate their suitability for palletization. Successively, rice straw and husk pellets were commercially produced. All the samples were characterized in terms of their main physical, chemical, and physico-chemical properties. In addition, axial/diametral compression and durability tests were performed to assess their mechanical performance. All the analyzed properties were compared with the established quality standards for non-woody pellets. In general, rice straw pellets presented suitable properties for their use as pelletized fuels. Rice husk pellets fell out of the standards in recommended size or durability and thus preliminary treatments might be required prior their use as fuels.

Keywords: rice harvest pellets; palletization; combustion; ash recovery; normative

1. Introduction

Rice is one of the most consumed crops worldwide, with an annual production of 700 million tons according to the Food and Agriculture Organization of the United Nations database [1]. The largest producers in Europe are Italy and Spain with an 80% of the total rice production. Italy (with a total cultivated surface of 220,000 ha) is the largest producer mainly in the Po basin (the Piedmont, Lombardy, Venetia, and the Romagna). The second-largest European rice producer is Spain, with 117,000 ha. Andalucia and Valencia are the main rice-producing regions, the latter harboring a more stable water supply which benefits the production.

The main by-products of rice harvest are rice straw and rice husk which represent an environmental and economic problem for the farmers. Rice straw, for example, is usually eliminated by uncontrolled burning with harmful consequences related to air, flora, and fauna pollution in wetlands. As an alternative, rice waste is abandoned or sunk, being decomposed in fields, causing major die-off of fish and other aquatic fauna in deeper areas [2]. Currently, different European moratoria permitting uncontrolled burning are being applied, since no reliable solutions are being implemented.

In this framework, the EU-funded project LIFE LIBERNITRATE [3] proposes a synergic application of efficient rice waste management to treat nitrate problems in over-cropping areas. This is achieved through the controlled combustion of rice straw and the successive production of silica-based adsorbents from the obtained ashes, which are applied as filters to decrease the high concentration of nitrates in waters.

Combustion efficiencies are optimized using rice straw in the form of pellets. Raw materials are light and irregular which results in difficulties for their collection, transport, storage, and operation.

Bulk density of rice straw and rice husk are around 60 kg/m^3 [4] and 90 kg/m^3 [5], respectively, with recommended pellet values in the range 600–800 kg/m^3 [6].

The quality of pellets depends both on the properties of the feeding and on the operational conditions for its production. For that reason, understanding the physical, chemical, and mechanical properties of the initial feedstock becomes essential to evaluate their behavior during combustion. The control of the moisture content plays an important role: Within the optimal percentage (10–14%) it serves as binding agent, facilitates heat transfer, and promotes self-bonding of individual particles in the pellet [7,8]. However, an increase of this optimal value could decrease inter-particle forces that might cause swelling and disintegration of pellets [9]. Also, the assessment of the applied load compression applied must be careful evaluated: Low void fractions inside the pellet can prevent oxygen diffusion into the inner part restricting the combustion only to the outer surface [10].

On the other hand, the chemical composition of the feedstock provides indications on the potential emissions and operational problems that could occur on the device helping in their prevention. In this sense, special attention has to be paid to sintering and fouling owing to the high content of silica present in rice straw (RS) and rice husk (RH): Their ash analysis show concentrations around 70 mg/kg and 100 mg/kg of silica, respectively [11]. This implies increased risks for sintering and dust emissions and a lowering on the heating value [12]. Also, emissions associated with biomass combustion, such as NO_X, need to be evaluated [13].

Finally, storage, transport, and processing of pellets can lead to a loss of their mechanical properties. A significant loss on mechanical strength might increase the level of dust and consequently, the risk of fire and explosions as well as being a health hazard for workers [14]. Moreover, high strength values can result in extremely hard pellets causing difficulties during solids feeding due to bridging.

Within this framework, the main physical, chemical, and mechanical properties of single press and commercially produced pellets using rice straw and rice husk were investigated. Their properties were evaluated by comparing them with the standard limits for non-woody pellets and their suitability as combustion feeding materials in the framework of LIFE LIBERNITRATE was discussed.

2. Materials and Methods

2.1. Production of Rice Straw Pellet with a Single Press

Rice straw was provided by Società Agricola Abbazia (45°22′59.34″ N 9°58′22.3″ E), Orzinuovi (Brescia, Italy). All tests were performed at the *Laboratorio di Ingegneria dei Materiali* (DICCA, UNIGE). Rice straw was initially cut to get short fibers and then grounded using a chopper with rotating blades for 1 min at 4000 rpm. Sieves with an aperture of 1 mm, 0.85 mm, 0.3 mm, and 0.15 mm were used to obtain the granulometric distribution of the straw powders. The obtained particle size distribution after grinding is presented in Figure 1. A particle size range between 0.5 to 0.7 mm is suggested in literature [15]. Sizes larger than 1 mm could act as preferential breaking points. On the contrary, small particles with large surface area could increase density values and lead to tougher pellets. In this work, only the fraction in the range 0.15–1 mm was used to provide a homogenous sample according to established criteria.

The pellet press consisted of a cylindrical 8 mm diameter die of steel, a fitted piston, and a hydraulic press. The compression of the material was obtained by pressing, with the hydraulic press, against a fixed backstop. A fixed quantity of 0.25 g of rice straw was used. A constant force detected by a 150 kN load cell was applied for one minute at room temperature. After compression, the pressure was released, the bottom of the cylinder removed, and the pellet pressed out of the cylinder. For determination of experimental precision, four replicates were carried out. Pellets had in general a cylindrical shape and are about 8 mm in diameter.

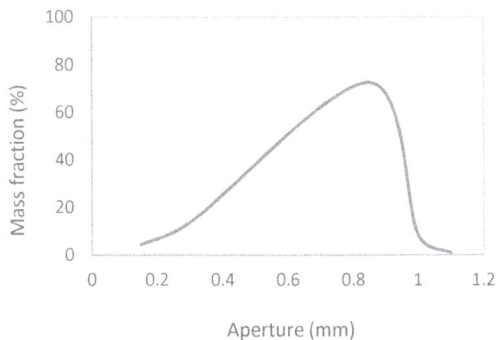

Figure 1. Particle size distribution of grounded rice straw.

Different compression forces (1–10 tons) and amount of added distilled water (40–60 mg) were used. Figure 2 shows examples of the produced pellets.

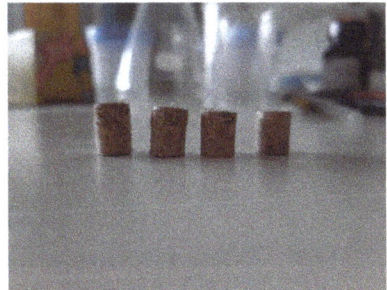

Figure 2. Pellets produced using the single press.

2.2. Production of Commercial Pellets

The available 10 kg of RS and 23 kg of RH were cut and reduced to a length less than 3–4 cm. This feedstock was sent to a commercial pellet producer, Novapellet (Novara, Italy) [16]. A size refinement was performed to obtain dimensions of 6 mm by 2 mm by 1 mm in order to ensure proper densification. The moisture content of both materials was within the range 10–14% wt. In accordance to the experimental procedure at lab-scale, no binder was used in the process. Pellets were produced at a temperature range of 60 °C to 70 °C with the commercial pelletizer N-PICO (Novapellet) with a maximum capacity of 200 kg/h and nominal power of 10 kW [17]. Figure 3 shows the pelletizer (a) and the pelletized RS (b).

(a)　　　　　　　　　　　　　　　(b)

Figure 3. N-Pico pelletizer (**a**) and pelletized rice straw (RS) (**b**).

2.3. Characterization of Pellets

Physical, chemical and mechanical properties of all samples were evaluated at the *Laboratorio di Ingegneria dei Materiali* (DICCA, UNIGE) and compared with standards [17]. The produced commercial pellets (6 kg of RS and 2 kg of RH) were stored for two weeks before characterization. Firstly, the specimens were sub-divided, and a representative sample was selected. The following characteristics were evaluated.

2.3.1. Physical Properties

Twenty randomly chosen pellets were used to determine the length (L) and the diameter (d) of the pellets (mm) using a caliper (precision 1/20 mm). Particle density (ρ_p) (UNE-EN 15,150) [18] was determined by weighing the individual pellet using a precision digital balance and calculating its volume based on the length and diameter previously determined. The pellet shape was assumed to be cylindrical. Bulk density (ρ_b) (UNE-EN 15103) [19] was calculated from the volume and weight of pellets using a laboratory balance and a graduated cylinder. Porosity (ε_0) was then calculated from particle and bulk density as:

$$\varepsilon_0(-) = 1 - \frac{\rho_b}{\rho_p} \quad (1)$$

2.3.2. Chemical Properties

Proximate analysis (PA) was calculated using a muffle furnace following the standards [20–22]. Ultimate analysis (UA) was calculated using a Vario MACRO Cube Elementar Analyzer. The heat release by the unit of mass during the combustion of RS and RH was evaluated through the higher heating value (HHV) using the empirical expression [23]:

$$HHV\ (MJ/Kg) = 0.3536\ FC + 0.1559\ VM - 0.0078\ A \quad (2)$$

With the fixed carbon (FC), volatiles (VM), and ash (A) content in mass percentage.

2.3.3. Mechanical Properties

Axial and diametral compression tests were carried out with a Hausfield universal mechanical tester. The equipment was set at a load cell capacity of 5 kN at a rate of 20 mm/min (Figure 4). The corresponding strength values were determined as the minimum force at which pellet broke. That way, the resistance of pellet to deformation and breakage when compressed vertically and horizontally were evaluated which are likely scenarios during transport and storage.

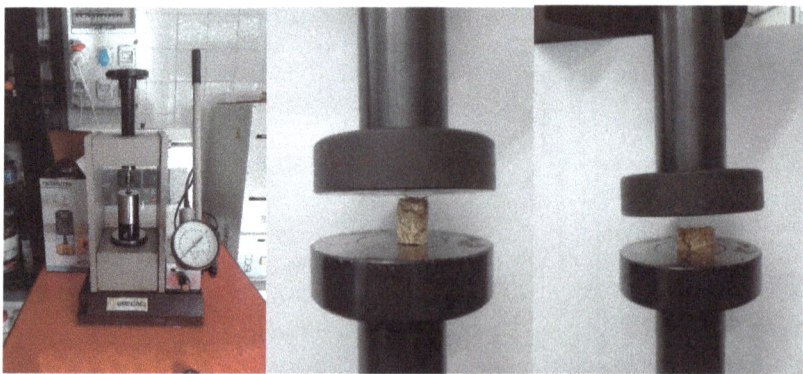

Figure 4. Axial compressive test.

The axial compressive strength (σ_{AC}) was calculated using:

$$\sigma_{AC}(MPa) = \frac{4P}{\pi d^2} \quad (3)$$

where d is the diameter of the pellet (mm) and P (N) is the failure force.

The elongation at break (ε) was defined as:

$$\sigma(\%) = \frac{s}{l} 100 \quad (4)$$

with s the length at break (mm) and l the initial length (mm).

The diametral compressive strength (σ_{DC}) was calculated as:

$$\sigma_{DC}(MPa) = \frac{2P}{\pi L d} \quad (5)$$

where L and d are the length (mm) and diameter (mm) of the pellet and P (N) is the failure force.

Durability (D) tests were performed on commercial RS and RH pellets to measure their friability and their possibility of breaking apart during processing. With this purpose, an initial quantity of material was weighed (W_i) and placed in a 0.25 L recipient. Then, the recipient was tumbled at 50 rpm for 10 min and sieved afterwards. Finally, the mass of pellets retained on the sieve after tumbling was measured (W_f). The final weight was compared against its initial value and the difference was expressed as:

$$D(\%) = \frac{W_f}{W_i} 100, \quad (6)$$

2.3.4. Structural Properties

Structural analyses were performed on lab-scale pellets using an optical microscope Nikon LV 100 at 50× in reflection.

All the described analyses were run in triplicate, except where indicated diversely. Averaged results were compared with the standard values established in the non-woody pellet norm (UNE-EN ISO 17225-6) [24]. Table 1 gathers the most representative properties used in this study.

Table 1. Guidelines for specifications of non-woody pellets.

Parameter	Guidelines
Length (mm)	3.15 < length < 40
Diameter (mm)	Die diameter ± 1
Bulk density (kg/m^3)	>600
Moisture content (wt %)	<10
Durability (%)	>97.5

3. Results

3.1. Lab-Scale Pellets

Palletization is a challenging task when non-conventional wood pellets are produced due to their particular properties in terms of bulk density, geometry, and composition. In order to evaluate the suitability of rice straw to produce adequate pellets, pellets were initially obtained using a single press as described in Section 2.1.

Two different operational conditions were varied: Compression force (from 1 to 10 tons) and percentage (in mass) of added water (40 mg and 60 mg, corresponding to a 14% and 20% wt moisture content, respectively). However, the highest moisture content presented processing difficulties and so only the moisture value of 14% wt was considered. Physical (dimensions and density), mechanical (resistance to axial compression), and structural (distribution of fibers and presence of silica) properties were evaluated.

3.1.1. Physical Properties

All the pellets presented a cylindrical shape with around 8 mm of diameter and 10 mm length. Samples produced at 6 and 7 tons showed the best performances and were reproduced to obtain a representative amount for their further characterization. Table 2 presents the averaged values of diameter, length, and density of all the repeated samples.

Table 2. Averaged results of physical properties of RS pellets produced at lab conditions.

Samples	l (mm)	d (mm)	ρ_p (kg/m^3)
RS_6 tons	10.1 ± 0.3	8.18 ± 0.02	980 ± 20
RS_7 tons	10.8 ± 0.6	8.14 ± 0.03	930 ± 49

3.1.2. Chemical Properties

Table 3 shows the PA, UA, and HHV of lab-scale RS pellets.

Table 3. Proximate analysis (PA), ultimate analysis (UA), and higher heating value (HHV) of lab-scale RS pellets.

Physical	RS
Proximate analysis (wt %)	
Moisture	7
Fixed Carbon	13
Volatile Matter	71
Ash	9
Ultimate analysis (wt %) (ash-free)	
Carbon	46.8
Hydrogen	5.1
Nitrogen	0.6
Oxygen	47.2
Sulfur	0.3
HHV (MJ/kg)	15.6

3.1.3. Mechanical Properties

Table 4 shows the averaged results of the axial compression tests. Samples obtained at 1, 8, 9, and 10 tons did not provide stable pellets and could not be used for their mechanical characterization.

Table 4. Averaged results of axial compression tests for RS pellets at lab conditions.

Samples	σ_{AC} (MPa)	ε (%)
RS_2 tons	56.5 ± 7.7	0.70 ± 0.04
RS_3 tons	67.7 ± 2.9	0.72 ± 0.02
RS_4 tons	58.2 ± 10.4	0.66 ± 0.04
RS_5 tons	61.2 ± 3.1	0.68 ± 0.05
RS_6 tons	58.4 ± 7.0	0.69 ± 0.06
RS_7 tons	54.4 ± 18.9	0.72 ± 0.15

3.1.4. Structural Properties

Microscopic analyses were performed on lab-scale pellets to evaluate the distribution of fibers in the matrix. Figure 5 shows the fractions 0.3–0.85 mm (a) and 0.15–0.3 mm (b) of the straw powders at the magnification of 50×. Particles appear as straight rods with diameters about 0.1–0.2 mm (fraction

0.3–0.85 mm) and as irregular needles with the presence of rectangular pieces in the fraction at lower dimensions (b). In both figures the presence of silica is clearly shown by the light reflection on the surface of the samples.

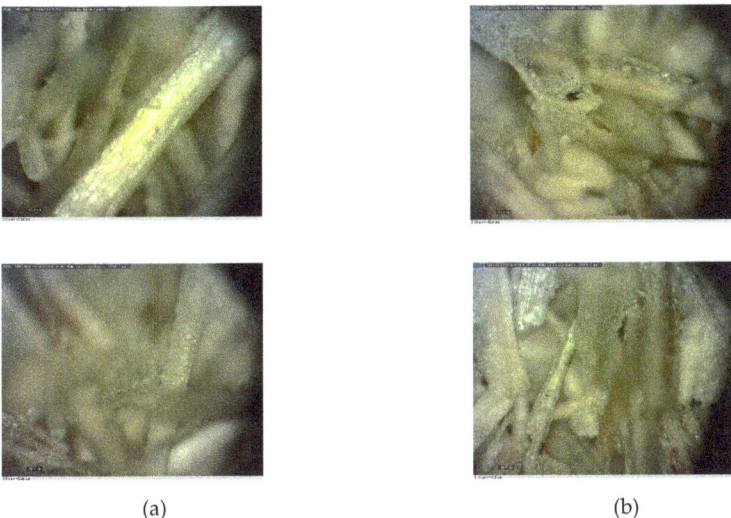

Figure 5. Fraction 0.3–0.85 mm (**a**) and 0.15–0.3 mm (**b**) of the straw powders at the magnification of 50× at optical microscope.

3.2. Commercial Rice Straw and Rice Husk Pellets

After the confirmation of the suitability of RS for palletization without the use of any binder, commercial pellets were produced as described in Section 2.2. Afterwards, their main properties were evaluated.

3.2.1. Physical Properties

The size of pellets greatly influences feeding and combustion processes. In general, shorter pellets promote a more continuous inflow, as solid obstructions are more unlikely to occur. Longer pellets can be more easily broken during storage, handling, and processing but can also provoke bridging problems. Thinner pellets allow a more uniform and efficient combustion rate.

As shown in Table 1, the norm [24] sets the diameter limit at die diameter ± 1 mm and a length range between 3.15 and 40 mm. Table 5 shows the dimensions of RS and RH pellets. Die diameter was 6 mm in all cases.

Table 5. Averaged dimensions of RS and rice husk (RH) pellets.

Dimensions	RS	RH
L (mm)	17.7 ± 1.0	8.5 ± 2.5
d (mm)	6.3 ± 0.1	6.4 ± 0.1
L/d	2.8 ± 0.1	1.3 ± 0.1

Visual observations of pellets showed differences in size (RS—Figure 6a; RH—Figure 6b). RH pellets showed lower length values than RS pellets and a more fragile behavior during its manipulation.

Figure 6. RS (a) and RH (b) pellets.

In any case, the dimensions of both pellets were within the range established by the norm. A slight increase in the diameter indicated a pellet expansion after its release from the die, an effect also shown in other materials [25]. RH pellets had the lowest L/d ratio and both materials presented values similar to other biomass pellets (e.g. black poplar = 3.64; holm oaks = 2.18; and leaves of olive trees = 1.76 [12]).

Bulk density (ρ_b) greatly influences storage and transport processes as well as combustion efficiencies. Particle density (ρ_p) might affect combustion efficiency as highly packed materials might prevent oxygen to access within the particles. The norm [24] suggests a value $\rho_b > 600$ kg/m^3. No guidelines are established for ρ_p with a recommended value of 1200 kg/m^3.

Porosity (ε_0) influences heat and mass transfer phenomena and thus the combustion parameters such as burning rate, conversion efficiency, and emissions [26]. Table 6 shows the densities and porosities of RS and RH pellets.

Table 6. Averaged density and porosity of RS and RH pellets.

Physical	RS	RH
ρ_b (kg/m^3)	606 ± 20	505 ± 15
ρ_p (kg/m^3)	1305 ± 34	1112 ± 38
ε_0 (−)	0.536	0.544

RS was just within the range established by the norm and above the recommended particle density value. On the contrary, RH fell out of the optimal ranges presenting lower values than those recommended in all cases. These results are in accordance with other studies in which RS pellets presented higher density values than RH pellets [11].

In summary and regarding physical properties of pellets, RS was considered as a suitable material according to the standards whereas RH presented length values close to the minimum permitted and a density out of the recommended range.

3.2.2. Chemical Properties

Chemical energy is stored in biomass in two forms: Volatiles and fixed carbon. VM is defined as the released gas by heating and FC is the mass remaining after this release, excluding the ash and moisture contents (MC).

The content of volatiles also influences the ignition temperature which is the temperature above which combustion reactions become self-sustaining. In general, the ignition temperature is lower for higher VM contents.

The composition of pellets described by their PA was carried out using a muffle furnace following the standards [20–22]. The energy developed during the combustion of RS and RH pellets was evaluated through the HHV. Table 7 lists the PA, UA, and HHV (calculated by Equation 2) of RS and RH pellets.

Table 7. PA, UA, and HHV of RS and RH pellets.

Physical	RS	RH
Proximate analysis (wt %)		
Moisture	15	17
Fixed Carbon	11	11
Volatile Matter	60	63
Ash	14	9
Ultimate analysis (wt %)—ash-free		
Carbon	43.9	52.1
Hydrogen	4.7	6.2
Nitrogen	0.6	2.2
Oxygen	50.5	39.3
Sulfur	0.3	0.2
HHV (MJ/kg)	13.2	13.8

Similar compositions were found for both feedstocks. A slightly higher value of volatile matter for RH pellets led to a slightly higher heating value. The ash content, main objective of LIFE LIBERNITRATE, was higher in RS pellets identifying them as a better choice in terms of material efficiency. Comparing the composition before and after densification, slightly higher volatile contents were found for pellets with respect to raw straw (VM_{raw} = 57.5% [27]). They also presented higher HHV (HHV_{raw} = 11.6 MJ/kg [27]) which indicated an increased energy emission during the combustion due to densification. On the contrary, the ash content slightly decreased with respect to raw materials (A_{raw} = 18.8% [27]). This suggested that a higher quantity of pellets will need to be processed in order to achieve the same quantitative efficiencies. HHV values were found to be lower than commercial wood pellets for both feedstocks (HHV_{wood} = 19.5 MJ/kg [28]). An increase in these values could be achieved by mixing them with other available residues ($HHV_{persimmon\ stems}$ = 17.3 MJ/kg [29] or $HHV_{appletrees}$ = 17.2 MJ/kg [30]). That way, the increase in the quality of the final pellet would permit using these residues with low calorific values in energy recovery processes. Sulfur and nitrogen contents remained in a very low percentage (below 0.3% wt and 2.5% wt, respectively). This low sulfur and nitrogen content results in low NO_X and SO_X emissions and contributing to minimize environmental damage.

3.2.3. Mechanical Properties

Axial and diametral compressive tests were performed on RS and RH samples at the Laboratorio di Ingegneria dei Materiali (DICCA, UNIGE). Ten specimens were tested except for the RH in the axial compression test where only one specimen was tested due to its extreme fragility. The load at fracture was obtained from the stress–strain curve (as shown in Figure 7a for RS and Figure 7b for RH) from which the main mechanical properties were calculated.

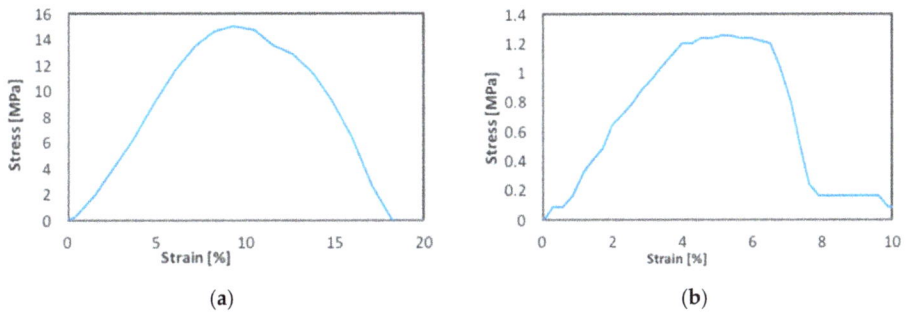

Figure 7. Axial compressive stress-strain curves for RS pellets (**a**) and RH pellets (**b**).

The mean values for axial compression strength (σ_{AC}, MPa), deformation at break (ε, %), and diametral compression strength (σ_{DC}, MPa) for RS and RH pellets are listed in Table 8. The high values of standard deviation were attributed to the heterogeneous nature of the biomass samples.

Table 8. Averaged results of axial and diametral compression tests for RS and RH pellets.

	RS	RH
σ_{AC} (MPa)	11.0 ± 2.0	1.17
ε (%)	7.5 ± 3.2	8.12
σ_{DC} (MPa)	7.8 ± 2.5	4.26 ± 2.57

RS pellets presented higher mechanical resistance in both directions in comparison to RH pellets. This fact indicated a stronger bonding as a result of a more efficient densification process. Mechanical strength varies significantly with direction of load and feedstock. Compressive stress was higher in the axial orientation for RS pellets whereas RH presented higher values in the diametral orientation. Because the diametral compression is an indirect measure of the tensile properties, a higher value of σ_{DC} with respect to σ_{AC} is probably due to the prevalent orientation of the fibers along the longitudinal direction of cylinder. Mechanical strength of pellets is highly related to their physical and chemical properties. The higher moisture content, lower density, and lower length of pellets resulted in lower mechanical strength values. This might be attributed to the lower adhesion forces between particles, in accordance with other works [11]. Compression strength values were consistent with other pellet values like eucalyptus pellets (σ_{AC} = 5.5 MPa; σ_{DC} = 9.9 MPa), wood pellets (σ_{AC} = 4.1 MPa; σ_{DC} = 9.5 MPa), or sunflower pellets (σ_{AC} = 7.5 MPa; σ_{DC} = 6.5 MPa) [31].

Following these results, it is concluded that RS pellets present an adequate mechanical strength whereas RH pellets were too fragile and not resistant to mechanical forces. This fact could lead to increased dust emissions and higher risks of fire and explosion and so they were not considered as a valid feedstock.

Durability is one of the most important factors defining the quality of pellets. Low values can lead to problems such as blocking in the feeding system, dust emissions, and higher risk of explosions during handling and storage [14]. The norm [24] recommends a value over 97.5%. The calculated durability values of RS and RH pellets were 99.8% and 91.8%, respectively. Durability of RS pellets was higher than RH pellets probably due to their different physical characteristics. RS is more flexible, and particles can be in closer contact than RH particles during palletization favoring a correct densification. The durability of RS pellets was well over the limit value according to the standards. On the contrary, the durability of RH pellets was slightly lower than the recommended value and so failed to accomplish the established requirements.

Overall, commercial RS pellets showed adequate properties according to the quality standards. However, the low HHV values suggested that mixing with other materials could improve their calorific value. On the contrary, commercial RH pellets fell out of the standards in several tests such as recommended size or durability. Different operational conditions during pellet production or mixing of rice husk with other biomass could permit using these residues for pellet production to take advantage of the discussed densification processes.

In summary, an adequate densification of biomass could reduce costs and operational difficulties with handling, transportation, storage, and utilization of low bulk density materials. However, the feedstock and the operational parameters for pellets production need to be carefully examined to determine the technical suitability of the produced materials and potential required improvements. In addition to these technical considerations, further economic and environmental studies will be performed in the context of LIFE LIBERNITRATE to evaluate the global impact on the project.

4. Conclusions

Pellets from residues from rice harvest (i.e., straw and husk) were produced and their main properties were evaluated. Rice straw pellets were initially produced at lab scale at varying operational conditions (i.e., load compression and wt % of feeding moisture content) to evaluate their suitability for palletization without the use of binders. Afterwards, pellets from rice straw and husk pellets were commercially produced without the use of a binder.

All the samples were characterized in terms of their main physical (density, dimension, porosity), physico-chemical (proximate and ultimate analysis, higher heating value), and structural properties. In addition, axial and diametral compression tests and durability tests were used to assess their mechanical performance. All the analyzed properties were compared with the established quality standards for non-woody pellets.

In summary, rice straw pellets presented suitable properties for their use as pelletized fuels in the context of the LIFE LIBERNITRATE project. Rice husk pellets fell out of the standards in several tests such as recommended size or durability and their pre-treatment might be necessary for their use as fuels.

Author Contributions: Conceptualization, methodology, writing draft, formal analysis C.M. and A.L. writing-review and editing, B.B., R.B. supervision, funding acquisition, E.A. All authors have read and agreed to the published version of the manuscript.

Funding: This work was funded through the LIFE LIBERNITRATE project (LIFE16 ENV/ES/000419).

Acknowledgments: Beatrice Ghinello, Matteo Toscanini and Alice Gabuti are acknowledged in the framework of their thesis work.

Conflicts of Interest: The authors declare no conflict of interest.

References

1. Food and Agriculture Organization of the United Nations. *World Agriculture: Towards 2015–2030*; Summary Report; Earthscan Publications Ltd: London, UK, 2002.
2. Moliner, C.; Badia, J.D.; Bosio, B.; Arato, E.; Kittikorn, T.; Strömberg, E.; Ribes-Greus, A. Thermal and thermo-oxidative stability and kinetics of decomposition of PHBV/sisal composites. *Chem. Eng. Commun.* **2018**, *205*, 226–237. [CrossRef]
3. Moliner, C.; Teruel-Juanes, R.; Primaz, C.; Badia, J.; Bosio, B.; Campíns-Falcó, P.; Morán, J. Reduction of Nitrates in Waste Water through the Valorization of Rice Straw: LiFE libernitrate Project. *Sustainability* **2018**, *10*, 3007. [CrossRef]
4. Moliner, C.; Curti, M.; Bosio, B.; Arato, E.; Rovero, G. Experimental Tests with Rice Straw on a Conical Square-Based Spouted Bed Reactor. *Int. J. Chem. React. Eng.* **2018**, *13*, 351–358. [CrossRef]
5. Mansaray, K.G.; Ghaly, A.E. Physical and Thermochemical Properties of Rice Husk. *Energy Sources* **1997**, *19*, 989–1004. [CrossRef]
6. Obernberger, I.; Thek, G. Physical characterisation and chemical composition of densified biomass fuels with regard to their combustion behaviour. *Biomass Bioenergy* **2004**, *27*, 653–669. [CrossRef]
7. Kaliyan, N.; Morey, R.V. Factors affecting strength and durability of densified biomass products. *Biomass Bioenergy* **2009**, *33*, 337–359. [CrossRef]
8. Larsson, S.H.; Thyrel, M.; Geladi, P.; Lestander, T.A. High quality biofuel pellet production from pre-compacted low density raw materials. *Bioresour. Technol.* **2008**, *99*, 7176–7182. [CrossRef] [PubMed]
9. Serrano, C.; Monedero, E.; Lapuerta, M.; Portero, H. Effect of moisture content, particle size and pine addition on quality parameters of barley straw pellets. *Fuel Process. Technol.* **2011**, *92*, 699–706. [CrossRef]
10. Ewida, K.T.; El-Salmawy, H.; Atta, N.N.; Mahmoud, M.M. A sustainable approach to the recycling of rice straw through pelletization and controlled burning. *Clean Technol. Environ. Policy* **2006**, *8*, 188–197. [CrossRef]
11. Yang, I.; Kim, S.H.; Sagong, M.; Han, G.S. Fuel characteristics of agropellets fabricated with rice straw and husk. *Korean J. Chem. Eng.* **2016**, *33*, 851–857. [CrossRef]
12. Zamorano, M.; Popov, V.; Rodríguez, M.L.; García-Maraver, A. A comparative study of quality properties of pelletized agricultural and forestry lopping residues. *Renew. Energy* **2011**, *36*, 3133–3140. [CrossRef]

13. Osman, A.I. Mass spectrometry study of lignocellulosic biomass combustion and pyrolysis with NOx removal. *Renew. Energy* **2020**, *146*, 484–496. [CrossRef]
14. Temmerman, M.; Rabier, F.; Jensen, P.D.; Hartmann, H.; Böhm, T. Comparative study of durability test methods for pellets and briquettes. *Biomass Bioenergy* **2006**, *30*, 964–972. [CrossRef]
15. ISO E. Solid Biofuels—Fuel Specifications and Classes—Part 1: General Requirements. 2014. Available online: https://www.iso.org/standard/59456.html (accessed on 18 January 2020).
16. Novapellet [online]. Available online: http://novapellet.it/inglese/index.htm (accessed on 3 May 2019).
17. Moliner, C.; Lagazzo, A.; Bosio, B.; Botter, R.; Arato, E. Production and characterisation of pellets from rice straw and rice husk. In Proceedings of the European Biomass Conference and Exhibition, Lisbon, Portugal, 27–30 May 2019; pp. 1023–1028.
18. UNE-EN 15150. Solid Biofuels. Determination of Particle Density. 2012. Available online: https://www.une.org/encuentra-tu-norma/busca-tu-norma/norma/?Tipo=N&c=N0049721 (accessed on 18 January 2020).
19. UNE-EN 15103. Solid Biofuels. Determination of Bulk Density. 2010. Available online: https://www.une.org/encuentra-tu-norma/busca-tu-norma/norma/?c=N0046378 (accessed on 18 January 2020).
20. UNE-EN 14774-3. Solid Biofuels—Determination of Moisture Content—Oven Dry Method—Part 3: Moisture in General Analysis Sample. 2010. Available online: https://www.une.org/encuentra-tu-norma/busca-tu-norma/norma/?c=N0045728 (accessed on 18 January 2020).
21. DIN-EN 15148. Solid Biofuels—Determination of Volatile Content—Oven Dry Method—Part 5: Volatiles in General Analysis 2010. Available online: https://infostore.saiglobal.com/en-us/Standards/DIN-EN-15148-2010-448122_SAIG_DIN_DIN_1010653/ (accessed on 18 January 2020).
22. DIN-EN 14775. Solid Biofuels—Determination of Ash Content—Oven Dry Method—Part 4: Ash in General Analysis 2010. Available online: https://www.beuth.de/en/standard/din-en-14775/165912452 (accessed on 18 January 2020).
23. Parikh, J.; Channiwala, S.A.; Ghosal, G.K. A correlation for calculating HHV from proximate analysis of solid fuels. *Fuel* **2005**, *84*, 487–494. [CrossRef]
24. UNE-EN ISO 17225-6. Solid Biofuels, Fuel Specifications and Classes. Part 6: Graded Non-Woody Pellets. 2014. Available online: https://www.iso.org/obp/ui/#iso:std:iso:17225:-6:ed-1:v1:en (accessed on 18 January 2020).
25. Said, N.; García-Maraver, A.; Zamorano, M. Influence of densification parameters on quality properties of rice straw pellets. *Fuel Process. Technol.* **2015**, *138*, 56–64. [CrossRef]
26. Igathinathane, C.; Tumuluru, J.S.; Sokhansanj, S.; Bi, X.; Lim, C.J.; Melin, S.; Mohammad, E. Simple and inexpensive method of wood pellets macro-porosity measurement. *Bioresour. Technol.* **2010**, *101*, 6528–6537. [CrossRef] [PubMed]
27. Moliner, C.; Bosio, B.; Arato, E.; Ribes, A. Thermal and thermo-oxidative characterisation of rice straw for its use in energy valorisation processes. *Fuel* **2016**, *180*, 71–79. [CrossRef]
28. Bove, D.; Moliner, C.; Curti, M.; Baratieri, M.; Bosio, B.; Rovero, G.; Arato, E. Preliminary tests for the thermo-chemical conversion of biomass in a spouted bed pilot plant. *Can. J. Chem. Eng.* **2018**, *97*, 57–66. [CrossRef]
29. Moliner, C.; Aguilar, K.; Bosio, B.; Arato, E.; Ribes, A. Thermo-oxidative characterisation of the residues from persimmon harvest for its use in energy recovery processes. *Fuel Process. Technol.* **2016**, *152*, 421–429. [CrossRef]
30. Bove, D. Experimental studies on the gasification of the residues from prune of apple trees with a spouted bed reactor. In Proceedings of the European Biomass Conference and Exhibition, Amsterdam, The Netherlands, 6–9 June 2016; pp. 858–862.
31. Williams, O.; Taylor, S.; Lester, E.; Kingman, S.; Giddings, D.; Eastwick, C. Applicability of Mechanical Tests for Biomass Pellet Characterisation for Bioenergy Applications. *Materials* **2018**, *11*, 1329. [CrossRef] [PubMed]

© 2020 by the authors. Licensee MDPI, Basel, Switzerland. This article is an open access article distributed under the terms and conditions of the Creative Commons Attribution (CC BY) license (http://creativecommons.org/licenses/by/4.0/).

Article

Simulating the Effect of Torrefaction on the Heating Value of Barley Straw

Dimitrios K. Sidiras *, Antonios G. Nazos, Georgios E. Giakoumakis and Dorothea V. Politi

Laboratory of Simulation of Industrial Processes, Department of Industrial Management and Technology, School of Maritime and Industrial Studies, University of Piraeus, 80 Karaoli & Dimitriou, GR 18534 Piraeus, Greece; anazos@yahoo.gr (A.G.N.); ggiakoum@unipi.gr (G.E.G.); doritapoliti@yahoo.gr (D.V.P.)
* Correspondence: sidiras@unipi.gr; Tel.: +30-210-4142360

Received: 29 December 2019; Accepted: 5 February 2020; Published: 7 February 2020

Abstract: Many recent studies focused on the research of thermal treated biomass in order to replace fossil fuels. These studies improved the knowledge about pretreated lignocellulosics contribution to achieve the goal of renewable energy sources, reducing CO_2 emissions and limiting climate change. They participate in renewable energy production so that sustainable consumption and production patterns can by ensured by meeting Goals 7 and 12 of the 2030 Agenda for Sustainable Development. To this end, the subject of the present study relates to the enhancement of the thermal energy content of barley straw through torrefaction. At the same time, the impact of the torrefaction process parameters, i.e., time and temperature, was investigated and kinetic models were applied in order to fit the experimental data using the severity factor, R_0, which combines the effect of the temperature and the time of the torrefaction process into a single reaction ordinate. According to the results presented herein, the maximum heating value was achieved at the most severe torrefaction conditions. Consequently, torrefied barley straw could be an alternative renewable energy source as a coal substitute or an activated carbon low cost substitute (with/without activation treatment) within the biorefinery and the circular economy concept.

Keywords: barley straw; torrefaction; higher heating value; severity factor; sustainable development; enhancement factor; energy yield

1. Introduction

Fossil fuels, such as oil, gas and coal, are the world's primary energy sources. However, these resources have limited reserves that will only be sufficient for the next 50 years [1,2]. Fossil fuels also make a significant contribution to the environmental impact of carbon dioxide emissions. Reductions in CO_2 emissions through the use of renewable energy aim to reduce greenhouse gas emissions from 1990 to 2030 by 40% and reduce greenhouse gas emissions by 80%–95% by 2050 [3–6]. It is imperative that we use natural resources to achieve the goals of the 2030 agenda so that the needs of the present situation and the satisfaction of future ones will be covered [7].

The use of renewable energy sources and particularly, of biomass, is important due to the economic factor, since the use of cheaper energy resources is more selective, enhancing the conservation of clean environment, as natural, abundant and reusable means of producing thermal energy are mostly in use [8]. Biomass is becoming more promising due to a set of features that allow fossil fuel substitution, thereby reducing greenhouse gas emissions [9,10]. Biomass is one of the major sources of renewable energy, accounting for about 10% of total primary energy and 78% of total renewable energy [11]. Thus, the need to utilize non-wood lignocellulosic biomass as a promising raw material for future renewable fuels is widely recognized, since the latter is in abundance [12,13]. Lignocellulosic biomass, while presenting several positive features, is, however, associated with various deficiencies, such as structural heterogeneity, non-uniform physical properties, low energy density, hygroscopic nature and low bulk density. All of these features create difficulties in transport, handling, storage, and conversion [14–18]. These features impede the use of biomass in the replacement of fossil fuels for energy production. Therefore, biomass must be pre-treated before it can be used in any thermochemical process. The torrefaction process is an appropriate such pretreatment method that removes many of the above limitations associated with crude biomass. The torrefaction process is the partial pyrolysis of the biomass which is carried out usually under atmospheric pressure over a small temperature range of 200–300 °C and under an inert environment [19–21]. The process is usually performed at a low heating rate, which gives a higher yield of solid product [22]. A great motivation for torrefaction is the maximization of solid performance, which is not achieved with pyrolysis. During the torrefaction process, three major phases, namely decomposition, rehabilitation and depolymerization, occur. The process releases concentrated hydrocarbons, hydrogen, oxygen and some of the carbon content of the biomass as carbon monoxide and carbon dioxide [23]. During the torrefaction process, drying is considered to be the most destructive between the intramolecular hydrogen bonds, C–O and C–H [24]. This results in the significant emissions of hydrophilic and oxygenated pollutants, and hydrophilic and oxygenated compounds, forming a black hydrophobic energy-dense product.

The main motivation of torrefaction is to improve the quality of biomass fuels and make it more suitable for energy use. The torrefied biomass can be applied in briquetting, pelletizing, gasification and thermal energy cogeneration [5,25,26]. Biomass torrefaction destroys biomass strength and fibrous structure and also increases energy density. Many studies have concluded that torrefied biomass can avoid many constraints associated with crude biomass because it produces moisture-free hydrophobic solid products [27], reduces the O/C ratio [13], decreases milling energy [15,28], increases energy density [29], increases bulk density and simplifies storage and transport [30]. It also improves particle size distribution [15], strengthens burning with less smoke [31], shifts the combustion zone to the high temperature zone in a gasifier [32] and increases resistance to biological decomposition [33]. Therefore, the torrefied biomass is more appropriate than the raw biomass for co-firing in the conventional coal power plants due to many of these improvements, as mentioned above. In addition, torrefied biomass is more appropriate than crude biomass for eligible fuel in conventional coal-fired power plants [20]. The removal of volatiles during torrefaction process leads to a decrease of the O/C ratio, and to an increase the energy density of the biomass [34].

In the present study, the process of torrefaction caused by muffle furnace on barley straw under different experimental conditions was studied, aiming at increasing the energy content of barley straw. Barley straw was placed in a porcelain capsule and was heated using a muffle furnace for various experiments with different sets of temperatures and residence times, allowing the critical parameters of the combustion process to be identified and affecting the energy content of the material. Furthermore, innovative kinetic models were applied to fit the experimental data using the severity factor (R_0), which combines the effect of temperature and time on the torrefaction process in a single reaction operator.

2. Materials and Methods

2.1. Material Development

Barley straw was collected from the Kapareli village of Thebes, Greece (38°14′8″ N 23°12′59″ E) and it was manually treated and prepared in small bunches. The specific fraction was considered to be suitable because in this way, homogeneity could be achieved when the torrefaction procedure was over. The untreated straw moisture was 6.0% w/w measured according to the procedure UNE-EN ISO 18134-1: 2015.

2.2. Torrefaction Process

The torrefaction process was applied to barley straw through a muffle furnace. The experimental setup is presented in Figure 1. The muffle furnace temperature was in room temperature at zero torrefaction time. The heat increase curve was from 20 °C up to 300 °C. Each experiment had a different reaction time. The time was increased by 2.5 min from 15 min to 50 min (see Table 1).

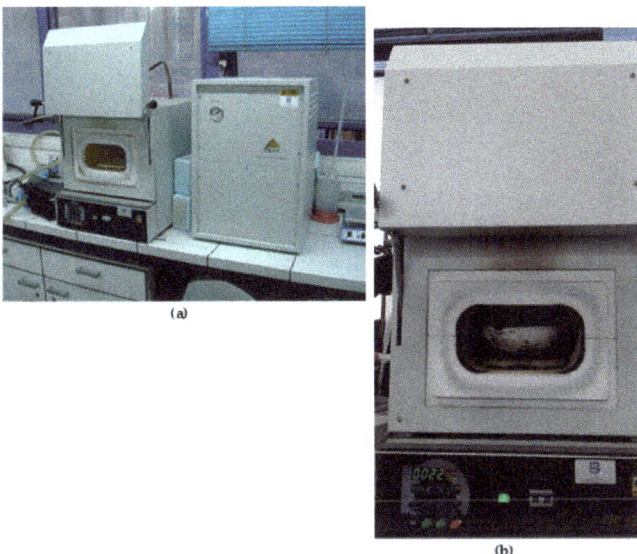

(a)

(b)

Figure 1. Experimental setup of the torrefaction process: (**a**) Nuve Muffle Furnace (Internal dimensions: 210W × 300D × 110Hmm) equipped with electric heating and a nitrogen generator that uses molecular sieves and (**b**) barley straw in a ceramic crucible placed in the cold furnace.

Table 1. Ranking, time, severity factor and logarithm of severity factor for barley straw torrefaction design of experiments.

Count	t (min)	R_0	$LogR_0$
1	15	36,010	4.56
2	17.5	332,910	5.52
3	20	1,724,755	6.24
4	22.5	5,098,025	6.71
5	25	8,922,535	6.95
6	27.5	8,949,697	6.95
7	30	13,005,480	7.11
8	32.5	10,659,469	7.03
9	35	15,435,573	7.19
10	37.5	16,313,425	7.21
11	40	16,032,904	7.21
12	42.5	22,099,405	7.34
13	45	18,959,522	7.28
14	47.5	19,904,975	7.30
15	50	16,903,655	7.23

Therefore, barley straw was placed in a porcelain capsule and after the process was completed, it was removed from the muffle furnace and finally, it was placed in a dryer for 15 min. Then, the char weight in the capsule was measured. The higher heating value was measured experimentally using an adiabatic bomb calorimeter, which measures the enthalpy change between reagents and products.

2.3. Bomb Calorimeter

In order to take the necessary measurements, a Parr 1341 Plain Jacket Bomb Calorimeter was used. In more detail, from the samples obtained after torrefaction for a given time, we accurately weighed quantities of about 0.5 g. Then, the 0.5 g was placed in the oxygen container with the pressure of 25 bar oxygen and a specific length of ignition wire. The combustion container was placed in the calorimeter's adiabatic tank, which contained 2000 g distilled water shaken at a steady speed while the temperature was measured per minute. Accordingly, two ignition lead wires were pushed into the terminal sockets on the bombs' head, the cover was set on the jacket and the stirrer was turned manually in order to ensure that it ran freely. Upon turning of the stirrer, the drive belt slipped onto the pulleys and the motor started operating.

With regards to temperature indications, they were measured manually with the help of a 6775 Parr Digital Thermometer each minute for 5 min in order to achieve equilibrium into the calorimeter. More specifically, at the start of the sixth minute, the ignition button was pushed and temperature measurements were taken each minute until the temperature became stable again. As such, the increase of the temperature was intense during the first minutes and slowed down when reaching the stage equilibrium. The energy equivalent of the calorimeter was determined by its standardization at 10,104 J/°C. The energy equivalent due to the formation of nitric acid and sulfuric acid was not included in the calculations while the moisture content of the samples was 6.0%.

Additionally, two ignition lead wires were pushed into the terminal sockets on the bombs' head, the cover was set on the jacket, and the stirrer was turned manually in order to ensure that it ran freely. Upon turning of the stirrer, the drive belt slipped onto the pulleys and the motor started operating. In this context, the diagram of Figure 2 depicts the temperature profile and how the latter is affected during the different stages described earlier.

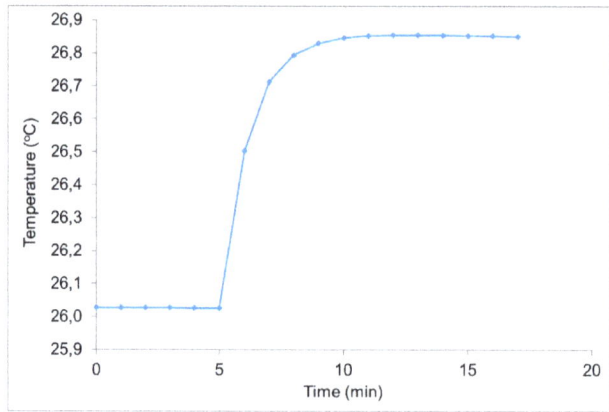

Figure 2. Temperature profile vs. time during the measurement of a typical sample's higher heating value of combustion.

2.4. Ultimate Analysis

Proximate and ultimate analysis provided information on the major categories that are important for the thermal conversion of biomass such as moisture, Volatile Matter (VM), ash, carbon etc. The moisture content of the samples was measured by drying according to the procedure UNE-EN ISO 18134-1: 2015; the nitrogen, carbon and hydrogen percentages were determined according to the UNE-EN ISO 16948: 2015 standard, the VM by the UNE-EN ISO 18123: 2015, the sulfur by UNE-EN ISO 16994: 2015, and the oxygen by difference. The measurement was conducted by the Centre for Research & Technology Hellas/Chemical Process and Energy Resources Institute (CERTH/CPERI),

Athens branch. The proximate and ultimate analysis data for untreated and torrefied barley straw are presented in Table 2. The torrefaction conditions were for the optimal heating value.

Table 2. Composition of untreated and torrefied barley straw.

Percentages (% wt. Dry Basis)	Untreated Barley Straw	Torrefied Barley Straw	Analytical Method
Proximate analysis (wt. %)			
Moisture Content	6.0	3.5	ISO 18134-1
Volatile Matter	74.3	62.5	ISO 18123
Ash	8.4	16.1	ISO 18122
Ultimate analysis (wt. %)			
Carbon	45.5	57.5	ISO 16948
Hydrogen	5.5	4.1	ISO 16948
Nitrogen	0.99	1.6	ISO 16948
Oxygen	47.9	36.4	by difference
Sulfur	0.11	57.5	ISO 16994

2.5. Scanning Electron Microscopy

The observation the surface morphology pattern changes of the untreated and torrefied barley straw was conducted by scanning electron microscopy (SEM) at the Institute of Nanoscience and Nanotechnology, National Centre for Scientific Research "Demokritos", Athens, using an FEI INSPECT SEM equipped with an EDAX super ultra-thin window analyzer for energy-dispersive X-ray spectroscopy (EDS).

3. Results and Discussion

The kinetics of higher heating value combustion of untreated and torrefied barley straw have been extensively studied using ISO 1716:2018 [35]. To this end, the well-known higher heating value of combustion equation is shown below:

$$H_g = (tW - e_1 - e_2 - e_3)/m \qquad (1)$$

where H_g represents the higher heating value of combustion, "m" stands for the mass of the sample in grams, e_1 refers to a correction coefficient concerning calories for heat of formation of nitric acid, e_2 to a correction coefficient concerning calories for heat of formation of sulfuric acid and e_3 to a correction coefficient for calories or heat of combustion of fuse wire. For the given case, both e_1 and e_2 are taken as being equal to zero since neither nitric acid nor sulfuric acid were used. Moreover, W is the energy equivalent of the calorimeter, which is determined under standardization and t is the net-corrected temperature increase, with equations following further analyzing the above variables.

$$t = t_c - t_a - r_1(b - a) - r_2(c - b) \qquad (2)$$

$$e_3 = 2.3 l_f \qquad (3)$$

$$W = 10{,}104 \text{ J/°C} \qquad (4)$$

To this end, a stands for the time of firing, b for the time when the temperature reaches 60% of the total rise and c for the time at the beginning of period in which the rate of temperature change is constant. Next, t_a corresponds to the temperature at firing time and t_c the temperature at time c, r_1 is the rate at which the temperature was rising until firing and r_2 is the rate at which the temperature is rising during the 5 min period after the time c. Finally, l_f is the size of the fuse wire consumed during the firing.

A severity factor was used in order to integrate the effects of reaction times and temperature into a single variable during torrefaction. In this context, a 'combined severity factor' for isothermal reactions was based on the 'P' factor, first introduced at 1965 by Brasch and Free [36], for the prehydrolysis-Kraft pulping of *pinus radiata*, and then at 1987 (under the name 'reaction ordinate') applied by Overend and Chornet [37] in the case of fractionation of lignocellulosics by steam-aqueous pretreatments (like wet torrefaction). The 'P' factor had units of time and was as follows:

$$('P' \text{ factor}) = [\exp(T - 100)/14.75] \cdot t \quad (5)$$

where t is the reaction time in min and T is the reaction temperature in degrees Celsius.

Moreover, in the case of torrefaction for high energy density solid fuel of fast-growing tree species, the following severity factor was used [38]:

$$SF = \log\left[t \cdot e^{\frac{T_h - T_R}{14.75}}\right] \quad (6)$$

where t is the reaction time of the torrefaction in min, T_h the reaction temperature and T_R the reference temperature, both in degrees Celsius.

In addition to the above, a combined severity factor for non-isothermal reaction conditions was also introduced in the case of the batch autohydrolysis of wheat straw [39],

$$R_0^* = 10^{-pH} \cdot \int_0^t e^{\frac{T_\theta - 100}{14.75}} dt \quad (7)$$

where T_θ is the reaction temperature in degrees Celsius.

At this point, it should be noted that since, in this work, the main variables used are time and temperature, pH was removed from the equation, with the simplified severity factor used for non-isothermal reaction conditions given in the following equation:

$$R_0 = \int_0^t e^{\frac{T_\theta - 100}{14.75}} dt \quad (8)$$

A similar severity factor was used by Aguado et al. [40] for wet torrefaction of almond-tree pruning. On the other hand, a severity index was used by Zhang et al. [41] for spend coffee grounds and microalga residue torrefaction. Several torrefaction severity reporting methods were reported by Campbell et al. [42], while the dry mass yield was suggested as an indicator for severity presuming that was the most reliable singular severity indicator for bench and pilot scale work.

Consequently, in the present work, the severity factor values according to Equation. (8) and for each of the experiments carried out are provided in Table 1. Therefore, the gradual reduction of the test sample mass from starting time (m_0) until the end of each experiment (m_t) is used, with the parameter of solid residue yield showing the percentage of the mass loss over torrefaction time.

In this context, the diagram of Figure 3 depicts the temperature profile and how the latter is affected during the time stages described earlier. An example of temperature profiles at different times of the muffle furnace during torrefaction of 300 °C is shown in this Figure. The preheating time in the case of 300 °C was around 25 min, because the initial temperature in the furnace was about 30 °C, i.e., the furnace was cold (not preheated). Similar temperature measurements were done at each torrefaction time for the muffle furnace.

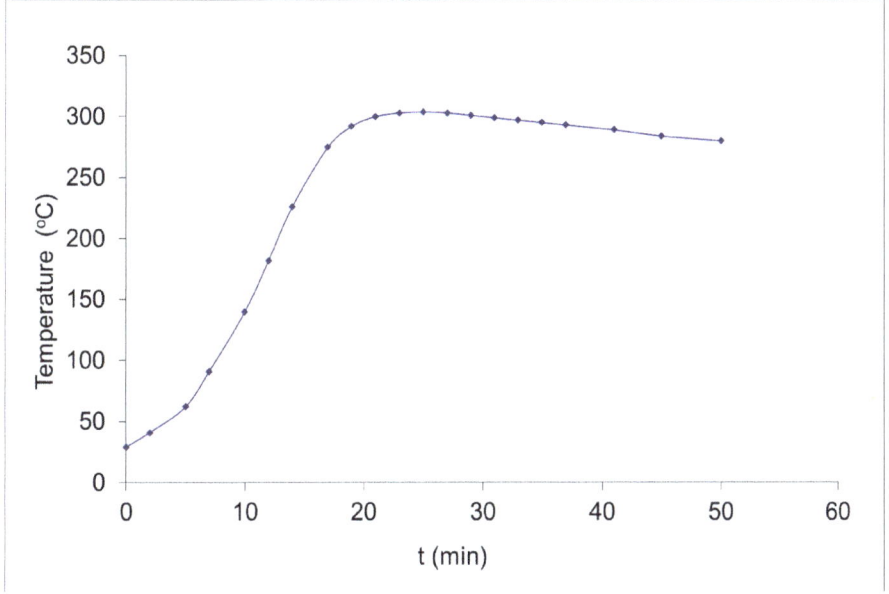

Figure 3. Experimental temperature profile for a typical barley straw torrefaction experiment.

Moreover, in Figure 4 the percentage of loss of mass during the torrefaction procedure following illustrates the impact of the time (Figure 4a), severity factor (Figure 4b), and logarithm of severity factor (Figure 4c) on the solid residue yield percentage decrease, with the latter showing a rapid reduction for small severity factor values which is gradually almost stabilized for higher severity factor values. Increased weight loss occurs when torrefaction temperature is also increased due to moisture removal and hemicellulose breakdown which produced H_2O, CO, CO_2 and other hydrocarbons. Finally, the following equations describe the exponential relation between the yield (y) and the time (t) or the severity factor (R_0 or $logR_0$), with the equation parameters given in Table 3.

$$\text{Model A1: } y = y_e + (y_0 - y_e)\exp(-kt) \tag{9}$$

where y_e is the value for y at infinite time, y_0 is the value for y at zero time, and k is the pseudo-first order kinetic constant.

$$\text{Model A2: } y = y_e + (y_0 - y_e)\exp(-kR_0) \tag{10}$$

Table 3. The parameters and standard error of estimate (SEE) of the three models for the solid residue yield (% w/w) of barley straw torrefaction.

	Model A1	Model A2	Model A3
y_0	101.10	80.23	70.58
k	0.0450	$1.739 \cdot 10^{-7}$	$7.778 \cdot 10^{-8}$
y_e	36.76	45.32	38.35
SEE	3.637	7.183	2.596

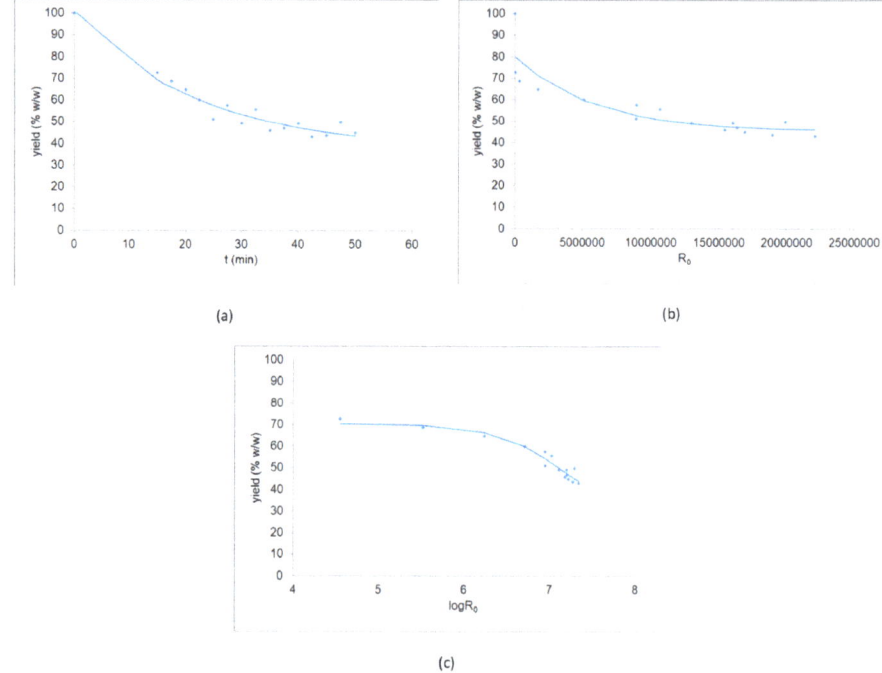

Figure 4. Torrefied barley straw solid residue yield vs. time (**a**), severity factor (**b**), and logarithm of severity factor (**c**).

It must be mentioned Model A3 is described by the same Equation (10) as Model A2, but its parameters were estimated without taking into account the experimental value for y at zero time. The standard error of estimate (SEE) values for these tree models are presented in Table 3, showing that the best fitting to the experimental data was for Model A3. The fitting of these three models is illustrated in Figure 4a,b and c for Model A1, A2 and A3, respectively.

Moreover, Figure 5 demonstrates the Higher Heating Value (H_g) of barley straw combustion vs. torrefying reaction time (Figure 5a), severity factor (Figure 5b), and logarithm of severity factor (Figure 5c). To this end, according to the experimental results obtained, the optimal time that gives the maximum output (H_g = 21.3 MJ.kg) was 47.5 min, where H_g increases by 21.7%. On the other hand, the gross heat of combustion for the untreated barley straw was measured a total of three times, with the average value found to be 17.5 MJ/kg and the standard deviation 0.17 (1.0%). Therefore there is an increase of H_g during conditions intensification. After all, the following equations describe the relation between the H_g and the time (t) or the severity factor (R_0 or $logR_0$) with the equation parameters given in Table 4.

$$\text{Model B1: } H_g = H_{ge} - [(H_{ge} - H_{g0})^{-1} + k_1 t]^{-1} \tag{11}$$

where H_{ge} is the value for H_g at infinite time, H_{g0} is the value for H_g at zero time, and k_1 is the pseudo-second order kinetic constant.

$$\text{Model B2: } H_g = H_{ge} - [(H_{ge} - H_{g0})^{-1} + k_1 R_0]^{-1} \tag{12}$$

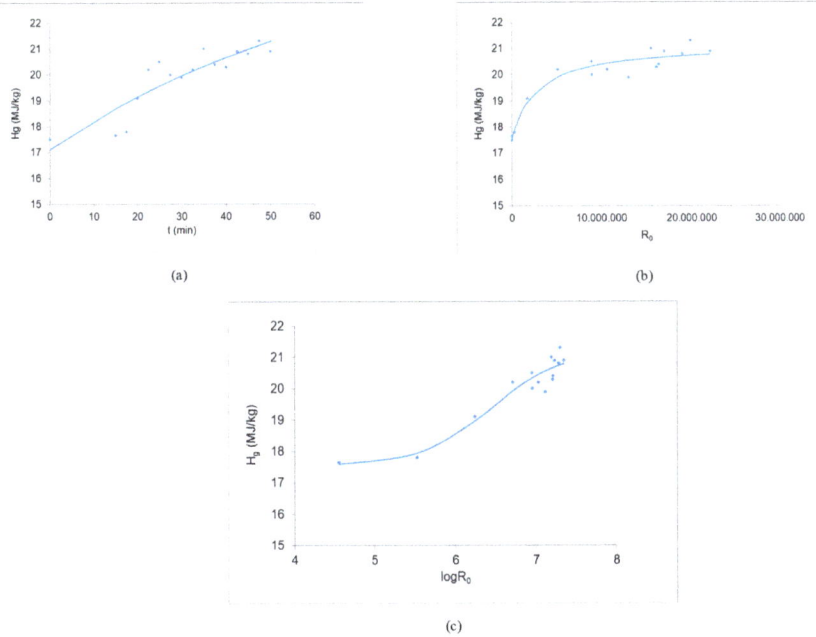

Figure 5. Torrefied barley straw Higher Heating Value of combustion vs. time (**a**), severity factor (**b**), and logarithm of the severity factor (**c**).

Table 4. The parameters and standard error of estimate (SEE) of the three models for the Higher Heating Value (MJ/kg) of the combustion of the torrefied barley straw.

	Model B1	Model B2	Model B3
H_{g0}	17.11	17.53	17.55
k_1	0.0006334	$1.020 \cdot 10^{-7}$	$1.000 \cdot 10^{-7}$
H_{ge}	30.87	21.18	21.19
SEE	0.6182	0.3344	0.3479

It must be mentioned the Model B3 is described by the same Equation (12) as Model B2, but its parameters were estimated without taking into account the experimental value for H_g at zero time. The SEE values for these tree models are presented in Table 4, showing that the best fitting to the experimental data was for Model B2. The fitting of these three models is illustrated in Figure 5a–c for Models B1, B2 and B3, respectively.

Figure 6 illustrates the relation between the Higher Heating Value of barley straw combustion and the material's mass loss percentage due torrefaction. The theoretical curve was estimated using Models A1 and B1 in combination. Moreover, Models A2 and B2 could successfully fit the experimental data. The maximum Higher Heating Value of the barley straw combustion is expected to be at the maximum material's mass loss percentage, i.e., at the most severe torrefaction conditions. Moderate torrefaction conditions could be chosen to reduce barley straw's mass loss but with a lower Higher Heating Value of the material combustion.

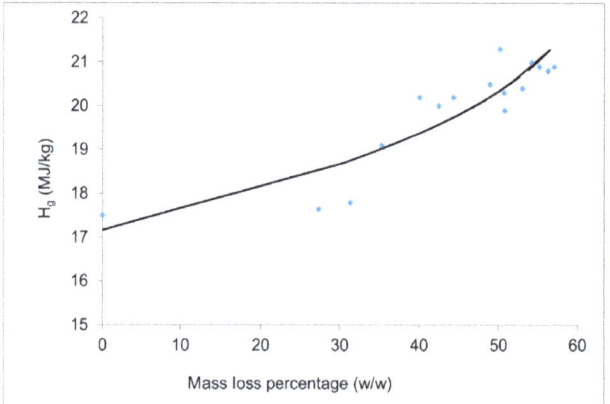

Figure 6. Torrefied barley straw Higher Heating Value of combustion vs. the mass loss percentage.

In Figure 7, the torrefied barley straw Enhancement Factor (EF) and Energy yield (EY) vs. the mass loss percentage are presented. The Enhancement Factor (EF) is given by

$$(EF) = H_{gt}/H_{gu} \qquad (13)$$

where H_{gt} is the HHV for torrefied straw and H_{gu} is the HHV for untreated straw. The Energy yield (EY) is given by the following equation:

$$(EY) = (EF) \cdot y \qquad (14)$$

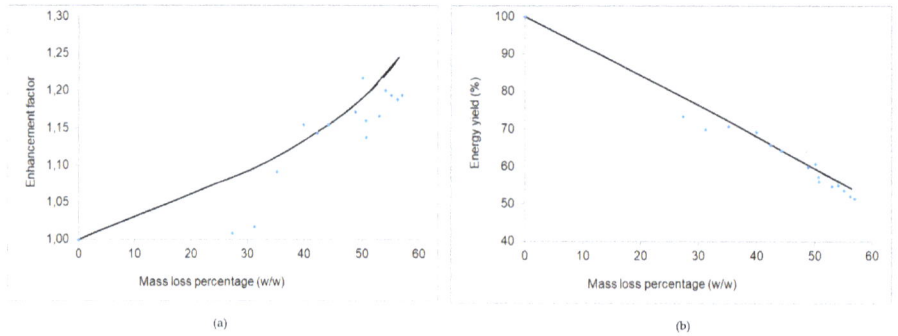

Figure 7. Torrefied barley straw (**a**) Enhancement Factor (EF) and (**b**) Energy yield (EY) vs. the mass loss percentage.

According to Figure 7a the Enhancement factor increases by mass loss decreasing, while according Figure 7b, Energy yield decreases almost linearly by mass loss decreasing. The theoretical curves are according to the same above-described Models A1 and A2, and Equations (10) and (12), respectively. There was no need for re-estimation of the models' parameters.

In Figure 8 are shown the Scanning Electron Microscopy (SEM) images of the untreated barley straw at (a) × 750, (c) × 7500 and (e) × 20,000 magnification, and torrefied barley straw (at optimal conditions) at (b) × 750, (d) × 7500 and (f) × 20,000 magnification. We observe that the effect of the torrefaction on the straw surface topology is the roughening of the surface. The effect might facilitate the use of torrefied barley straw for the production of adsorbents (low-cost activated carbon substitute). This could be an alternative use to the torrefied straw as energy production material (coal substitute).

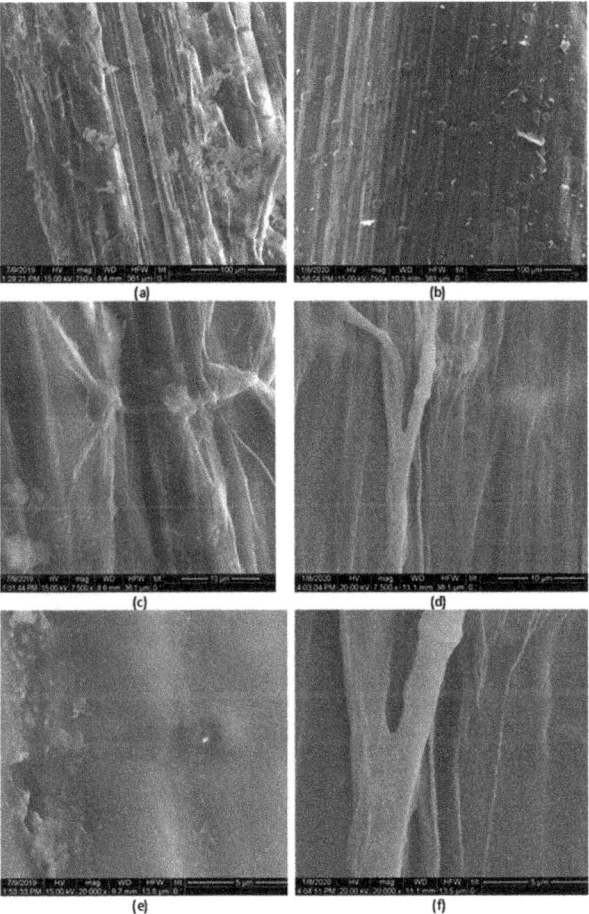

Figure 8. Scanning Electron Microscopy (SEM) images of untreated barley straw at (**a**) ×750, (**c**) ×7500 and (**e**) ×20,000 magnification, and torrefied barley straw at (**b**) ×750, (**d**) ×7500 and (**f**) ×20,000 magnification.

In Table 5, the Higher Heating Value of combustion, the Solid residue yield, the Enhancement factor and the Energy yield for some untreated and torrefied lignocellulosic residues according to the recent literature are presented.

Table 5. Higher Heating Value of combustion, Solid residue yield, Enhancement factor and Energy yield for some untreated and torrefied lignocellulosic residues.

Materials	HHV (MJ/kg)	Solid Residue Yield, y (% wt.)	Enhancement Factor, EF	Energy Yield, EY (%)	References
Almond-tree pruning	17.6				[28]
Almond-tree pruning pretreated by wet torrefaction	24	57.1	1.36	77.9	[28]
Barley straw	17.7				[11]
Barley straw torrefied	21.5	44.1	1.21	53.6	[11]
Barley straw	17.5				This study
Barley straw torrefied	21.3	49.9	1.22	60.7	This study
Eucalyptus grandis	20.1				[43]
Eucalyptus grandis torrefied	25.0	65.2	1.24	81.0	[43]
Herbal medicine wastes	19				[44]
Herbal medicine wastes torrefied	20.3	82.1	1.07	87.7	[44]
Microalga residue	12.7				[41]
Microalga residue torrefied	17.3	68.0	1.36	92.6	[41]
Spent coffee grounds	21.8				[41]
Spent coffee grounds torrefied	29.8	72.4	1.37	98.9	[41]
Spruce	20.3				[42]
Spruce char	21.2	91.5	1.04	95.6	[42]
Wheat straw	17.8				[11]
Wheat straw torrefied	20.5	64	1.15	73.7	[11]
Wheat straw	19				[42]
Wheat straw char	20.1	84.8	1.06	89.7	[42]
Willow	20.1				[42]
Willow char	21.2	87.1	1.05	91.9	[42]

These HHV values are comparable to the values found in the present study with regards to untreated and torrefied barley straw [34], but there are significant differences when another lignocellulosic material was used. Moreover, the EF value of torrefied barley straw [11] was similar to the findings of the present work, while the EY value [11] was lower compared to that of the present work. On the other hand, most of the other lignocellulosic materials presented in Table 5 have higher EY values (73.7–98.9%) compared to the 60.7% found herein. The high EY values were found due to high EY and/or high solid residue yield.

The higher heating values (HHV) of the barley straw samples in the present work can be calculated from their C, H and N contents (see Table 2) in a dry basis, using the following expression, as derived by Friedl et al. [45] for biomass from plant origin:

$$HHV = 3.55C^2 - 232C - 230H + 51.2C \cdot H + 131N + 20600 \quad (15)$$

The values calculated according to Equation (15) for untreated and torrefied barley straw of this work were 18.1 and 22.4 MJ/kg, respectively. This is very close to the experimental values shown in Table 5. The EF was 1.24 very close to 1.22, i.e., the experimental one.

Lignocellulosic biomass torrefaction (dry or wet, in the absence of oxygen or not, under atmospheric pressure or not) is a pretreatment process used to overcome the disadvantages of using biomass as a fuel such as low energy density, high moisture, and oxygen contents [46,47]. The torrefaction increases energy density, hydrophobicity, and reduces grinding energy requirement of biomass. The environmental and economic aspects of the torrefaction process and torrefied product, and various applications of torrefaction products have been taken into account by various researchers. The cost competitiveness of torrefied materials is one of the major concerns of the torrefaction process.

Integrating the torrefaction with other processes makes it economically more viable than as a standalone process [47].

4. Conclusions

Torrefaction of biomass is a promising process for improving the characteristics of biomass, as an alternative renewable energy source over the use of fossil fuels. This process provokes the interest of investors in this sector. From this perspective, the potential development of biomass heat conversion technologies, such as combustion, is promising as far as the use of new forms of biomass is concerned, i.e., more eco-friendly, more abundant and more economical, as is the case of barley straw. The fact that publications on this topic have significantly increased indicates the strong academic relevance and industrial interest in this subject in recent years. In the current study, torrefaction conditions were investigated for increasing the Higher Heating Value of combustion for barley straw. An integrated methodology was applied to this end, with the main focus given on the impact of the temperature and time parameters and with the results presented herein eventually indicating that severe treatment conditions are the optimum ones in order to maximize the heating value of barley straw combustion. According to these experimental results, the optimal time that gives the maximum output equal to 21.3 MJ/kg was 47.5 min where H_g increases by 21.7%., for $R_0 = 1.99 \cdot 10^7$ and consequently $\log R_0 = 7.3$. On the other hand, according to the developed models, the maximum Higher Heating Value of the barley straw combustion is expected to be at the maximum material's mass loss percentage, i.e., at the most severe torrefaction conditions. More or less, moderate torrefaction conditions could be chosen to reduce barley straw's mass loss but with enhanced Higher Heating Value of the material combustion compared to the untreated material but lower value compared to the optimal one.

Author Contributions: Investigation, A.G.N.; Methodology, G.E.G.; Software, D.V.P.; Supervision, D.K.S. All authors have read and agreed to the published version of the manuscript.

Funding: This research received no external funding.

Acknowledgments: This work has been partly supported by the University of Piraeus Research Center.

Conflicts of Interest: The authors declare no conflict of interest.

References

1. Fragkos, P.; Tasios, N.; Paroussos, L.; Capros, P.; Tsani, S. Energy system impacts and policy implications of the European Intended Nationally Determined Contribution and low-carbon pathway to 2050. *Energy Policy* **2017**, *100*, 216–226. [CrossRef]
2. Corradini, M.; Costantini, V.; Markandya, A.; Paglialunga, E.; Sforna, G. A dynamic assessment of instrument interaction and timing alternatives in the EU low-carbon policy mix design. *Energy Policy* **2018**, *120*, 73–84. [CrossRef]
3. Sung, B.; Park, S.-D. Who Drives the Transition to a Renewable-Energy Economy? Multi-Actor Perspective on Social Innovation. *Sustainability* **2018**, *10*, 448. [CrossRef]
4. Chen, W.H.; Kuo, P.C. Torrefaction and co-torrefaction characterization of hemicellulose, cellulose and lignin as well as torrefaction of some basic constituents in biomass. *Energy* **2011**, *36*, 803–811. [CrossRef]
5. Van der Stelt, M.J.C.; Gerhauser, H.; Kiel, J.H.A.; Ptasinski, K.J. Biomass upgrading by torrefaction for the production of biofuels: A review. *Biomass Bioenergy* **2011**, *35*, 3748–3762. [CrossRef]
6. Vassilev, S.V.; Vassileva, C.G.; Vassilev, V.S. Advantages and disadvantages of composition and properties of biomass in comparison with coal: An overview. *Fuel* **2015**, *158*, 330–350. [CrossRef]
7. Ntanos, S.; Kyriakopoulos, G.; Chalikias, M.; Arabatzis, G.; Skordoulis, M. Public Perceptions and Willingness to Pay for Renewable Energy: A Case Study from Greece. *Sustainability* **2018**, *10*, 687. [CrossRef]
8. Muench, S. Greenhouse gas mitigation potential of electricity from biomass. *J. Clean. Prod.* **2015**, *103*, 483–490. [CrossRef]
9. Chen, W.; Peng, J.; Bi, X.T. A state-of-the-art review of biomass torrefaction, densification and applications. *Renew. Sustain. Energy Rev.* **2015**, *44*, 847–866. [CrossRef]

10. Dietrich, R.-U.; Albrecht, F.G.; Maier, S.; König, D.H.; Estelmann, S.; Adelung, S.; Bealu, Z.; Seitz, A. Cost calculations for three different approaches of biofuel production using biomass, electricity and CO_2. *Biomass Bioenergy* **2018**, *111*, 165–173. [CrossRef]
11. Satpathy, S.K.; Tabil, L.G.; Meda, V.; Naik, S.N.; Prasad, R. Torrefaction of wheat and barley straw after microwave heating. *Fuel* **2014**, *124*, 269–278. [CrossRef]
12. Bridgeman, T.G.; Jones, J.M.; Shield, I.; Williams, P.T. Torrefaction of reed canary grass, wheat straw and willow to enhance solid fuel qualities and combustion properties. *Fuel* **2008**, *87*, 844–856. [CrossRef]
13. Prins, M.J.; Ptasinski, K.J.; Janssen, F.J.J.G. Torrefaction of wood: Part 2: Analysis of products. *J. Anal. Appl. Pyrol.* **2006**, *77*, 35–40. [CrossRef]
14. Arias, B.; Pevida, C.; Fermoso, J.; Plaza, M.G.; Rubiera, F.; Pis, J.J. Influence of Torrefaction on the Grindability and Reactivity of Woody Biomass. *Fuel Process. Technol.* **2008**, *89*, 169–175. [CrossRef]
15. Phanphanich, M.; Mani, S. Impact of Torrefaction on the Grindability and Fuel Characteristics of Forest Biomass. *Bioresour. Technol.* **2011**, *102*, 1246–1253. [CrossRef] [PubMed]
16. Medic, D.; Darr, M.; Shah, A.; Potter, B.; Zimmerman, J. Effects of Torrefaction Process Parameters on Biomass Feedstock Upgrading. *Fuel* **2011**, *91*, 147–154. [CrossRef]
17. Uemura, Y.; Omar, W.N.; Tsutsui, T.; Yusup, S.B. Torrefaction of Oil Palm Wastes. *Fuel* **2011**, *90*, 2585–2591. [CrossRef]
18. Wannapeera, J.; Fungtammasan, B.; Worasuwannarak, N. Effects of Temperature and Holding Time during Torrefaction on the Pyrolysis Behaviors of Woody Biomass. *J. Anal. Appl. Pyrol.* **2011**, *92*, 99–105. [CrossRef]
19. Bergman, P.C.A.; Boersma, A.R.; Kiel, J.H.A.; Prins, M.J.; Ptasinski, K.J.; Janseen, F.J.J.G. Torrefaction for Entrained Flow Gasification of Biomass. In *Biomass for Energy, Industry and Climate Protection: Second World Biomass Conference; Proceedings of the World Conference, Rome, Italy, 10–14 May 2004*; van Swaaij, W.P.M., Ed.; ECN-RX; ETA-Renewable Energies: Florence, Italy, 2004; Volume 04-046, pp. 679–682.
20. Clausen, L.R.; Houbak, N.; Elmegaard, B. Techno-Economic Analysis of a Low CO_2 Emission Dimethyl Ether (DME) Plant Based on Gasification of Torrefied Biomass. *Energy* **2010**, *35*, 4831–4842. [CrossRef]
21. Prins, M.J.; Ptasninski, K.J.; Janssen, F.J.J.G. More Efficient Biomass Gasification via Torrefaction. *Energy* **2006**, *31*, 3458–3470. [CrossRef]
22. Deng, J.; Wang, G.J.; Kuang, J.H.; Zhang, Y.-L.; Luo, Y.H. Pretreatment of Agriculture Residues for Co-Gasification via Torrefaction. *J. Anal. Appl. Pyrol.* **2009**, *86*, 331–337. [CrossRef]
23. Pach, M.; Zanzi, R.; Bjornbom, E. Torrefied Biomass a Substitute for Wood and Charcoal. In Proceedings of the 6th Asia-Pacific International Symposium on Combustion and Energy Utilization, Kuala Lumpur, Malaysia, 20–22 May 2002.
24. Tumuluru, J.S.; Sokhansanj, S.; Hess, J.R.; Wright, C.T.; Boardman, R.D. A Review on Biomass Torrefaction Process and Product Properties for Energy Applications. *Ind. Biotechnol.* **2011**, *7*, 384–401. [CrossRef]
25. Bridgeman, T.G.; Jones, J.M.; Williams, A.; Waldron, D.J. An Investigation of the Grindability of the Torrefied of Energy Crops. *Fuel* **2010**, *89*, 3911–3918. [CrossRef]
26. Felfli, F.F.; Luengo, C.A.; Suarez, J.A.; Beaton, P.A. Wood Briquette Torrefaction. *Energy Sustain. Dev.* **2005**, *9*, 19–22. [CrossRef]
27. Acharjee, T.C.; Coronella, C.J.; Vasquez, V.R. Effect of Thermal Pretreatment on Equilibrium Moisture Content of Lignocellulosic Biomass. *Bioresour. Technol.* **2011**, *102*, 4849–4854.
28. Repellin, V.; Govin, A.; Rolland, M.; Guyonment, R. Energy Required for Fine Grinding of Torrefied Wood. *Biomass Bioenergy* **2010**, *34*, 923–930. [CrossRef]
29. Yan, W.; Acharjee, T.C.; Coronella, C.J.; Vasquez, V.R. Thermal Pretreatment of Lignocellulosic Biomass. *Environ. Prog.* **2009**, *28*, 435–440. [CrossRef]
30. Mobini, M.; Meyer, J.C.; Trippe, F.; Sowlati, T.; Fröhling, M.; Schultmann, F. Assessing the integration of torrefaction into wood pellet production. *J. Clean. Prod.* **2014**, *78*, 216–225. [CrossRef]
31. Pentananunt, R.; Rahman, A.N.M.M.; Bhattacharya, S.C. Upgrading of Biomass by Means of Torrefaction. *Energy* **1990**, *15*, 1175–1179. [CrossRef]
32. Ge, L.; Zhang, Y.; Wang, Z.; Zhou, J.; Cen, K. Effects of Microwave irradiation Treatment on Physicochemical Characteristics of Chinese Low-Rank Coals. *Energy Convers. Manag.* **2013**, *71*, 84–91. [CrossRef]
33. Chaouch, M.; Petrissans, M.; Petrissans, A.; Gerardin, P. Use of Wood Elemental Composition to Predict Heat Treatment Intensity and Decay Resistance Different softwood and Hardwood Species. *Polym. Degrad. Stab.* **2010**, *95*, 2255–2259. [CrossRef]

34. Nhuchhen, D.R.; Basu, P.; Acharya, B. A Comprehensive Review on Biomass Torrefaction. *Int. J. Renew. Energy Biofuels* **2014**, *2014*, 1–56. [CrossRef]
35. International Organization for Standardization ISO 1716:2018 Reaction to Fire Tests for Products—Determination of the Gross Heat of Combustion (Calorific Value). 2018. Available online: https://www.iso.org/standard/70177.html (accessed on 20 December 2019).
36. Brasch, D.J.; Free, K.W. Prehydrolysis-kraft pulping of Pinus radiata grown in New Zealand. *Tappi* **1965**, *48*, 245–248.
37. Overend, R.; Chornet, E. Fractionation of lignocellulosics by steam-aqueous pretreatments. *Philos. Trans. R. Soc. Lond. B Biol. Sci.* **1987**, *321*, 523–536.
38. Kim, Y.H.; Na, B.I.; Ahn, B.J.; Lee, H.W.; Lee, J.W. Optimal condition of torrefaction for high energy density solid fuel of fast growing tree species. *Korean J. Chem. Eng.* **2015**, *32*, 1547–1553. [CrossRef]
39. Sidiras, D.; Batzias, F.; Ranjan, R.; Tsapatsis, M. Simulation and optimization of batch autohydrolysis of wheat straw to monosaccharides and oligosaccharides. *Bioresour. Technol.* **2011**, *102*, 10486–10492. [CrossRef] [PubMed]
40. Aguado, R.; Cuevas, M.; Perez-Villarejo, L.; Martínez-Cartas, M.L.; Sanchez, S. Upgrading almond-tree pruning as a biofuel via wet torrefaction. *Renew. Energy* **2020**, *145*, 2091–2100. [CrossRef]
41. Zhang, C.; Ho, S.-H.; Chen, W.-H.; Xie, Y.; Liu, Z.; Chang, J.S. Torrefaction performance and energy usage of biomass wastes and their correlations with torrefaction severity index. *Appl. Energy* **2018**, *220*, 598–604. [CrossRef]
42. Campbell, W.A.; Coller, A.; Evitts, R.W. Comparing severity of continuous torrefaction for five biomass with a wide range of bulk density and particle size. *Renew. Energy* **2019**, *141*, 964–972. [CrossRef]
43. Silveira, E.A.; Gustavo, L.; Galva, O.; Sa, I.A.; Silva, B.F.; Macedo, L.; Rousset, P.; Caldeira-Pires, A. Effect of torrefaction on thermal behavior and fuel properties of *Eucalyptus grandis* macro-particulates. *J. Therm. Anal. Calorim.* **2019**, *138*, 3645–3652. [CrossRef]
44. Xin, S.; Huang, F.; Liu, X.; Mi, T.; Xu, Q. Torrefaction of herbal medicine wastes: Characterization of the physicochemical properties and combustion behaviors. *Bioresour. Technol.* **2019**, *287*, 121408. [CrossRef] [PubMed]
45. Friedl, A.E.; Padouvas, H.R.; Varmuza, K. Prediction of heating values of biomass fuel from elemental composition. *Anal. Chim. Acta* **2005**, *544*, 191–198. [CrossRef]
46. Bach, Q.-V.; Skreiberg, Ø. Upgrading biomass fuels via wet torrefaction: A review and comparison with dry torrefaction. *Renew. Sustain. Energy Rev.* **2016**, *54*, 665–677. [CrossRef]
47. Cahyanti, M.N.; Doddapaneni, T.R.K.C.; Kikas, T. Biomass torrefaction: An overview on process parameters, economic and environmental aspects and recent advancements. *Bioresour. Technol.* **2020**, *301*, 122737. [CrossRef]

© 2020 by the authors. Licensee MDPI, Basel, Switzerland. This article is an open access article distributed under the terms and conditions of the Creative Commons Attribution (CC BY) license (http://creativecommons.org/licenses/by/4.0/).

Article

The Influence of pH on the Combustion Properties of Bio-Coal Following Hydrothermal Treatment of Swine Manure

Aidan Mark Smith [1,*], Ugochinyere Ekpo [2] and Andrew Barry Ross [2]

1 School of Biological and Chemical Engineering, Aarhus University, 8200 Aarhus, Denmark
2 School of Chemical and Process Engineering, University of Leeds, Leeds LS2 9JT, UK; cn10une@leeds.ac.uk (U.E.); a.b.ross@leeds.ac.uk (A.B.R.)
* Correspondence: aidan.smith@eng.au.dk; Tel.: +45-93-52-12-15

Received: 29 November 2019; Accepted: 6 January 2020; Published: 9 January 2020

Abstract: The application of excessive amounts of manure to soil prompted interest in using alternative approaches for treating slurry. One promising technology is hydrothermal carbonisation (HTC) which can recover nutrients such as phosphorus and nitrogen while simultaneously making a solid fuel. Processing manure under acidic conditions can facilitate nutrient recovery; however, very few studies considered the implications of operating at low pH on the combustion properties of the resulting bio-coal. In this work, swine manure was hydrothermally treated at temperatures ranging from 120 to 250 °C in either water alone or reagents including 0.1 M NaOH, 0.1 M H_2SO_4, and finally 0.1 M organic acid (CH_3COOH and $HCOOH$). The influence of pH on the HTC process and the combustion properties of the resulting bio-coals was assessed. The results indicate that pH has a strong influence on ash chemistry, with decreasing pH resulting in an increased removal of ash. The reduction in mineral matter influences the volatile content of the bio-coal and its energy content. As the ash content in the final bio-coal reduces, the energy density increases. Treatment at 250 °C results in a more "coal like" bio-coal with fuel properties similar to that of lignite coal and a higher heating value (HHV) ranging between 21 and 23 MJ/kg depending on pH. Processing at low pH results in favourable ash chemistry in terms of slagging and fouling. Operating at low pH also appears to influence the level of dehydration during HTC. The level of dehydration increases with decreasing pH, although this effect is reduced at higher temperatures. At higher-temperature processing (250 °C), operating at lower pH increases the yield of bio-coal; however, at lower temperatures (below 200 °C), the reverse is true. The lower yields obtained below 200 °C in the presence of acid may be due to acid hydrolysis of carbohydrate in the manure, whereas, at the higher temperatures, it may be due to the acid promoting polymerisation.

Keywords: HTC; bio-coal; manure; slagging; fouling; corrosion; process chemistry; combustion; waste to energy

1. Introduction

Historically, animal manures were returned to land and used in agriculture to increase soil organic matter and provide plant nutrients. The expansion of concentrated animal husbandry over the latter half of the 20th century, however, resulted in thousands of animals often being concentrated into small geographical areas, overwhelming the nutrient needs and soil-absorbing capacity of the nearby land. Excessive nutrients can then leach into groundwater, potentially leading to surface and groundwater pollution. As such, disposal of animal manures is a problem [1]. Hydrothermal treatment, including hydrothermal carbonisation (HTC), is an emerging technology which is well suited to processing wet wastes such as manure, and it has potential for recovery of nutrients from biomass and wastes such as phosphorus and nitrogen while simultaneously producing a solid fuel for energetic purposes [2].

Hydrothermal treatment involves the processing of biomass in water at temperatures above 100 °C at elevated pressure to ensure the water is in the liquid phase. HTC typically uses a temperature range of 180 to 250 °C, while temperatures below 180 °C are typically regarded as thermal hydrolysis. Under hydrothermal conditions, water acts as both reagent and medium for a series of aqueous and solid-phase reactions to take place, leading to the carbonisation of biomass, resulting in a hydrochar or bio-coal, which has similar properties to a low-rank coal. [3]. Animal manures are typically composed of faeces, urine, discarded bedding, and waste feed. They have high moisture content, and they are, therefore, well suited for conversion by HTC [1]. During the hydrothermal processing of plant biomass, a number of key plant nutrients, including potassium and phosphorus from soluble phosphates, can be extracted into the aqueous phase and subsequently precipitated and recovered [4,5]. The extent to which the phosphorus is extracted is feedstock-dependent, with the inorganic content of the feedstock, particularly calcium content, often a key variable [6].

Previous studies investigating the HTC of manures found that phosphorus within manures is not easily extracted, leading to the immobilisation of the phosphorus in the bio-coal. This prompted the application of acids in HTC to aid phosphorus extraction [2,7,8]. The addition of acids during HTC was widely investigated, and it is thought to improve the overall rate of reaction in HTC [9–13]. In a study reported by Reza et al. [14], the influence of feedwater pH on the HTC of wheat straw was investigated using acetic acid and potassium hydroxide. The results indicated that the feedwater pH influences carbon density and higher heating value (HHV) in wheat straw, with higher carbon densities associated with lower pH.

At present, the application of acid catalysis for the processing of manures is primarily focused on increasing the extraction of phosphorus. Ekpo et al. [8] and Dai et al. [7] investigated the influence of acids on the recovery of phosphorus and nitrogen in swine and cattle manures, respectively. Ekpo et al. [8] investigated the addition of sodium hydroxide, sulphuric acid, acetic acid, and formic acid at 0.1 molar concentration and demonstrated that the presence of acidic additives improves the extraction of phosphorus and nitrogen. This study showed that phosphorus extraction is pH- and temperature-dependent and enhanced under acidic conditions. Phosphorus was most readily extracted using sulphuric acid, reaching 94% at 170 °C, while largely retained in the residue for all other conditions [8]. Dai et al. [7] performed HTC at 190 °C for 12 h using hydrochloric acid at varying concentrations. The results indicated that HTC in 2% hydrochloric acid extracted almost 100% phosphorus and 63% nitrogen. Decreasing the pH results in a small increase in carbon content and a large decrease in oxygen content, which will increase energy content (not stated). Decreasing the pH, however, also impacts the yields of bio-coal, which reduces from 70% (db) to 53% (db), and this reduction appears to be predominantly associated with the removal of oxygen. Fuel volatile matter is also seen to decrease, corresponding to an increase in fixed carbon at low pH. Ghanim et al. [15] investigated the HTC of poultry litter at 250 °C for 2 h at different pH using acetic and sulphuric acid. Once again, the results indicate that operation at low pH increases the carbon content and HHV of the bio-coal. Increasing sulphuric acid content appears to both increase yield of bio-coal and reduce ash content. These results suggest that performing HTC in dilute acid can simultaneously facilitate nutrient recovery from manure while upgrading the manure to a higher-quality bio-coal.

The studies performed to date did not consider the implications of operating at low pH on the inorganic chemistry, and how this affects the combustion behaviour of the bio-coal. The presence of inorganics and heteroatoms is a particular issue during thermochemical conversion of biomass and feedstocks that contain large amounts of potassium, sodium, sulphur, and chlorine; it can result in corrosion and slagging, or fouling in furnace and retorts [16]. Slagging is a process that occurs when ash deposits melt due to exposure to radiant heat, such as flames in a furnace. As this ash begins to melt, it starts to fuse, becomes sticky, and eventually forms a hard glassy slag known as a clinker, making ash removal difficult. A high ash melting temperature is desirable as most furnaces are designed to remove ash as a powdery residue [16]. Fouling occurs when potassium and sodium chlorides within the fuel partially evaporate on exposure to radiant heat and then condense on cooler surfaces such as heat exchangers forming alkali chloride deposits, which reduces their efficiency on heat exchangers. These alkali chlorides can also play a role in the corrosion as they can react with sulphur in the flue gas, forming alkali sulphates and liberating chlorine within the deposit. This chlorine then catalyses the active oxidation and corrosion of the steel on which the deposit is formed [16,17].

To reduce the chance of a fuel slagging or fouling, it is important to minimise the alkali metal content in the ash along with chlorine. Leaching of alkaline metals and chlorine during HTC was demonstrated in a number of studies that concluded that slagging, fouling, and corrosion can be reduced by reducing alkali metals [18–24]. This reduction in slagging and fouling propensity following HTC was first demonstrated by Reza et al. [25] and later developed by Smith et al. [26], who demonstrated the effect of alkali metal removal on ash melting temperatures for HTC bio-coal using ash fusion analysis. This was later validated by subsequent studies [27–30].

The work presented in Smith et al. [26], however, demonstrated that the reduction in slagging and fouling propensity of HTC bio-coal is only partially due to a reduction of alkali metals, and it is also influenced by the retention of calcium and phosphorus within the bio-coal. Calcium and phosphorus are important as, while alkali metals, such as potassium and sodium, act as a flux for alumina–silicate ash, alkaline earth metals, such as calcium and magnesium, tend to increase melting temperatures [16]. In addition to the alkali and alkaline earth metals, the presence of phosphorus can prevent alkali metals forming low-melting-temperature alkali silicates, instead forming thermally stable phosphate compounds [31]. Phosphorus is also important from a fouling perspective, as potassium and sodium chlorides present within the ash can bind with calcium-rich phosphates to produce potassium or sodium phosphates, which then further react with calcium oxides. The resulting calcium potassium phosphate/calcium sodium phosphate complexes are stable and remove the potassium/sodium available to form low-melting-temperature potassium silicates [31,32]. Calcium oxide, calcium carbonate, and calcium hydroxide would otherwise dissolve into potassium/sodium silicate melts, bringing about the release of the potassium or sodium into the gas phase [33,34]. The removal of inorganics by the addition of acids during HTC may have a profound effect on the ash chemistry and affect the properties of the bio-coal during subsequent thermochemical processing.

In this work, swine manure was hydrothermally treated between 120 and 250 °C in water or 0.1 M NaOH, 0.1 M H_2SO_4, or 0.1 M organic acid (CH_3COOH and $HCOOH$). The influence of pH on the on the HTC process was assessed, and the combustion properties of the resulting fuels were assessed.

2. Materials and Methods

2.1. Materials

The swine manure was collected from the University of Leeds farm. Prior to processing and characterisation, the manure was dried in an oven at 60 °C for several days and homogenised in an Agate Tema barrel.

2.2. Hydrothermal Processing

Hydrothermal processing of the swine manure was performed in an unstirred 600-mL Parr reactor (Parr, Moline, IL, USA). For each experiment, the reactor was filled with 24 g of swine manure and either 220 mL of de-ionised water (pH ≈ 6) or solutions of 0.1 M NaOH, 0.1 M H_2SO_4 (pH 13 and 1, respectively), or 0.1 M organic acid (CH_3COOH and $HCOOH$) (pH 2.88 and 2.38, respectively) to form a slurry. Hydrothermal processing was performed at 120 °C, 170 °C, 200 °C, and 250 °C for 1 h. The heating rate was approximately 10 °C·min^{-1}, and the residence time was taken from when the reactor reached the desired temperature. After one hour, the reactor was removed from the heating jacket and then allowed to cool to room temperature before the products were separated.

Bio-coal samples were dried in an oven at 60 °C for a minimum of 24 h, and yields were taken as dry bio-coal mass compared with the original dry mass of unprocessed manure.

2.3. Analysis

2.3.1. Inorganic Analysis

For analysis, dried samples were ground and homogenised to below 100 µm. Samples were ashed in a muffle furnace (Nabertherm, Lilenthal, Germany) to a final temperature of 550 °C, with a hold at 250 °C to minimise volatile metal loss, as directed in BS EN ISO 18122-2. The ash was then mixed with a lithowax binder at a 10:1 ratio and palletised using a laboratory press (Spex, Stanmore, UK). The elemental composition of the ash was then determined using wavelength-dispersive X-ray fluorescence (WD-XRF) (Rigaku, Tokyo, Japan) using a metal oxide method. To correct for residual carbon within the ash, the carbon content was determined using a CHNS analyser (Thermo Scientific, Waltham, MA, USA), and carbon content was manually input to the XRF component list.

2.3.2. Organic Analysis and Determination of Combustion Properties

The carbon, hydrogen, nitrogen, and oxygen contents were determined using a Flash 2000 CHNS-0 analyser (Thermo Scientific, Waltham, MA, USA), calibrated using both chemical standards and certified biomass reference materials (Elemental Microanalysis, Okehampton, UK). Hydrogen and oxygen values were corrected to account for residual moisture, and figures are given on a dry free basis, in accordance with ASTM D3180-15. The error stated is based on the calculated standard error. The higher heating value (HHV) was calculated by bomb calorimetry (Parr, Moline, IL, USA). Proximate analysis was undertaken using a thermogravimetric analyser (TGA) (Mettler Toledo, Columbus, OH, USA). To obtain the residual moisture, 10 µg of homogenised sample was heated under nitrogen to 105 °C, where the temperature was held for 10 min before heating at 25 °C·min^{-1} to 900 °C to determine the volatile carbon content. Fixed carbon was determined by holding the temperature at 900 °C and switching to air. Burning profiles, ignition, flame stability, and burnout temperature were obtained by temperature-programmed oxidation (TPO) in a TGA (Mettler Toledo, Columbus, OH, USA). Then, 10 µg of homogenised sample was heated at a rate of 10 °C·min^{-1} in air to 900 °C, and the first derivative of the weight loss was calculated.

2.3.3. Prediction of Slagging and Fouling Propensity

The propensity of the fuels to slag and foul was assessed using both predictive slagging and fouling indices in the ash fusion test (AFT). Slagging and fouling indices are numerical indices based on the ash composition, as determined in Section 2.3.1. The equations for the alkali index (AI), bed agglomeration index (BAI), and acid base ratio (R^b_a) are given in Table 1, along with the key values indicative of the onset of issues. The AFT is a qualitative method of assessing the propensity of a fuel to slag, and it works by heating an ash test piece and analysing the transitions in the ash chemistry in accordance with DD CEN/TS 15370-1:2006. Cylindrical test pieces are formed using 550 °C ash and a dextrin binder (Sigma-Aldrich, USA). The sample is then heated from 550 °C to 1570 °C in an ash fusion furnace (Carbolite, UK) using an airflow of 50 mL·min^{-1} to give an oxidising atmosphere. The key transitions are as follows: (i) shrinkage, which predominantly represents the decomposition of carbonates in hydrothermally derived chars, (ii) deformation temperature, essentially representing the onset point at which the powdery ash starts to agglomerate and starts to stick to surfaces, (iii) hemisphere, whereby ash is agglomerating and is sticky, and (v) flow, whereby the ash melts [29]. The temperature for each transition is given to the nearest 10 °C in accordance with the standard. For most power stations, slagging becomes problematic between the deformation and hemisphere temperature [29]; thus, the deformation temperature is taken as the onset temperature for slag related issues.

Table 1. Predictive slagging and fouling indices.

Slagging/Fouling Index	Expression	Limit	
Alkali Index (AI)	$AI = \frac{Kg\ (K_2O+Na_2O)}{GJ}$	AI < 0.17, safe combustion AI > 0.17 < 0.34, probable slagging and fouling AI > 0.34, almost certain slagging and fouling [35]	Equation (1)
Bed Agglomeration Index (BAI)	$BAI = \frac{\%(Fe_2O_3)}{\%(K_2O+Na_2O)}$	BAI < 0.15, bed agglomeration likely [36]	Equation (2)
Acid Base Ratio (R_a^b)	$R_a^b = \frac{\%(Fe_2O_3+CaO+MgO+K_2O+Na_2O)}{\%(SiO_2+TiO_2+Al_2O_3)}$	R_a^b < 0.5, low slagging risk [36]	Equation (3)

3. Results

3.1. Influence of pH on Bio-Coal Composition

The yields and ultimate analysis of the bio-coals derived from the swine manure are given in Table 2. The results show that, with decreasing pH, there is increased removal of ash, with lower-pH treatments having the lowest ash content for their respective temperature. The exception is acetic acid (pH 2.88), which appears to have slightly lower ash content than formic acid (pH 2.38). Yields on a dry basis also appear to be influenced by pH, with higher yields associated with lower pH. The exception to this is sulphuric acid at lower temperatures, with lower yields observed at 120 °C than that for the higher-pH treatments. This lower yield is in part because of the reduced ash content; however, the reduction in ash alone does not account for the reduced yield, and the results suggest that the lower yields appear to be due to enhanced oxygen removal. Due to the variation between ash contents with differing pH, Table 3 gives the ultimate analysis on a dry ash free (daf) basis, to enable direct comparisons between the organic chemistry of the different treatments. The results show that, for the 120 °C treatment in sulphuric acid, there is a higher carbon density and lower oxygen content for this treatment than the higher-pH treatments at this temperature.

Table 2. Ultimate analysis and yields of fuels on a dry basis; n/a—not applicable.

Sample Name	Dry Basis						Ash (wt.%)
	Yield (%)	C (wt.%)	H (wt.%)	N (wt.%)	S (wt.%)	O (wt.%)	
Unprocessed Pig Manure	n/a	44.0 ± 0.5	5.1 ± 0.1	2.9 ± 0.1	0.2 ± 0.0	35.7 ± 0.7	12.6
Sodium Hydroxide 120 °C	84	42.7 ± 0.3	5.0 ± 0.0	2.4 ± 0.0	0.1 ± 0.0	37.1 ± 0.5	12.6
Sodium Hydroxide 170 °C	64	45.0 ± 0.2	5.1 ± 0.1	2.1 ± 0.0	0.1 ± 0.0	32.6 ± 0.9	15.2
Sodium Hydroxide 200 °C	59	48.4 ± 0.2	5.2 ± 0.1	2.4 ± 0.0	0.2 ± 0.0	27.8 ± 0.0	16.0
Sodium Hydroxide 250 °C	41	49.8 ± 0.0	4.8 ± 0.0	2.5 ± 0.0	0.3 ± 0.0	15.0 ± 0.1	27.6
Water 120 °C	86	45.4 ± 0.5	5.8 ± 0.4	2.4 ± 0.1	0.1 ± 0.0	35.5 ± 1.0	10.7
Water 170 °C	62	47.3 ± 0.3	5.0 ± 0.0	2.3 ± 0.1	0.2 ± 0.0	31.2 ± 0.4	14.0
Water 200 °C	59	50.8 ± 0.3	5.2 ± 0.0	2.4 ± 0.0	0.2 ± 0.0	27.0 ± 0.1	14.3
Water 250 °C	43	53.5 ± 3.4	4.9 ± 0.4	2.8 ± 0.2	0.3 ± 0.0	16.7 ± 0.3	21.9
Acetic Acid 120 °C	83	45.0 ± 0.2	5.1 ± 0.1	2.6 ± 0.1	0.1 ± 0.0	36.2 ± 0.4	11.0
Acetic Acid 170 °C	61	47.9 ± 0.7	5.2 ± 0.1	2.5 ± 0.1	0.1 ± 0.0	31.1 ± 0.6	13.0
Acetic Acid 200 °C	59	50.1 ± 0.8	5.1 ± 0.1	2.3 ± 0.1	0.1 ± 0.0	27.9 ± 1.9	14.5
Acetic Acid 250 °C	44	56.6 ± 0.6	5.2 ± 0.1	2.9 ± 0.0	0.2 ± 0.0	15.4 ± 0.1	19.7
Formic Acid 120 °C	83	45.0 ± 1.1	5.4 ± 0.2	2.5 ± 0.1	0.2 ± 0.0	35.9 ± 1.4	11.0
Formic Acid 170 °C	61	49.8 ± 0.1	5.7 ± 0.2	2.6 ± 0.0	0.2 ± 0.0	29.0 ± 0.9	12.7
Formic Acid 200 °C	58	50.8 ± 0.2	5.0 ± 0.1	2.3 ± 0.1	0.2 ± 0.0	27.1 ± 0.3	14.6
Formic Acid 250 °C	44	56.0 ± 1.4	5.1 ± 0.1	2.9 ± 0.1	0.3 ± 0.0	14.9 ± 0.3	20.8
Sulphuric Acid 120 °C	75	47.9 ± 0.5	5.6 ± 0.0	2.2 ± 0.0	1.0 ± 0.0	35.2 ± 2.3	8.1
Sulphuric Acid 170 °C	58	50.5 ± 0.6	5.7 ± 0.0	2.4 ± 0.1	1.7 ± 0.1	29.8 ± 0.7	10.0
Sulphuric Acid 200 °C	57	52.4 ± 0.2	5.5 ± 0.0	2.2 ± 0.0	1.7 ± 0.0	28.1 ± 0.2	10.1
Sulphuric Acid 250 °C	47	56.4 ± 0.7	5.0 ± 0.0	2.7 ± 0.0	3.4 ± 0.0	17.3 ± 0.3	15.1

Table 3. Ultimate analysis proximate analysis and yields of fuels on a dry ash free basis.

Sample Name	Dry Ash Free Basis								
	Yield (%)	C (wt.%)	H (wt.%)	N (wt.%)	S (wt.%)	O (wt.%)	H/C	O/C	Volatile Matter (%)	Fixed Matter (%)
Unprocessed Pig Manure	n/a	50.1	5.8	3.3	0.2	40.6	1.39	0.61	74	26
Sodium Hydroxide 120 °C	84	48.9	5.7	2.8	0.2	42.5	1.40	0.65	80	20
Sodium Hydroxide 170 °C	62	53.0	6.0	2.4	0.1	38.4	1.36	0.54	80	20
Sodium Hydroxide 200 °C	57	57.7	6.2	2.8	0.2	33.1	1.29	0.43	77	23
Sodium Hydroxide 250 °C	34	68.8	6.6	3.4	0.4	20.8	1.15	0.23	68	32
Water 120 °C	88	50.9	6.5	2.7	0.2	39.7	1.53	0.59	82	18
Water 170 °C	61	54.9	5.9	2.7	0.3	36.2	1.28	0.49	78	22
Water 200 °C	58	59.3	6.1	2.8	0.3	31.5	1.24	0.40	76	24
Water 250 °C	38	68.5	6.2	3.6	0.4	21.3	1.09	0.23	67	33
Acetic Acid 120 °C	85	50.5	5.8	2.9	0.2	40.7	1.37	0.60	80	20
Acetic Acid 170 °C	61	55.1	6.0	2.9	0.2	35.8	1.30	0.49	79	21
Acetic Acid 200 °C	58	58.6	5.9	2.7	0.2	32.7	1.21	0.42	75	25
Acetic Acid 250 °C	40	70.5	6.5	3.7	0.3	19.1	1.10	0.20	67	33
Formic Acid 120 °C	85	50.6	6.0	2.8	0.2	40.4	1.43	0.60	81	19
Formic Acid 170 °C	61	57.0	6.5	3.0	0.2	33.2	1.38	0.44	79	21
Formic Acid 200 °C	57	59.5	5.9	2.6	0.2	31.7	1.19	0.40	75	25
Formic Acid 250 °C	40	70.7	6.5	3.6	0.3	18.9	1.10	0.20	68	32
Sulphuric Acid 120 °C	79	52.1	6.1	2.4	1.1	38.3	1.40	0.55	83	17
Sulphuric Acid 170 °C	60	56.1	6.3	2.6	1.9	33.1	1.35	0.44	83	17
Sulphuric Acid 200 °C	59	58.3	6.1	2.4	1.9	31.3	1.26	0.40	76	24
Sulphuric Acid 250 °C	46	66.4	5.9	3.2	4.0	20.4	1.07	0.23	66	34

A van Krevelen plot of the bio-coals from different pH and temperatures is presented in Figure 1. The results indicate that, with increasing temperature, the bio-coal has a more coal-like property, indicating increasing levels of dehydration with increasing temperature and decreasing pH. The pH appears to have a significant influence on dehydration, particularly at lower temperatures (120–170 °C), with sulphuric acid (pH 1) promoting the greatest levels of dehydration and sodium hydroxide (pH 13) showing the least dehydration. This effect of pH on dehydration, however, becomes less as the temperature is increased.

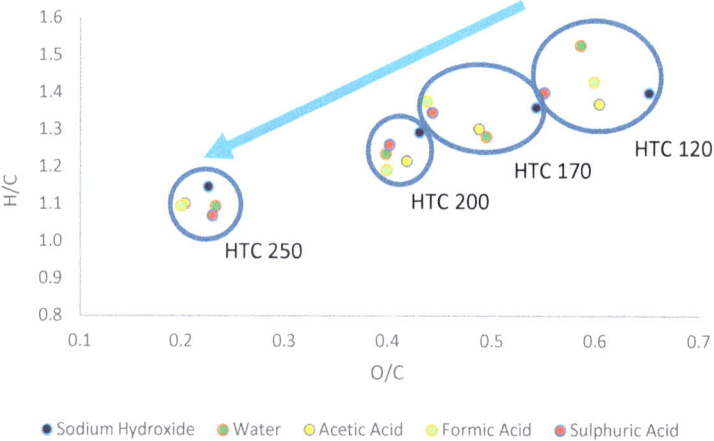

Figure 1. Van Krevelen plot of the bio-coals from different pH and temperature.

It should also be noted that the most dehydrated/coalified fuel for the 250 °C treatment is actually formic acid, followed by acetic acid (pH 2.38 and 2.88, respectively). The dry ash free ultimate analysis data presented in Table 3 also indicate that these two acid-treated bio-coals have the highest carbon density with 71% (daf). This higher carbon content could, however, be due to formic and acetic acid adding to the carbon in the bio-coal, lowering the O/C ratio. The use of a mineral acid at pH 2.38 and 2.88 could give an O/C ratio similar to that of water in Table 3 and Figure 1. This is because temperatures above 200 °C bring about high dissociation of H$^+$ and OH$^-$ in the water, leading to the decomposition of monosaccharides to organic acids, rapidly dropping process water pH to approximately 3 [37,38]. The results in Table 3 suggest that the bio-coal from sulphuric acid at 250 °C has the lowest carbon content at 66% (daf). It should, however, be noted that the sulphuric acid samples acquired sulphur from the acid, which makes up 4% (daf) of the fuel, while the other samples are low in sulphur; this would reduce the relative carbon content of the fuel when compared to the other treatments, even when correcting to a dry ash free basis. The analysis of the 250 °C bio-coals in Table 3 would suggest that the bio-coal is similar in property to lignite A coal, as described in Smith et al. [39], for all pH treatments.

When the yields are corrected on a dry ash free basis, as shown in Table 3, it indicates that decreasing pH increases yields at 250 °C but decreases yields below 200 °C. This is likely due to the lower pH enhancing the rate of hydrolysis. The generation of hydronium ions due to the presence of acids is known to catalyse the hydrolysis of hemi-cellulose and cellulose into monosaccharides in lignocellulosic biomass [40].

Figure 2 shows the combustion profiles of the bio-coals following different treatments with different pH and temperature, and it shows a distinct volatile burn peak at around 300 °C, which is normally consistent with the presence of cellulose within a fuel [29,30]. Following HTC at 250 °C, this peak is absent, suggesting that the cellulose was removed. The lower yields associated with lower pH in the 120 °C, 170 °C, and 200 °C treatments are due to hydrolysis of the cellulose or "cellulose-like" component within the feedstock, resulting in its removal from the resulting bio-coal. For the 250 °C treatment, the increased yield at lower pH would suggest that pH is catalysing repolymerisation.

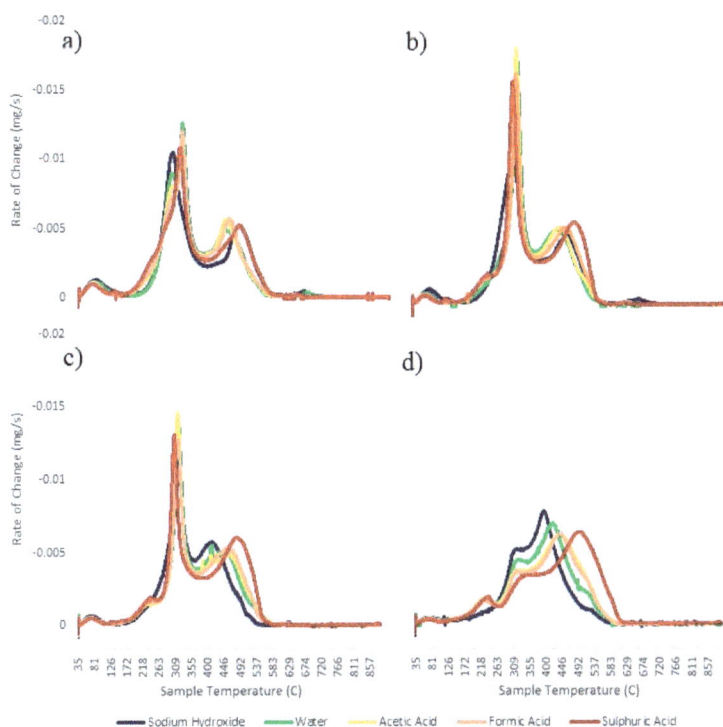

Figure 2. Derivative thermogravimetric (DTG) burning profiles for (**a**) 120 °C, (**b**) 170 °C, (**c**) 200 °C, and (**d**) 250 °C treatments.

In HTC, the hydrolysis of components such as hemi-cellulose and cellulose results in the formation of monosaccharides within the process water, which then undergo dehydration and fragmentation processes, giving rise to different soluble products such as furfural and hydroxymethylfurfural (HMF), benzenetrol, carboxylic acids, and aldehydes (acetaldehyde, acetonitrilacetone) [40,41]. These decomposition products undergo polymerisation and condensation to form insoluble polymers, often known as humins, which make up a portion of the bio-coal [42,43]. The pH should play a key role in this polymerisation due to the influence pH has on the zeta potential. Zeta potential, or, as it is more correctly known, electrokinetic potential, is a key indicator of the stability of colloidal dispersions, and the magnitude of the zeta potential indicates the degree of repulsion of like-charged particles in that dispersion. A high zeta potential will result in the solution or dispersion resisting agglomeration and flocculation. Bio-coals have a negative zeta potential. This is because they have oxygenated functional groups on their surface, making them behave like weak acids [44]. When a base is added to a suspension with a negative zeta potential, the particles tend to acquire a more negative charge; this reduces the chance of the suspension flocculating and, in the case of hydrothermal suspensions, polymerising to form bio-coal [44]. This is demonstrated in the sodium hydroxide 250 °C bio-coal, which has the lowest yield. If an acid is added to the suspension, a point is reached where the negative charge is neutralised [45]. At this point, the zeta potential is at zero, and it is called the isoelectric point. At this moment, the suspension is most likely to flocculate. For the water- and acid-treated samples, the lower pH increases the yields for the 250 °C treatments, which would suggest that lowering pH reduces the electrokinetic potential of the decomposition products within the aqueous phase, enabling them to polymerise and increase char yield.

The hypothesis that lower pH catalyses flocculation and polymerisation would also support the findings of Ghanim et al. [15], who performed HTC of poultry litter at 250 °C for 2 h using various initial pH with sulphuric acid and found that increasing sulphuric acid content increased bio-coal yield and decreased ash content. This, however, does not agree with the findings in Chen et al. [46], who used sulphuric acid and bagasse and found the reverse. Shorter retention times were, however, used in Chen et al. [46] (5 min, 15 min, and 30 min) and, while the samples underwent polymerisation in Ghanim et al. [15], the extent of repolymerisation was less in Chen et al. [46], as repolymerisation and aromatisation are considerably slower than hydrolysis, decarboxylation, and dehydration reactions which initially occur [3].

When looking at the bulk properties, excluding sulphur content (carbon, hydrogen, nitrogen, oxygen, fixed carbon, and volatile carbon), there is only limited difference between the compositions of the bio-coals produced at 250 °C. This would suggest that, under these conditions, temperature is a more important parameter than pH. The bio-coals produced using the addition of formic and acetic acid result in an increased carbon content in the bio-coal; however, this could be due to the addition of carbon in the form of organic acid, as opposed to an influence of pH. A major increase in sulphur content (4%) is observed in the bio-coal produced using sulphuric acid, indicating incorporation of sulphur, most likely though Maillard chemistry [47].

3.2. Influence of pH on Fuel Inorganic Chemistry

The metal analysis of the bio-coals and unprocessed pig manure is given in Table 4. The results show that, despite apparent increases in ash content, the overall composition of mineral matter in the bio-coal changes depending on the temperature and the pH. For all treatments, most of the potassium, sodium, and strontium is removed into the aqueous phase. The 120 °C treatment with water demonstrates that, for all three metals, around 11% of the original metal is retained within the char, with the remaining 10% gradually reducing with increasing temperature and almost complete removal at 250 °C, regardless of pH. The exception is for the sodium hydroxide treatment, where, at 120 °C, additional sodium is added from the solution. This sodium addition, however, diminishes as the temperature increases and, by the 250 °C treatment, the bio-coal has the equivalent of only 12% of the original sodium in the feedstock along with almost complete removal of the potassium.

Table 4. Main inorganics present within the unprocessed pig manure and derived bio-coals; n/d—not determined.

Sample Name	mg/kg Fuel (db)											
	Na	K	Mg	Ca	Al	Si	P	Mn	Fe	Cu	Zn	Sr
Unprocessed Pig Manure	1426	12,100	9395	27,035	434	3654	15,197	314	1090	140	634	367
Sodium Hydroxide 120 °C	2933	4126	9869	30,212	413	3447	15,431	445	1278	153	604	39
Sodium Hydroxide 170 °C	2020	2927	10,002	41,385	620	4067	19,254	633	1712	230	864	61
Sodium Hydroxide 200 °C	1121	1371	7056	45,915	715	3324	24,748	531	1797	254	963	51
Sodium Hydroxide 250 °C	404	577	17,221	74,240	1174	10,311	41,254	819	3177	394	1771	93
Water 120 °C	193	1493	6232	31,645	546	3710	13,139	411	1464	195	701	45
Water 170 °C	92	1043	5388	41,790	674	3249	21,806	599	1632	227	788	61
Water 200 °C	156	411	6544	38,053	660	3203	25,178	452	1224	197	657	40
Water 250 °C	n/d	392	10,323	61,727	1008	4533	36,468	766	2287	380	1444	80
Acetic Acid 120 °C	199	1920	4201	30,840	572	4709	15,020	433	1447	220	701	45
Acetic Acid 170 °C	37	469	4201	38,345	783	3699	20,876	499	1527	262	758	45
Acetic Acid 200 °C	105	425	4596	43,490	721	3321	23,494	573	1555	271	853	55
Acetic Acid 250 °C	53	397	7123	57,455	941	3504	34,106	590	2252	324	1489	71
Formic Acid 120 °C	114	1048	3685	30,815	576	5344	15,808	403	1523	208	653	42
Formic Acid 170 °C	85	644	3624	37,304	750	3995	20,297	462	1336	234	642	48
Formic Acid 200 °C	125	508	4542	43,529	809	3646	24,236	500	1411	286	803	55
Formic Acid 250 °C	161	743	8307	60,023	1023	4893	34,543	901	2528	317	1516	89
Sulphuric Acid 120 °C	171	1312	1490	19,342	609	5182	2566	56	518	195	234	27
Sulphuric Acid 170 °C	160	1252	1170	25,280	670	4559	1728	n/d	593	222	226	28
Sulphuric Acid 200 °C	125	1213	2152	24,507	709	4287	3815	138	734	203	281	32
Sulphuric Acid 250 °C	28	406	4083	34,688	877	3472	6670	414	1295	336	923	55

The results of the 120 °C treatment with water are significant, as autoclaving fuel samples at 120 °C for one hour in water is a method provided in BS EN ISO 16995-2 as a methodology for determining free ionic salts within a biofuel. Consequently, metals removed in this treatment can, under BS EN ISO 16995-2, be regarded as in free ionic form. This result would suggest that around 90% of the sodium, potassium, and strontium is in free ionic form within the pig manure, along with 40% of the magnesium and 25% of the phosphorus. The calcium, aluminium, silicon, manganese, iron, copper, and zinc do not appear to be in a water-soluble ionic form. In higher plants, typically over 90% of potassium and sodium is in ionic form, while 60–90% and 30–85% of magnesium and calcium is ionic [48]. The results for the 120 °C treatment with water would suggest that, for pig manure, there is a similar proportion of free ionic potassium and sodium in the fuel, but a greater proportion of the calcium, magnesium, and phosphorus is either organically associated or present in mineral form than in lignocellulosic biomass [16,49]. The relatively low extraction of phosphorus in water at 120 °C would suggest that phosphorus is present as low-solubility salts such as calcium and magnesium phosphate.

The behaviour of silicon is particularly notable in the results as, for all pH treatments, it appears to undergo increasing removal with increasing temperature, with between 45% and 60% retained, depending on pH, at 250 °C. In previous work, it was generally found that silicon is reasonably recalcitrant, being largely retained in the bio-coal, as, to become water-soluble silicon, it has to be hydrated and become silicic acid (H_4O_4Si), which, unless kept buffered within certain boundaries, readily degrades back to insoluble silicon dioxide (SiO_2) [29]. Magnesium retention appears to be strongly influenced by pH, with the highest retentions seen with the sodium hydroxide (pH 13) and reduced retention with decreasing pH. Temperature is also critical, with the highest removal of magnesium typically being observed at 170 °C and increasing retention at 200 °C and 250 °C at all pH.

The metal retention behaviour for magnesium, whereby there is an initial reduction in metal retention up to 170 °C followed by increased retention of metals with increasing temperature between 200 °C and 250 °C, is also observed for calcium, magnesium, iron, and zinc, along with phosphorus. For these elements, there appears to be almost complete retention within the bio-coal at 250 °C for the sodium hydroxide, water, acetic acid, and formic acid treatments. The exception is sulphuric acid, which extracts 50% of the calcium between 120 °C and 200 °C, with 60% retained at 250 °C. Most of the phosphorus is extracted by the sulphuric acid with 87% extracted at 120 °C and 200 °C and over 90% extracted at 170 °C. Additionally, 80% of the original phosphorus is extracted at 250 °C, leaving 20% of the original phosphorus within the bio-coal. Similar ratios of magnesium are extracted as seen for the phosphorus using sulphuric acid. Aluminium and copper appear to remain within the bio-coal irrespective of temperature and pH.

The additional removal of the metals with the addition of sulphuric acid is due to the high dissociation constant of sulphuric acid, which is a strong acid and forms hydronium ions in two stages, with the initial loss of a hydrogen (Equation (4)) and the subsequent decomposition of the bi-sulphate (Equation (5)). Here, 0.1 molar sulphuric acid gives a pH of 1, but the same pH is possible using the same concentration of other strong mineral acids such as hydrochloric acid, or higher concentrations of weaker acids such as acetic acid which only partially dissociate (hence, pH 2.88 at the same molar concentration). The justification for using acid catalysts in HTC, according to Ekpo et al. [8] and Dai et al. [2,7], is to principally mobilise the phosphorus into the aqueous phase for subsequent recovery. The relatively low extraction of phosphorus in water at 120 °C would suggest that phosphorus is present as low-solubility salts such as calcium and magnesium phosphate. The high concentrations of hydronium ions generated by the sulphuric acid are required for the acid leaching of the phosphorus, with the mechanism for acid leaching of calcium given in Equation (6). With sulphuric acid, the calcium is converted to calcium sulphate (see Equation (7)) liberating the phosphorus as phosphoric acid [50]. Similar phosphorus extraction is possible using hydrochloric acid, although, when leaching iron ores, slightly higher efficiency is observed for sulphuric acid [50]. Greater calcium extraction may be possible using hydrochloric acid, as calcium chloride is more water-soluble than calcium sulphate (745 g/L as opposed to 2.6 g/L at standard temperature and pressure (STP)).

This could explain the higher retention of calcium in the bio-coal when compared with magnesium, calcium, phosphorus, manganese, iron, and zinc, as the process water may be saturated once cooled to room temperature.

$$H_2SO_4 + H_2O \rightarrow H_3O^+ + HSO_4^- \quad K1 = 2.4 \times 10^6 \text{ (strong acid)} \tag{4}$$

$$HSO_4^- + H_2O \rightarrow H_3O^+ + HSO_4^- \quad K2 = 1.0 \times 10^{-2} \tag{5}$$

$$Ca_{10}(PO_4)_6X(s) + 20H^+ = 6H_3PO_4 + 10Ca^{2+} + H_2X \tag{6}$$

$$Ca^{2+} + SO_4^{2-} = CaSO_4. \tag{7}$$

3.3. Influence of pH on Fuel Combustion Chemistry

Table 5 gives the energy content, the volatile content, and the results of slagging and fouling indices derived from the inorganic chemistry are given in Table 4. The results show that the reaction temperature has the biggest impact on the energy content of the fuel, irrespective of the pH. The results indicate that HTC with sodium hydroxide gives the lowest energy density at all temperatures, and that undertaking HTC in the presence of acids at decreasing pH increases the HHV of the bio-coal, as demonstrated in Ghanim et al. [15]. The highest HHV is observed for acetic and formic acid due to their higher carbon contents, due to the increased carbon within the hydrothermal reaction, as previously discussed; however, there still appears a trend of increasing HHV with decreasing pH.

Table 5. Energy content, volatile content, and slagging and fouling indices for the bio-coals. HHV—higher heating value.

Sample Name	Dry Basis					
	HHV (MJ/kg)	Volatile Matter (%)	Fixed Matter (%)	AI	BAI	R_a^b
Unprocessed Pig Manure	15.8	64.8	22.6	1.17	0.08	8.54
Sodium Hydroxide 120 °C	14.9	70.9	18.1	0.86	0.14	9.05
Sodium Hydroxide 170 °C	16.7	67.9	16.7	0.54	0.27	8.76
Sodium Hydroxide 200 °C	18.9	62.8	18.8	0.25	0.55	9.91
Sodium Hydroxide 250 °C	21.0	48.3	22.4	0.09	2.55	5.76
Water 120 °C	17.3	76.9	16.9	0.13	0.90	6.63
Water 170 °C	17.6	68.3	18.9	0.09	1.55	8.73
Water 200 °C	19.8	65.7	20.3	0.05	1.92	8.31
Water 250 °C	22.1	54.0	27.0	0.02	6.92	9.32
Acetic Acid 120 °C	16.1	69.9	17.1	0.18	0.73	4.97
Acetic Acid 170 °C	18.1	67.3	17.8	0.04	3.29	6.81
Acetic Acid 200 °C	19.2	63.2	21.0	0.04	2.80	8.52
Acetic Acid 250 °C	23.8	53.4	26.2	0.03	5.19	10.44
Formic Acid 120 °C	16.5	73.5	16.9	0.10	1.39	4.27
Formic Acid 170 °C	19.8	69.3	18.1	0.05	1.90	6.19
Formic Acid 200 °C	19.6	63.4	21.1	0.05	2.12	7.72
Formic Acid 250 °C	23.6	55.4	25.8	0.06	2.72	8.35
Sulphuric Acid 120 °C	17.9	77.7	16.4	0.11	0.36	2.66
Sulphuric Acid 170 °C	19.8	74.3	14.8	0.10	0.44	3.67
Sulphuric Acid 200 °C	20.6	66.7	20.9	0.09	0.58	3.91
Sulphuric Acid 250 °C	23.2	56.3	28.7	0.02	3.28	6.41

The volatile matter content of the bio-coal is also given in Table 5. The volatile matter content is important for predicting combustion behaviour, as, during combustion, the volatiles prevent oxygen from oxidising the carbon, hydrogen, and sulphur present within the fuel particle, bringing about two-stage combustion within a furnace [51]. Moreover, the escaping volatiles burn much more quickly than the char (the fraction remaining after devolatilisation); therefore, understanding the devolatilisation behaviour of a fuel is important in terms of flame ignition, flame stability, flammability limits, and the formation of pollutants such as nitrogen oxides [52]. The volatile content is also useful when determining the equivalent coal rank, with coals with higher ranks having lower volatile contents.

The volatile content appears to be most strongly influenced by reaction temperature, with lower volatile content with increasing reaction temperature. Figure 2 shows the combustion profiles of the bio-coals following different treatments, with the 120 °C, 170 °C, and 200 °C treatments giving a distinct volatile burn peak at around 300 °C. By the 250 °C treatments, this peak is all but removed from the profiles of all fuel. For lignocellulosic biomass combustion, the volatile burn is often closely associated with the thermal decomposition of the hemi-cellulose and cellulose [53,54]. Hemi-cellulose and cellulose are readily degraded at hydrothermal temperatures of 180 °C and 200 °C, respectively [55], and the distinct volatile burn peak at around 300 °C is normally consistent with the presence of cellulose within a fuel [29,30]. This would then suggest the presence, and removal, of fibrous material within the pig manure in a similar manor to that seen for lignocellulosic biomass.

The results presented in Ghanim et al. [15] stated that low pH with sulphuric acid brought higher volatile matter contents in the bio-coal; however, this contradicts the findings in Chen et al. [46], who used sulphuric acid and bagasse and found the inverse. The volatile contents presented in Table 5 would initially support the findings of Ghanim et al. [15]; however, the reason that the volatile matter appears to increase is due to the decrease in ash content of the fuel, increasing the relative amount of volatile matter present. Due to the variation between ash contents with differing pH, the volatile contents are calculated on a dry ash free basis in Table 3 to enable direct comparisons between the volatile chemistry of the different treatments. These results show there is little change between pH, perhaps even suggesting a small decrease in the volatile matter content of the 250 °C treatments with decreasing pH. In the work presented in Chen et al. [46], the fuel was considerably lower in ash than the samples used in Ghanim et al. [15]; thus, there would not be the apparent increase due to the decrease in ash content of the fuel. Consequently, the results presented here would support the findings of both studies.

Combustion of these HTC bio-coals is considered as a coal substitute for coal-powered power plants, enabling utilisation of pre-existing infrastructure. Coal-powered power plants are usually designed to burn a specific type of coal, normally a coal obtained in and around the locality of the plant. Consequently, power stations have a design fuel specification that sets out, amongst others, the ash content, energy content, particle size, and slagging and fouling properties of the fuel. When changing from a design fuel specification, care is required to ensure that the new fuel achieves a stable flame, required to ensure safe boiler operation [56]. Biomass often has different combustion characteristics to that of coal, principally due to a higher proportion of volatile carbon and a much smaller char fraction than those seen in coals [57]. When burning high volatile fuels, combustion starts with the ignition of volatile gases surrounding the fuel particle and prevents oxygen from reaching and igniting the char, resulting in a two-phase combustion called homogeneous ignition, whereas, for typical coals used in pulverised fuel applications, devolatilisation, ignition, and combustion of the volatiles and char combustion occur almost simultaneously [58]. Issues with two-phase combustion occur when you start getting two areas of burning within the furnace that can then draw the flame from the burner and higher in the furnace, bringing about flame instability [52]. This can be a particular issue when co-firing biomass fuels with coal or two fuels with different burning characteristics, as mismatched burning characteristics can result in two fuels burning independently within a furnace. In this instance, the rate of burning (flame velocity) may not match the rate of material feed, leading to the flame either blowing out or flashing back [56]. Thermogravimetric analysis (TGA) is one method originally developed to compare and evaluate fuel burning characteristics using the first derivative thermogravimetric (DTG) curve [52]. When undertaking this test, five key characteristic temperatures are taken, which were developed from the Babcock and Wilcox TGA method for coal and adapted to biomass. The first temperature is the volatile initiation temperature, where the weight loss begins. The second temperature is peak volatile burn, where you get the highest rate of mass loss during devolatilisation. The third temperature is the char initiation temperature, where the rate of combustion changes due to the onset of char combustion. The fourth temperature is peak char burn temperature. The fifth temperature is the burn-out temperature where the weight is constant, indicating the completion of combustion [52].

Figure 2 shows the DTG combustion profiles of the bio-coals following different treatments. For the 120 °C treatment, the acids reduce the first initiation temperature, with weight loss starting at 160 °C for the three acid treatments but 200 °C for the water and alkali treatment. For the water, acetic acid, and formic acid treatments, there appears to be an initial peak at 300 °C, which can be constant with the presence of a fibrous component, such as hemi-cellulose, followed by a larger second peak at 325 °C, which is typically associated with cellulose in lignocellulosic biomass [29,30]. This peak is the peak volatile burn. For the sulphuric acid 120 °C sample, there does not appear to be a distinct peak where the "hemi-cellulose-like" material decomposes, but the presence of a shoulder at 300 °C. A higher mass loss is observed between 160 °C and 300 °C than any other sample, which may suggest that this component is still present but partially hydrolysed, resulting in it thermally decomposing earlier. The sodium hydroxide treatment has only one distinct volatile peak, peaking at about 300 °C. This result is most likely a consequence of the strong basic conditions degrading the "hemi-cellulose-like" material observed with the strong acid conditions, but the high sodium and potassium contents of the fuel (see Table 4) could also be catalysing the volatile burn, giving a different combustion profile [59]. Both the sodium hydroxide and the sulphuric acid treatments have higher char burn temperatures with temperatures of 500 °C, as opposed to 450 °C for the other treatments. The burnout temperature is similar at 580 °C for all treatments.

As the process severity increases, the profiles for 170 °C, 200 °C, and 250 °C retain the first initiation temperature of 160 °C, and a new initial peak begins to arise at 235 °C for the water and acids, initially starting in the 170 °C profile but becoming increasingly pronounced in the 250 °C profile. This peak represents potentially hydrolysed or repolymerised structures chemisorbed to oxygen functional groups on the char surface, yet to dehydrate to form the stable ether or pyrone functional groups required to fix the carbon in the fixed carbon [41]. This peak does not appear in the sodium hydroxide profiles, which has a different combustion profile to that of the water and acid samples, particularly with regard to the volatile burn in the 170 °C and 200 °C profiles, potentially due to the influence of sodium (alkali metals) on catalysing devolatilisation and combustion [59–61]. For all the 170 °C and 200 °C combustion profiles, the main volatile peak becomes increasingly dominant at 325 °C, although, upon reducing its dominance in the 200 °C profiles, the volatile content within the fuel decreases and the fixed matter increases (see Table 5). The increasing dominance of this peak is most likely due to the removal of hydroxyl, carboxyl, and carbonyl groups present within the biomass, along with any structural components that are hydrolysed at lower temperatures such as hemi-cellulose in any cellulosic material present [3,54]. The peak at 325 °C is constant with the presence of residual cellulose and the breaking of its glyosidic linkages [30]. By the 250 °C treatments, the samples adopt a "coal-like" single-stage combustion profile [29], whereby the transition between the volatile release and initiation of char burn (char initiation temperature) is marked more by a "shoulder" as opposed to a distinct peak. This shoulder becomes less distinct with lower pH and correlates with a modest reduction in volatile matter.

The benefit of creating a more "coal-like" burning profile is that it aids flame stability. As previously discussed, when there is a two-stage burn, as seen in the burning profiles for the 120 °C treatments in Figure 2, you get homogeneous combustion, whereby the volatile burn and char burn occur in isolation. In this case, upon drying and devolatilisation, the fuel particle can become entrained in the gas stream and move higher in the furnace while still burning, drawing the flame upward and promoting flame instability [52]. With the single-stage "coal-like" profiles seen in the 250 °C treatments in Figure 2, the devolatilisation, ignition, and combustion of the volatiles do not occur in isolation; instead, the char should oxidise/combust at the same time (heterogeneous reaction), promoting a simultaneous combustion of the fuel mass and a stable flame [57].

In all the profiles, with increasing reaction severity, lower pH increases char burnout temperature. This is particularly notable at 250 °C as the peak temperature (char burn) increases from 400 °C for the sodium hydroxide (pH 13) to 500 °C for the sulphuric acid (pH 1). This result would suggest that pH influences reactivity of the char. There is strong consensus in the literature that the alkali metals, potassium and sodium, catalyse char reactivity [59,60,62,63], although the concentrations of alkali metals appears similar for the 250 °C bio-coals (see Table 4). The alkaline earth metals, calcium and magnesium, are understood to catalyse char reactivity but to a lesser extent than the alkali metals [63–65], with iron known to behave using similar mechanisms [66]. Given that pH appears to more strongly influence alkaline earth metal content, with higher calcium and magnesium contents, along with iron, associated with higher pH (see Table 4), it is possible that the higher concentrations at higher pH catalyse the thermal decomposition.

Figure 3 displays the ash transition temperatures obtained from the ash fusion test for the four different temperatures and five different pH values. Table 6 gives the transition temperatures for all samples, along with the unprocessed sample, and their standard errors. The results suggest that the unprocessed swine manure has a reasonably high deformation temperature of 1320 °C, compared to between 980 °C and 1140 °C for the conventional *Miscanthus* [26,29]. Despite this result, the shagging and fouling indices suggest almost certain flagging and fouling for these samples. The results of the ash fusion test certainly would suggest a low slagging fuel, possibly due to the high calcium and phosphorus content of the fuel. This would strongly suggest that calcium potassium phosphate complexes and calcium sodium phosphate complexes remove the potassium and sodium available to form low-melting-temperature potassium and sodium silicates [31,32]. A high ash melting temperature is usually indicative of a low alkali metal content and, thus, low fouling propensity temperatures [16], although caution is required here. The results of the 120 °C treatment with water would suggest that around 90% of the sodium and potassium in the fuel is in the form of free ionic salts. Potassium and sodium, when in the form of free ionic salts, are more readily released into the vapour phase and likely to bring about issues with fouling [16,67,68]. The high calcium and phosphorus content of the fuel may, however, prevent this release, instead forming the stable calcium potassium phosphate complexes and calcium sodium phosphate complexes [69]. The slagging and fouling indices used in this paper would not consider such mechanisms when predicting the fuel propensity to slag and foul.

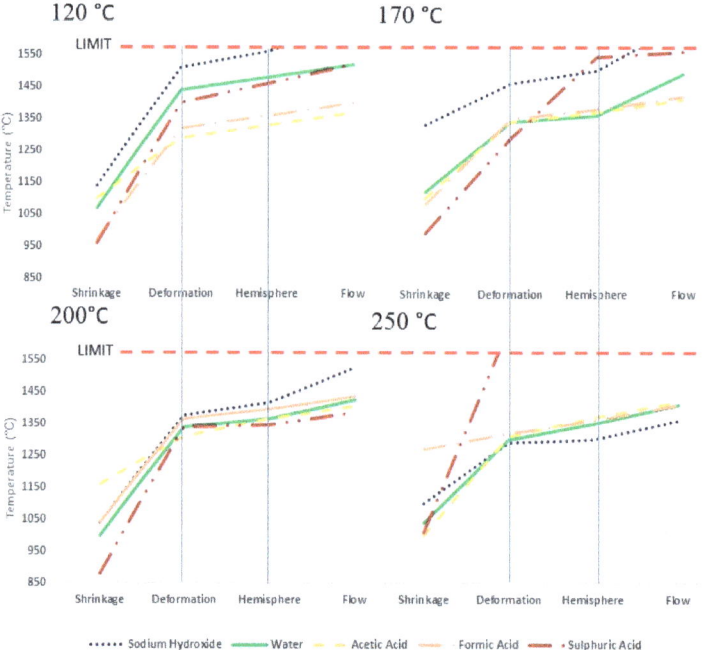

Figure 3. Ash transition temperatures from the ash fusion test for the different treatments at differing hydrothermal treatment temperatures.

Table 6. Ash transition temperatures from the ash fusion test.

Sample Name	Ash Transition Temperature (°C)			
	Shrinkage	Deformation	Hemisphere	Flow
Unprocessed	1260 ± 0	1320 ± 0	1380 ± 0	1500 ± 0
Sodium Hydroxide 120 °C	1140 ± 0	1510 ± 0	1560 ± 10	>1570
Sodium Hydroxide 170 °C	1330 ± 0	1460 ± 20	1500 ± 10	>1570
Sodium Hydroxide 200 °C	1040 ± 0	1380 ± 10	1420 ± 0	1530 ± 0
Sodium Hydroxide 250 °C	1100 ± 0	1290 ± 0	1300 ± 0	1360 ± 0
Water 120 °C	1070 ± 0	1440 ± 0	1480 ± 0	1520 ± 0
Water 170 °C	1120 ± 0	1340 ± 0	1360 ± 0	1490 ± 0
Water 200 °C	1000 ± 0	1340 ± 0	1370 ± 0	1430 ± 0
Water 250 °C	1040 ± 0	1300 ± 0	1350 ± 0	1410 ± 0
Acetic Acid 120 °C	1100 ± 0	1290 ± 0	1330 ± 0	1370 ± 0
Acetic Acid 170 °C	1100 ± 0	134 ± 00	1370 ± 0	1410 ± 0
Acetic Acid 200 °C	1160 ± 0	1310 ± 0	1370 ± 0	1410 ± 0
Acetic Acid 250 °C	1000 ± 0	1310 ± 0	1370 ± 0	1420 ± 0
Formic Acid 120 °C	970 ± 0	1320 ± 0	1360 ± 0	1400 ± 0
Formic Acid 170 °C	1080 ± 0	1350 ± 0	1380 ± 0	1420 ± 0
Formic Acid 200 °C	1040 ± 0	1370 ± 0	1400 ± 0	1440 ± 0
Formic Acid 250 °C	1270 ± 0	1320 ± 0	1360 ± 0	1410 ± 0
Sulphuric Acid 120 °C	960 ± 0	1400 ± 80	1460 ± 60	1520 ± 20
Sulphuric Acid 170 °C	990 ± 0	129 ± 00	1545 ± 5	1560 ± 0
Sulphuric Acid 200 °C	880 ± 0	1340 ± 0	1350 ± 0	1390 ± 0
Sulphuric Acid 250 °C	1010 ± 0	>1570		

For the hydrothermally treated samples, the results show that ash shrinkage temperature is reduced; however, this is believed to be due to formation of carbonates during the hydrothermal process [26,28,29]. The deformation and hemisphere temperatures, however, appear to change reasonably little, indicating limited change to slagging propensity. The exception to this is the low-temperature sodium hydroxide-treated samples and the 250 °C treated sulphuric acid sample. The 120 °C, 170 °C, and 200 °C sodium hydroxide samples are suppressing, given that these are the only hydrothermal samples where the alkali index (given in Table 5) suggests almost certain slagging and fouling. Sodium and potassium contents are, however reduced, when compared to the starting feedstock, which may explain this increase.

In the 250 °C treated sulphuric acid sample, there is the greatest improvement in ash behaviour with the sample not undergoing deformation within the test conditions (test limit 1570 °C). This is predominantly due to the ash becoming a highly stable magnesium calcium phosphate silicate complex. It should be noted that the sulphuric acid samples acquire sulphur from the acid, which makes up 3.4% (db) of the fuel. Sulphur can play both a positive and a negative role during combustion in large combustion plants. During combustion, sulphur is predominantly oxidised to sulphur dioxide (SO_2) (>95%), but some sulphur trioxide (SO_3) is also formed [16]. While the sulphur dioxide plays an undesirable role in terms of corrosion, active oxidation of furnace components, and emissions of sulphur dioxide to the atmosphere (unless abated using flue gas desulphurisation) [17], sulphur trioxide plays an important role in the abatement of particulate emissions. This is due to the thermally derived sulphur trioxide forming sulphuric acid in the flue gas, which then adsorbs onto the fly ash particulates [70]. This affects the surface electrical conductivity of the particulate, greatly increasing the efficiency of the electrostatic pacificators [71]. This could be particularly advantageous for the 250 °C treated sulphuric acid sample, given that potential to emit (PTE) metals (strontium, copper, and zinc) are present within the swine manure and the derived hydrothermal fuels. Precipitation of these in the fly ash would be required to avoid issues with emissions.

When combusting fuel in pulverised applications, fuel sulphur is desirable as, while sulphur emissions are largely in the form of sulphur dioxide, when a fuel with high sulphur content is combusted, there is generally enough sulphur trioxide formed to bring the electrical resistivity of the fly ash into a range which results in good precipitator operation [72]. This can be a particular issue with biomass, which is typically low in sulphur and results in a very low collection efficiency of electrostatic pacificators [72]; consequently, biomass-fuelled furnaces, such as Drax (UK), add sulphur to overcome this. The sulphur content of the 250 °C treated sulphuric acid sample is typically too high for single-fuel combustion; however, if blended with low sulphur biofuels or coal, this sulphur content could be brought within "design fuel" specification. Moreover, the magnesium calcium phosphate silicate ash of the fuel would have an additive influence and could improve the slagging and fouling propensity of the blended biomass and coal. Consequently, with blending, the 250 °C treated sulphuric acid fuel could be safely combusted within the pulverised fuel plant if appropriately blended.

4. Conclusions

The influence of pH most strongly influences ash chemistry, with decreasing pH increasing the removal of ash. This reduction in ash has the biggest influence on the volatile carbon and energy content of the fuel, with lower ash contents bringing about higher energy densities when calculated on a dry basis for a given temperature. The pH also influences dehydration, with fuel dehydration increased with decreasing pH, although, with increasing temperature, the influence pH has on dehydration becomes less. The pH and temperature appear to influence yield, with lower pH increasing yields above 250 °C but decreasing yields below 200 °C. The lower yields below 200 °C appear due to the acids catalysing hydrolysis of "cellulose-like" fibres within the swine manure, whereas the higher yields at 250 °C could be due to the low pH catalysing polymerisation due to its influence on the electrokinetic potential of the hydrothermal suspension.

The water experiments at 120 °C would suggest that around 90% of the sodium and potassium is in free ionic form within the pig manure, along with 40% of the magnesium and 25% of the phosphorus. This free ionic sodium and potassium are more readily released into the vapour phase, and they are likely to bring about issues with fouling if combusted. Slagging and fouling indices suggest that they cannot be safely combusted without treatment. Nonetheless, the ash fusion test suggests reasonably high deformation temperatures, suggesting low slagging and fouling. This paradox is brought about through the high calcium and phosphorus content of the fuel forming the stable calcium potassium/sodium phosphate complexes. Hydrothermally treating the fuels achieves almost complete removal of sodium and potassium and their associated issues with fouling. Increasing reaction temperature appears to immobilise calcium, magnesium, iron, zinc, and phosphorus within the bio-coal unless treated at low pH, which enables mobilisation of the phosphorus and alkaline earth metals. Treatment at 250 °C results in a more coal-like combustion fuel, with fuel properties similar to that of lignite coal and an HHV between 21 and 23 MJ/kg depending on pH. The removal of the alkaline earth metals and iron reduces the reactivity of the fuel treated at pH 1. Despite the mobilisation of calcium and phosphorus using strong acid, sufficient calcium and phosphorus is retained within the ash to give very favourable ash chemistry in terms of slagging and fouling. The use of sulphuric acid does result in residual sulphur within the fuel; however, this sulphur may be beneficial due to the influence that thermally derived sulphur trioxide has on the collection efficiency of electrostatic precipitators and particulate removal, if appropriately blended with another low-sulphur fuel.

Author Contributions: Conceptualisation, A.M.S., U.E., and A.B.R.; methodology, A.M.S.; validation, A.M.S.; formal analysis, A.M.S.; investigation, A.M.S.; resources, A.B.R.; data curation, A.M.S. and U.E.; writing—original draft preparation, A.M.S.; writing—review and editing, A.M.S. and A.B.R.; visualisation, A.M.S.; supervision, A.B.R. All authors have read and agreed to the published version of the manuscript.

Funding: This research was funded by the Engineering and Physical Sciences Research Council (EPSRC) Doctoral Training Centre in Low Carbon Technologies (EP/G036608/1), the Niger Delta Development Commission (NDDC), the European Commission ERDF Interreg IVb NEW "Biorefine" project and Grønt Udviklings- og Demonstrationsprogram (GUDP) (34009-18-1435).

Acknowledgments: The authors would like to thank the University of Leeds farm for the supply of swine manure and would also like to thank Simon Lloyd, Karine Alves Thorne, and Adrian Cunliffe from the University of Leeds for their technical assistance.

Conflicts of Interest: The authors declare no conflicts of interest.

References

1. Szogi, A.A.; Vanotti, M.B.; Ro, K.S. Methods for Treatment of Animal Manures to Reduce Nutrient Pollution Prior to Soil Application. *Curr. Pollut. Rep.* **2015**, *1*, 47–56. [CrossRef]
2. Heilmann, S.M.; Molde, J.S.; Timler, J.G.; Wood, B.M.; Mikula, A.L.; Vozhdayev, G.V.; Colosky, E.C.; Spokas, K.A.; Valentas, K.J. Phosphorus Reclamation through Hydrothermal Carbonization of Animal Manures. *Environ. Sci. Technol.* **2014**, *48*, 10323–10329. [CrossRef]
3. Funke, A.; Ziegler, F. Hydrothermal carbonization of biomass: A summary and discussion of chemical mechanisms for process engineering. *Biofuels Bioprod. Biorefining* **2010**, *4*, 160–177. [CrossRef]
4. Heilmann, S.M.; Davis, H.T.; Jader, L.R.; Lefebvre, P.A.; Sadowsky, M.J.; Schendel, F.J.; von Keitz, M.G.; Valentas, K.J. Hydrothermal carbonization of microalgae. *Biomass Bioenergy* **2010**, *34*, 875–882. [CrossRef]
5. Heilmann, S.M.; Jader, L.R.; Sadowsky, M.J.; Schendel, F.J.; von Keitz, M.G.; Valentas, K.J. Hydrothermal carbonization of distiller's grains. *Biomass Bioenergy* **2011**, *35*, 2526–2533. [CrossRef]
6. Ekpo, U.; Ross, A.B.; Camargo-Valero, M.A.; Williams, P.T. A comparison of product yields and inorganic content in process streams following thermal hydrolysis and hydrothermal processing of microalgae, manure and digestate. *Bioresour. Technol.* **2016**, *200*, 951–960. [CrossRef]
7. Dai, L.; Tan, F.; Wu, B.; He, M.; Wang, W.; Tang, X.; Hu, Q.; Zhang, M. Immobilization of phosphorus in cow manure during hydrothermal carbonization. *J. Environ. Manag.* **2015**, *157*, 49–53. [CrossRef] [PubMed]
8. Ekpo, U.; Ross, A.B.; Camargo-Valero, M.A.; Fletcher, L.A. Influence of pH on hydrothermal treatment of swine manure: Impact on extraction of nitrogen and phosphorus in process water. *Bioresour. Technol.* **2016**, *214*, 637–644. [CrossRef] [PubMed]
9. Titirici, M.-M.; Thomas, A.; Antonietti, M. Back in the black: hydrothermal carbonization of plant material as an efficient chemical process to treat the CO_2 problem? *New J. Chem.* **2007**, *31*, 787–789. [CrossRef]
10. Hu, B.; Yu, S.-H.; Wang, K.; Liu, L.; Xu, X.-W. Functional carbonaceous materials from hydrothermal carbonization of biomass: an effective chemical process. *Dalton Trans.* **2008**, 5414–5423. [CrossRef] [PubMed]
11. Lu, X.; Zhang, Y.; Angelidaki, I. Optimization of H_2SO_4-catalyzed hydrothermal pretreatment of rapeseed straw for bioconversion to ethanol: focusing on pretreatment at high solids content. *Bioresour. Technol.* **2009**, *100*, 3048–3053. [CrossRef]
12. Demir-Cakan, R.; Baccile, N.; Antonietti, M.; Titirici, M.-M. Carboxylate-rich carbonaceous materials via one-step hydrothermal carbonization of glucose in the presence of acrylic acid. *Chem. Mater.* **2009**, *21*, 484–490. [CrossRef]
13. Lynam, J.G.; Coronella, C.J.; Yan, W.; Reza, M.T.; Vasquez, V.R. Acetic acid and lithium chloride effects on hydrothermal carbonization of lignocellulosic biomass. *Bioresour. Technol.* **2011**, *102*, 6192–6199. [CrossRef] [PubMed]
14. Reza, M.T.; Rottler, E.; Herklotz, L.; Wirth, B. Hydrothermal carbonization (HTC) of wheat straw: Influence of feedwater pH prepared by acetic acid and potassium hydroxide. *Bioresour. Technol.* **2015**, *182*, 336–344. [CrossRef] [PubMed]
15. Ghanim, B.M.; Kwapinski, W.; Leahy, J.J. Hydrothermal carbonisation of poultry litter: Effects of initial pH on yields and chemical properties of hydrochars. *Bioresour. Technol.* **2017**, *238*, 78–85. [CrossRef] [PubMed]
16. Koppejan, J.; Van Loo, S. *The Handbook of Biomass Combustion and Co-Firing*; Earthscan: London, UK, 2012.
17. Riedl, R.; Dahl, J.; Obernberger, I.; Narodoslawsky, M. Corrosion in fire tube boilers of biomass combustion plants. In Proceedings of the China International Corrosion Control Conference, Beijing, China, 9–13 October 1999.
18. Volpe, M.; Goldfarb, J.L.; Fiori, L. Hydrothermal carbonization of Opuntia ficus-indica cladodes: Role of process parameters on hydrochar properties. *Bioresour. Technol.* **2018**, *247*, 310–318. [CrossRef] [PubMed]
19. Mäkelä, M.; Kwong, C.W.; Broström, M.; Yoshikawa, K. Hydrothermal treatment of grape marc for solid fuel applications. *Energy Convers. Manag.* **2017**, *145*, 371–377. [CrossRef]
20. Lane, D.J.; Truong, E.; Larizza, F.; Chiew, P.; de Nys, R.; van Eyk, P.J. Effect of Hydrothermal Carbonization on the Combustion and Gasification Behavior of Agricultural Residues and Macroalgae: Devolatilization Characteristics and Char Reactivity. *Energy Fuels* **2017**. [CrossRef]

21. Petrovil, J.; Perisil, N.; Maksimovil, J.D.; Maksimovil, V.; Kragovil, M.; Stojanovil, M.; Lauševil, M.; Mihajlovil, M. Hydrothermal conversion of grape pomace: Detailed characterization of obtained hydrochar and liquid phase. *J. Anal. Appl. Pyrolysis* **2016**, *118*, 267–277. [CrossRef]
22. Bach, Q.-V.; Tran, K.-Q.; Skreiberg, Ø. Accelerating wet torrefaction rate and ash removal by carbon dioxide addition. *Fuel Process. Technol.* **2015**, *140*, 297–303. [CrossRef]
23. Benavente, V.; Calabuig, E.; Fullana, A. Upgrading of moist agro-industrial wastes by hydrothermal carbonization. *J. Anal. Appl. Pyrolysis* **2014**, *113*, 89–98. [CrossRef]
24. Broch, A.; Jena, U.; Hoekman, S.; Langford, J. Analysis of Solid and Aqueous Phase Products from Hydrothermal Carbonization of Whole and Lipid-Extracted Algae. *Energies* **2013**, *7*, 62. [CrossRef]
25. Reza, M.T.; Lynam, J.G.; Uddin, M.H.; Coronella, C.J. Hydrothermal carbonization: Fate of inorganics. *Biomass Bioenergy* **2013**, *49*, 86–94. [CrossRef]
26. Smith, A.M.; Singh, S.; Ross, A.B. Fate of inorganic material during hydrothermal carbonisation of biomass: Influence of feedstock on combustion behaviour of hydrochar. *Fuel* **2016**, *169*, 135–145. [CrossRef]
27. Mäkelä, M.; Yoshikawa, K. Ash behavior during hydrothermal treatment for solid fuel applications. Part 2: Effects of treatment conditions on industrial waste biomass. *Energy Convers. Manag.* **2016**, *121*, 409–414. [CrossRef]
28. Smith, A.M.; Ross, A.B. Production of bio-coal, bio-methane and fertilizer from seaweed via hydrothermal carbonisation. *Algal Res.* **2016**, *16*, 1–11. [CrossRef]
29. Smith, A.M.; Whittaker, C.; Shield, I.; Ross, A.B. The potential for production of high quality bio-coal from early harvested Miscanthus by hydrothermal carbonisation. *Fuel* **2018**, *220*, 546–557. [CrossRef]
30. Smith, A.M.; Ross, A.B. The Influence of Residence Time during Hydrothermal Carbonisation of Miscanthus on Bio-Coal Combustion Chemistry. *Energies* **2019**, *12*, 523. [CrossRef]
31. Grimm, A.; Skoglund, N.; Boström, D.; Ohman, M. Bed agglomeration characteristics in fluidized quartz bed combustion of phosphorus-rich biomass fuels. *Energy Fuels* **2011**, *25*, 937–947. [CrossRef]
32. Lindström, E.; Sandström, M.; Boström, D.; Öhman, M. Slagging Characteristics during Combustion of Cereal Grains Rich in Phosphorus. *Energy Fuels* **2007**, *21*, 710–717. [CrossRef]
33. Thy, P.; Jenkins, B.M.; Grundvig, S.; Shiraki, R.; Lesher, C.E. High temperature elemental losses and mineralogical changes in common biomass ashes. *Fuel* **2006**, *85*, 783–795. [CrossRef]
34. Thy, P.; Lesher, C.E.; Jenkins, B.M. Experimental determination of high-temperature elemental losses from biomass slag. *Fuel* **2000**, *79*, 693–700. [CrossRef]
35. Jenkins, B.; Baxter, L.; Miles, T. Combustion properties of biomass. *Fuel Process. Technol.* **1998**, *54*, 17–46. [CrossRef]
36. Bapat, D.; Kulkarni, S.; Bhandarkar, V. *Design and Operating Experience on Fluidized Bed Boiler Burning Biomass Fuels with High Alkali Ash*; American Society of Mechanical Engineers: New York, NY, USA, 1997.
37. Antal, M.J.; Mok, W.S.L.; Richards, G.N. Mechanism of formation of 5-(hydroxymethyl)-2-furaldehyde from d-fructose and sucrose. *Carbohydr. Res.* **1990**, *199*, 91–109. [CrossRef]
38. Jin, F.; Zhou, Z.; Moriya, T.; Kishida, H.; Higashijima, H.; Enomoto, H. Controlling hydrothermal reaction pathways to improve acetic acid production from carbohydrate biomass. *Environ. Sci. Technol.* **2005**, *39*, 1893–1902. [CrossRef]
39. Smith, K.L.; Smoot, L.D.; Fletcher, T.H.; Pugmire, R.J. *The Structure and Reaction Processes of Coal*; Springer Science & Business Media: New York, NY, USA, 2013.
40. Garrote, G.; Domínguez, H.; Parajó, J.C. Hydrothermal processing of lignocellulosic materials. *Holz als Roh-und Werkst.* **1999**, *57*, 191–202. [CrossRef]
41. Sevilla, M.; Fuertes, A.B. Chemical and Structural Properties of Carbonaceous Products Obtained by Hydrothermal Carbonization of Saccharides. *Chem. Eur. J.* **2009**, *15*, 4195–4203. [CrossRef]
42. Kei-ichi, S.; Yoshihisa, I.; Hitoshi, I. Catalytic Activity of Lanthanide(III) Ions for the Dehydration of Hexose to 5-Hydroxymethyl-2-furaldehyde in Water. *Bull. Chem. Soc. Jpn.* **2001**, *74*, 1145–1150. [CrossRef]
43. Patil, S.K.R.; Lund, C.R.F. Formation and Growth of Humins via Aldol Addition and Condensation during Acid-Catalyzed Conversion of 5-Hydroxymethylfurfural. *Energy Fuels* **2011**, *25*, 4745–4755. [CrossRef]
44. Baccile, N.; Antonietti, M.; Titirici, M.-M. One-Step Hydrothermal Synthesis of Nitrogen-Doped Nanocarbons: Albumine Directing the Carbonization of Glucose. *ChemSusChem* **2010**, *3*, 246–253. [CrossRef]
45. Yu, L.; Falco, C.; Weber, J.; White, R.J.; Howe, J.Y.; Titirici, M.-M. Carbohydrate-Derived Hydrothermal Carbons: A Thorough Characterization Study. *Langmuir* **2012**, *28*, 12373–12383. [CrossRef] [PubMed]

46. Chen, W.-H.; Ye, S.-C.; Sheen, H.-K. Hydrothermal carbonization of sugarcane bagasse via wet torrefaction in association with microwave heating. *Bioresour. Technol.* **2012**, *118*, 195–203. [CrossRef] [PubMed]
47. Wohlgemuth, S.-A.; Vilela, F.; Titirici, M.-M.; Antonietti, M. A one-pot hydrothermal synthesis of tunable dual heteroatom-doped carbon microspheres. *Green Chem.* **2012**, *14*, 741–749. [CrossRef]
48. Marschner, H.; Marschner, P. *Marschner's Mineral Nutrition of Higher Plants*; Academic Press: Cambridge, MA, USA, 2012.
49. Korbee, R.; Kiel, J.; Zevenhoven, M.; Skrifvars, B.; Jensen, P.; Frandsen, F. Investigation of biomass inorganic matter by advanced fuel analysis and conversion experiments. In *Proceedings of the Power Production in the 21st Century: Impacts of Fuel Quality and Operations*; United Engineering Foundation Advanced Combustion Engineering Research Center: Snowbird, UT, USA, 2001.
50. Jin, Y.-S.; Jiang, T.; Yang, Y.-B.; Li, Q.; Li, G.-H.; Guo, Y.-F. Removal of phosphorus from iron ores by chemical leaching. *J. Cent. South Univ. Technol.* **2006**, *13*, 673–677. [CrossRef]
51. Miller, B.G.; Tillman, D.A. *Combustion Engineering Issues for Solid Fuel Systems*; Elsevier: Amsterdam, The Netherlands, 2008.
52. Tillman, D.A.; Duong, D.N.B.; Harding, N.S. Chapter 4—Blending Coal with Biomass: Cofiring Biomass with Coal. In *Solid Fuel Blending*; Tillman, D.A., Duong, D.N.B., Harding, N.S., Eds.; Butterworth-Heinemann: Boston, MA, USA, 2012; pp. 125–200. [CrossRef]
53. Dahlquist, E. *Technologies for Converting Biomass to Useful Energy*, 1st ed.; CRC Press: Boca Raton, FL, USA, 2013.
54. Yang, H.; Yan, R.; Chen, H.; Lee, D.H.; Zheng, C. Characteristics of hemicellulose, cellulose and lignin pyrolysis. *Fuel* **2007**, *86*, 1781–1788. [CrossRef]
55. Peterson, A.A.; Vogel, F.; Lachance, R.P.; Fröling, M.; Antal, M.J., Jr.; Tester, J.W. Thermochemical biofuel production in hydrothermal media: A review of sub-and supercritical water technologies. *Energy Environ. Sci.* **2008**, *1*, 32–65. [CrossRef]
56. Su, S.; Pohl, J.H.; Holcombe, D.; Hart, J.A. Techniques to determine ignition, flame stability and burnout of blended coals in p.f. power station boilers. *Prog. Energy Combust. Sci.* **2001**, *27*, 75–98. [CrossRef]
57. Williams, A.; Jones, J.; Ma, L.; Pourkashanian, M. Pollutants from the combustion of solid biomass fuels. *Prog. Energy Combust. Sci.* **2012**, *38*, 113–137. [CrossRef]
58. Yarin, L.P.; Hetsroni, G.; Mosyak, A. *Combustion of Two-Phase Reactive Media*; Springer Science & Business Media: Heidelberg, Germany, 2013.
59. Saddawi, A.; Jones, J.M.; Williams, A. Influence of alkali metals on the kinetics of the thermal decomposition of biomass. *Fuel Process. Technol.* **2012**, *104*, 189–197. [CrossRef]
60. Jones, J.M.; Darvell, L.I.; Pourkashanian, M.; Williams, A. The Role of Metals in Biomass Char Combustion. In Proceedings of the European Combustion Meeting, Louvain-la-Neuve, Belgium, 3–6 April 2005.
61. Nowakowski, D.J.; Jones, J.M.; Brydson, R.M.D.; Ross, A.B. Potassium catalysis in the pyrolysis behaviour of short rotation willow coppice. *Fuel* **2007**, *86*, 2389–2402. [CrossRef]
62. Huang, H.Y.; Yang, R.T. Catalyzed Carbon–NO Reaction Studied by Scanning Tunneling Microscopy and ab Initio Molecular Orbital Calculations. *J. Catal.* **1999**, *185*, 286–296. [CrossRef]
63. Backreedy, R.I.; Jones, J.M.; Pourkashanian, M.; Williams, A. Burn-out of pulverised coal and biomass chars. *Fuel* **2003**, *82*, 2097–2105. [CrossRef]
64. Kannan, M.P.; Richards, G.N. Gasification of biomass chars in carbon dioxide: dependence of gasification rate on the indigenous metal content. *Fuel* **1990**, *69*, 747–753. [CrossRef]
65. Stojanowska, G.; Jones, J. Influence of added calcium on thermal decomposition of biomass, lignite and their blends. *Arch. Combust.* **2006**, *26*, 91.
66. Backreedy, R.I.; Jones, J.M.; Pourkashanian, M.; Williams, A. Modeling the reaction of oxygen with coal and biomass chars. *Proc. Combust. Inst.* **2002**, *29*, 415–421. [CrossRef]
67. Skrifvars, B.-J.; Laurén, T.; Hupa, M.; Korbee, R.; Ljung, P. Ash behaviour in a pulverized wood fired boiler—A case study. *Fuel* **2004**, *83*, 1371–1379. [CrossRef]
68. Miles, T.R.; Miles, T., Jr.; Baxter, L.; Bryers, R.; Jenkins, B.; Oden, L. *Alkali Deposits Found in Biomass Power Plants: A preliminary Investigation of Their Extent and Nature*; National Renewable Energy Lab.: Golden, CO, USA; Miles Thomas R.: Portland, OR, USA; Sandia National Labs.: Livermore, CA, USA; Foster Wheeler Development Corp.: Livingston, NJ, USA; California University: Davis, CA, USA; Bureau of Mines, Albany Research Center: Albany, OR, USA, 1995; Volume 1.

69. Wang, L.; Hustad, J.E.; Skreiberg, Ø.; Skjevrak, G.; Grønli, M. A Critical Review on Additives to Reduce Ash Related Operation Problems in Biomass Combustion Applications. *Energy Procedia* **2012**, *20*, 20–29. [CrossRef]
70. Dahlin, R.S.; Vann Bush, P.; Snyder, T.R. *Fundamental Mechanisms in Flue-Gas Conditioning. Topical Report No. 1, Literature Review and Assembly of Theories on the Interactions of Ash and FGD Sorbents*; Southern Research Inst.: Birmingham, AL, USA, 1992.
71. Parker, K.R. *Applied Electrostatic Precipitation*; Springer Science & Business Media: Dordrecht, The Netherlands, 2012.
72. Shanthakumar, S.; Singh, D.N.; Phadke, R.C. Flue gas conditioning for reducing suspended particulate matter from thermal power stations. *Prog. Energy Combust. Sci.* **2008**, *34*, 685–695. [CrossRef]

© 2020 by the authors. Licensee MDPI, Basel, Switzerland. This article is an open access article distributed under the terms and conditions of the Creative Commons Attribution (CC BY) license (http://creativecommons.org/licenses/by/4.0/).

Article

High-Temperature, Dry Scrubbing of Syngas with Use of Mineral Sorbents and Ceramic Rigid Filters

Mateusz Szul *, Tomasz Iluk and Aleksander Sobolewski

Institute for Chemical Processing of Coal, 41-803 Zabrze, Zamkowa 1, Poland; tiluk@ichpw.pl (T.I.); asobolewski@ichpw.pl (A.S.)
* Correspondence: mszul@ichpw.pl

Received: 28 February 2020; Accepted: 22 March 2020; Published: 24 March 2020

Abstract: In this research, the idea of multicomponent, one-vessel cleaning of syngas through simultaneous dedusting and adsorption is described. Data presented were obtained with the use of a pilot-scale 60 kW$_{th}$ fixed-bed GazEla reactor, coupled with a dry gas cleaning unit where mineral sorbents are injected into raw syngas at 500–650 °C, before dedusting at ceramic filters. The research primarily presents results of the application of four calcined sorbents, i.e., chalk (CaO), dolomite (MgO–CaO), halloysite (AlO–MgO–FeO), and kaolinite (AlO–MgO) for high-temperature (HT) adsorption of impurities contained in syngas from gasification of biomass. An emphasis on data regarding the stability of the filtration process is provided since the addition of coating and co-filtering materials is often necessary for keeping the filtration of syngas stable, in industrial applications.

Keywords: adsorption; biomass; ceramic filter; gasification; hot-gas cleaning

1. Introduction

For most small- and medium-scale applications of gasification technology, the gasified feedstock is either bio-based or, more preferably, derived from waste feedstocks such as solid recovered fuel (SRF), lignin, or sewage sludge [1,2]. In the case of most waste-derived feedstocks, the step of their thermochemical conversion is more demanding than commonly encountered for conventional, clean biomass. The problems are mostly encountered in the operation of reactors; however, the use of waste materials also introduces major changes in the way the syngas needs to be purified. Importantly though, where the control and efficiency of a gasifier is the most important aspect for the operation of any gasification installation, it is the possibility to clean the produced syngas that renders the whole operation economically and environmentally sound. With the increasing complexity of syngas application, the complexity of its purification also increases. In general, direct combustion requires only marginal gas cleaning, which can often be limited merely to dedusting in cyclones, while, for reciprocating or gas-turbine engines, complete removal of fines, deep removal of condensing species (tars, light organics, and water), and careful control of the content of acidic species are demanded. By far, the most advanced applications of syngas are chemical synthesis (methanation, Fisher-Tropsch) and fuel cells, both of which demand almost complete removal of any contaminants [3]. It is compulsory, then, to devise such gas cleaning systems so that three major goals are simultaneously obtained. The first is to achieve the required cleanliness of the gas. The second is to perform the process with the least environmental impact, while the third connects with the economic aspect of the investment. Thus, the third issue not being technologically limiting remains crucial when developing new technologies.

Due to the progress achieved in gasification and gas cleaning technologies over the past 30 years, today, we most often see that it is the economy that holds back the development of many gasification installations rather than their technical or technological limitations. The fact is that, depending on a given case, from technological and technical standpoints, different feedstocks perform better in different types of reactors. Thus, the choice of gasifier can determine the whole installation's success or failure. Furthermore, gas cleaning units are developed for a given gasifier and feedstock, not vice versa. It is also well established that gas cleaning with a combination of the following methods provides the best performance in the widest area of syngas cleaning applications: catalytic upgrading, high-temperature (HT) dedusting on barrier filters, wet scrubbing, and final polishing with dry adsorption methods [4]. These advanced solutions are compulsory for high-value end products obtained from syngas. However, at the same time, they tend to lower the profitability of installations in applications where low-value products, such as heat and power, are produced. Currently, in many developing countries such as Poland, it is the eco-friendly, renewable generation of electrical power that shapes the scene in terms of political decision-making and market development initiatives much more firmly than the drive toward, e.g., production of liquid biofuels. To make progress in the amount of waste biomass, SRF, or sewage sludge utilized for power generation, it is vital then to take advantage of small- and medium-scale distributed gasification installations, which offer better efficiencies, flexibilities, costs, and environmental impact than conventional combustion methods. For these reasons, development of gas cleaning methods that provide the above-mentioned benefits needs to be done.

The article describes the concept of an adsorption technology tested for HT cleaning of syngas derived from the gasification of biomass. The concept is based on the use of a multi-component cleaning method performed in one reactor vessel, where a hot-gas ceramic candle filter is precoated with naturally occurring cheap and abundant minerals. Multi-component cleaning solutions greatly improve the reliability and efficiency of thermochemical conversion plants, while also lowering their footprint. HT, raw syngas coming out from a reactor contains the highest amount of contaminants, which makes its cleaning most efficient. Furthermore, because HT dedusting is most of the time the first step of syngas cleaning, it is interesting to develop a method for simultaneous removal of not only dust particles but also other contaminants. For this reason, it is proposed to inject adsorbents upstream of ceramic filters and, thus, to conduct the adsorption process on a continuously regenerated bed of filter cake composed of char, ash, and the adsorbent material itself. This concept has multiple potential merits. The most important ones are the complete removal of solids and the possibility to reduce the amount of Cl, S, and tar species.

Firstly, it is noteworthy that, after the gasification process, both chars and ashes entrained in syngas have substantially activated surface area and, thus, can act as a sorbent for contaminants. This phenomenon is currently thoroughly researched since major improvements in fixed-bed gasifiers are possible this way. Nakamura et al. [4] proposed the use of gasification by-products as a means for process gas cleaning. In the system, the use of water–tar condensate as a washing medium in a scrubber and char in the form of a fixed-bed adsorber was tested. The results showed that the scrubber efficiency for tar removal reached its optimum at 50%; however, the efficiency of the char adsorber reached over 81%. Furthermore, Yafei et al., in their work, reviewed and studied the concept of closing the management cycle for tar, char, and heavy-metal gasification by-products through integrated concepts [5]. In the study, the pathway for using gasification- or pyrolysis-based chars as support for adsorption of heavy metals and their subsequent use (after deactivation) for catalytic elimination of tars was presented. Finally, the closure of the tar/char management cycle was proposed through the use of catalytic gasification of the spent catalyst. In this way, it is possible to recover C from the catalyst as syngas and the heavy metals as bottom ash. However, the most straightforward and visual way of using creative design of thermochemical processes to use the characteristics of gasification char was presented by Obernberger et al. [6]. The study showed that it is possible to use a two-stage gasification–combustion method for the energetic use of waste biomasses of low ash melting temperatures for highly efficient generation of heat with unprecedentedly low emissions of CO, total organic carbon, and particulate matter. Here, the process layout resembles a common updraft gasifier with a gas combustion chamber that is fitted just above the fixed bed. However, the obtained in-bed temperature profile is distinctly different for the char/biomass bed to act as adsorber/filter and not to produce excessive amounts of tars, as is often encountered in conventional updraft gasifiers.

Furthermore, process conditions where most syngas ceramic filters are commonly operated induce the need to use auxiliary co-filtering materials in the first place. The inert materials increase the filter's particle collection efficiency, as well as keeping the pressure-drop low. Such inerts are mostly based on fine powders composed of SiO or CaO. Through this "co-filtration" effect, it is often possible to keep dedusting of hot syngases stable, where filtration of char alone leads to a constant rise of pressure drop on the filter. Due to its higher thermal stability and flowability, the inert material eases pulse-back regeneration of the filter, while collecting part of the polymerizing tar and condensing mineral matter, thus preventing the clogging of filter pores. By changing the inert solid for materials that exhibit chemical activity in the process condition, it is possible to perform high-temperature, dry scrubbing of syngas.

In the proposed system, the sorbent is injected upstream of the ceramic filter through a Venturi nozzle. This method assures good mixing of the solid and syngas in the turbulent region of the filter intake. In the first step, the adsorption takes place in a diluted two-phase system, where contact time is extended by the design of the dirty plenum of the filter. Importantly, the particle size distribution of the sorbents needs to be controlled to avoid disengagement of the sorbent from syngas before reaching the surface of the filter cake. The adsorption process is finalized in thorough cleaning on a fixed-bed layer composed of the filter cake which collects on the surface of the ceramic filters. When adsorption on the filter cake is considered, a few differences render the process distinctive from a conventional adsorption set-up. Firstly, filters applied in the filtration of gases are mostly developed to operate at as high linear gas velocities as possible (face velocity). However, in HT applications, to keep them stable, the units need to be run in the range of face velocities much lower than nominal for low-temperature bag filters. Values from 0.5 to 3.0 cm/s often provide the lowest operational and investment costs for hot gas filters, while keeping the process stable. In adsorption processes, hourly spaced gas velocity (HSGV) can be thought of as a parameter comparable to face velocity in filters because, here, the filter cake is the active bed composed of the sorbent material. For adsorption to be efficient, the HSGV in fixed-beds should be kept in the range from 0.2 to 0.5 m/s with a contact time of approximately 3 s [7]. To get close to the above-mentioned standards, it is necessary to lower the filtration velocity to minimal values and increase the thickness of filter cake. In practice, reaching the benchmark with the use of ceramic filters is technically impossible.

For the pilot installation used in the research, the ceramic filter integrated with the pilot GazEla gas generator was designed to be operated at 0.5 to 1.0 cm/s. Depending on the operational conditions of the filter, the thickness of its filter cake should range from 1 mm for freshly pre-coated new filters up to 6 to 8 mm when the filter is run with candles not regenerated regularly. Realistically speaking, this set-up may provide from a minimum of 0.2 s up to a maximum of 1 s of contact time between the gas phase and solid phase of the filter cake with an average of 0.5 s. Such residence times should be considered as low values for fixed-bed adsorbers; however, they are more than reasonable for many fluidized bed (FB) units [7]. In the proposed system, the first stage of adsorption resembles FB adsorption in a diluted circulating fluidized bed (CFB), as the adsorption process starts from the moment the sorbent is injected into syngas. For the pilot installation, the contact time for the first stage of adsorption can reach up to a few seconds depending on the reactor power.

The second reason why the HT filtration/adsorption system cannot be directly compared to any of the two above-mentioned systems is the particle size of sorbents that build up the filter cake. Preferably, for the filter, the sorbent particle size distribution should range from 15 to 50 μm, which makes it 10-fold smaller than sorbents used in FB reactors and more than 1000-fold smaller in comparison to fixed-bed units.

Moreover, it is important that, after a regeneration cycle takes place, part of the surface of the filters is stripped of the filter cake. If the regeneration is too harsh, the thickness of the cake may be reduced to the point where both particles and other contaminants slip through the filter, thus impairing the efficiency of cleaning. Furthermore, as mentioned above, the filter cake only partially consists of the dedicated adsorbent, whereas the rest of the cake is composed of char and ash filtered from the syngas. Both ash and char are transported through a gasifier; thus, their surface is activated to some extent, which has a positive effect on the adsorption efficiency. Sorption in the filter cake also brings forward another relation. The thickness of the cake positively influences the efficiencies of both dedusting and adsorption, while increasing pressure drop (dP) across the vessel. Thus, the importance of precise control of the regeneration process is vital here.

Research presented in this paper focused on the collection of process data and operational experience. However, to assess the efficiency of the method, it also leaned into the determination of adequate analytical procedures. In contrast to flue gas cleaning systems, the problem of qualitative and quantitative determination of cleaning efficiencies of syngas cleaning units is still demanding and by no means should be regarded as trivial. This issue is even more pronounced when research is performed in real conditions and on a small scale, which brings forward many technical limitations that are not present in lab-scale installations.

The extent to which the syngas needs to undergo cleaning is determined by the type of feedstock, the type of gasification reactor, and the of the final application. Generally speaking, when the reactor is treated as an equilibrium black box, its products are always the same and depend only on process efficiency, temperature, pressure, type of gasifying agent used, etc. Thus, models often do not take into account the recognition of the type of gasifier used. However, in reality, differences between the gasification processes run at fixed, fluidized, or entrained beds are substantial and decide the composition of generated syngas. As far as contaminants of syngas are concerned, their generation in gasifiers takes place mainly due to volatilization from the solid phase and subsequent gas–gas and gas–solid reactions. Due to different process conditions taking place in gasifiers, differences in syngas composition are pronounced. For fixed-bed reactors, a readsorption of contaminants on activated char and ash present in the bed is observed [8]. For FB, on the other hand, the amount of char in the bed is small, but the bed is often composed of a material which can influence the process through catalysis or sorption. For example, in FB reactors, olivine is a frequently used as a bed material for in situ tar reforming, and CaO beds are used for CO_2 removal from syngas [9,10]. Finally, entrained flow reactors are operated at very high temperatures that can lead to very clean syngases through the almost complete conversion of tars and collection of containments in the form of vitrified slag [11]. For the Institute for Chemical Processing of Coal (IChPW), the choice of fixed-bed reactor systems was done based on the scale of solution sought by the market and the intrinsic characteristics of fixed-bed units, meaning their ease of operation, robustness, flexibility, and ability to use the bed for active removal of contaminants from syngas. In fixed-bed reactors, char is activated by process conditions and acts as the sorbent material. For example, in laboratory conditions, it was shown that, inside gasifiers, chlorine can be almost fully volatilized (>90%) from fuel and, thus, is a constituent of raw syngas. In the case of S, it rarely undergoes complete volatilization in the reactor and, on average, 50% of S from biomass remains in the bottom ash from the reactor even at temperatures reaching 1000 °C [12]. On the other hand, alkalis, such as potassium, almost exclusively volatilize with correlation to Cl. Therefore, if Cl content of the biomass is low, K mostly remains in the bottom ash. Other elements that tend to volatilize under gasification conditions are mainly Na, Ca, Si, Mg, P, and Al; thus, in this research, their content in the filter cake was the subject of examination. Okuno et al. [13] suggested that alkali and alkali earth metals (AAEM) leave gasifier mostly in correlation with water-insoluble tar and, thus, with aromatic compounds that are derivatives of benzene, xylene, furfural, and naphthalene in the syngas [14,15]. These compounds remain gaseous under gasification and HT gas cleaning conditions; hence, their enhanced recovery in filer cake is a sign of adsorption from the gas phase. Moreover, Sonoyama et al. [16] showed that AAEM species have high affinity to bonding with char in fixed-bed conditions unless the flow of gas is significant enough to force them to pass through the bed. In their research, helium was used as the carrier gas; however, the principle also applies to the research presented in this paper. These results showed that volatile AAEM species undergo repeated adsorption/desorption cycles on the surface of the char bed; thus, their readsorption on filter cake should also be visible. In the study, the authors experimentally verify the theoretical possibility of removing the following contaminants from syngas with the use of a combined process of HT filtration and sorption: hydrogen sulfide, hydrogen halides, ammonia, AAEMs, heavy metals, and tars on the surface particles collected in the filter cake.

2. Materials and Methods

2.1. Characteristics of Gasified Feedstock

For all testing, conventional wood chips from alder trees obtained from local wood mills were used. For adsorption tests, the wood chips were not dried or sieved before feeding into the gasifier. The particle size of the chips was consistent (ca. 20 × 20 × 8). Their moisture content was checked regularly. Due to similar storage conditions, the fuel was air-dried and had on average 18% of moisture content in the working state. Table 1 presented below summarizes the proximate and ultimate analysis of the fuel (analytical state). Later in the study, experiments connected with the stability of filtration were also conducted, in which the moisture content of the feedstock was regulated.

Table 1. Proximate and ultimate analysis of wood chips gasified in GazEla reactor (a—analytical state, ar—as received = working state).

C_a wt.%	H_a wt.%	N_a wt.%	S_a wt.%	Cl_a wt.%	F_a wt.%	H_2O_a wt.%	H_2O_{ar} wt.%	A_a wt.%	V_a wt.%	HHV_a J/g	LHV_a J/g
47.6	5.29	0.15	0.04	0.08	0.003	8.9	18	0.5	76.66	18,536	17,164

2.2. Chemical Composition of Analyzed Sorbents

Four groups of naturally occurring minerals were selected for testing. The first adsorbent was chalk, which is rich in calcium carbonate. This material is widely available and used for the protection of filter elements (pre-coat), as well as a sorbent in high- and low-temperature cleaning of flue gases from combustion plants. It consists of ca. 94 wt.% clean $CaCO_3$, which, under the filter conditions, undergoes calcination to CaO. The calcined form actively takes part in the adsorption of acidic compounds. However, it is also suggested that CaO can take an active role in catalytic recombination and degradation of tar components [14]. Chalk is used here as a representative of the Ca-rich family of sorption materials. The fine chalk for testing was obtained from LabTar sp. z o.o., and the product is commonly available on the market as a dietary additive for cattle.

The next two sorbents used in the research are closely related because they are derived from a similar family of sedimentary rocks. In nature, they are mostly distinguished by their tertiary structure and content of impurities. Thus, the second applied sorbent was halloysite. It is widely applied in the industry as a sorbent for petrochemicals and adapts a form of hollow tubes, similar to nanotubes. It is the form of halloysite that predisposes it for adsorption purposes. Due to these characteristics, halloysite also finds use as a feedstock for ceramic industry, as a dietary supplement for cattle, as a catalyst, and as a material for FB. The third sorbent was kaolinite, which mostly finds its application as a basic material for ceramic masses for the production of roof tiles, in chemical or paper industries, as well as in cosmetics. It mostly takes up the form of flat wafers of rhombic structure. From the chemical standpoint, both are hydrated alumina silicate and can have the following formula: $Al_4[Si_4O_{10}](OH)_8 \cdot 4H_2O$. Both were obtained from mines located in the southwest regions of Poland (Bolesławiec). In Poland, calcinated, fine halloysite is a market product of Intermark sp. z o.o., while kaolinite is marketed by Surmin-Kaolin S.A.

The last analyzed sorbent was dolomite, which is found in nature as a sedimentary, carbonate rock. In composition, dolomite mostly consists of a mixture of Ca and Mg (ca. 90 wt.%); thus, in this research, it was used as received for its Mg content, to judge its affinity for sorption processes. Similarly to chalk, fine dolomite used in the study is a commonly available fertilizer product manufactured by ZChSiarkopol sp. z o.o.

Even though, in filtration process conditions, all of the sorbents should take up their calcined forms, during the research, the sorbents were pre-calcined.

Regarding the handling of the sorbents, kaolinite showed a tendency to agglomerate when its moisture content increased even in an air-dried state. The same phenomenon was noticed for chalk, although to a smaller extent. The sorbent which by far was the easiest to flow and handle was dolomite, followed by halloysite.

Tables 2 and 3 presented below summarize the content of AAEMs and heavy metals in gasified biomass, as well as in sorbents, used for the gas cleaning research. As the adsorption tests were done in the order starting from chalk (T1), through halloysite (T2), kaolinite (T3), and ending up with dolomite (T4), the same numbering is used in the article to present the obtained data. For halloysite, two additional tests were also performed to assess its ability to stabilize filtration/regeneration cycles. The two are, thus, abbreviated T2.1 and T2.2.

Table 2. Overview of the chemical composition of alkali and alkali earth metals (AAEMs) present in the gasified fuel and applied sorbents.

	Unit	T1	T2	T3	T4	Wood Chips
SiO_2	%	2.79	35.75	65.68	2.79	44.49
Al_2O_3	%	0.85	25.77	27.11	1.61	4.22
Fe_2O_3	%	0.49	21.37	1.17	1.49	3.99
CaO	%	50.69	0.68	0.18	27.18	20.18
MgO	%	0.35	0.58	0.14	17.89	4.27
P_2O_5	%	0.02	0.77	0.03	0.30	2.48
SO_3	%	0.87	0.07	0.07	1.22	1.02
Mn_3O_4	%	0.02	0.56	0.02	0.17	1.21
TiO_2	%	0.04	3.78	0.92	0.09	0.43
BaO	%	0.01	0.11	0.01	0.01	0.11
SrO	%	0.05	0.02	0.01	0.01	0.08
Na_2O	%	0.06	0.22	0.15	0.08	1.58
K_2O	%	0.19	0.45	0.42	0.59	7.54

Table 3. Overview of the content of heavy metals present in the applied sorbents (d—dry state).

	Unit	T1	T2	T3	T4	Wood Chips
As^d	mg/kg	<1.7	<1.70	<1.70	7.26	-
Cd^d	mg/kg	<0.27	0.8	2.06	0.47	-
Co^d	mg/kg	1.08	2.28	280	3.53	-
Cr^d	mg/kg	5.83	15.4	790	53.8	-
Cu^d	mg/kg	3.16	8.28	70.9	13.8	-
Mn^d	mg/kg	95.6	1.0	3749	25.2	-
Mo^d	mg/kg	1.32	1.54	4.98	1.28	-
Ni^d	mg/kg	4.46	11.8	558	10.2	-
Pb^d	mg/kg	10.1	16.4	26.5	273	-
Sb^d	mg/kg	<2.0	<2.00	13.1	4.54	-
V^d	mg/kg	7.15	15.3	340	35.1	-
Zn^d	mg/kg	43.8	84.3	288	17.0	-

Due to the unknown degree of volatilization of heavy metals from raw biomass, for determination of their content in filter cake, a separate analysis of filter cake obtained from sole gasification and HT filtration of the char/ash was performed. Here, no sorbent was used to coat the filters or to support the filtration process.

Figure 1 presented below compares ratios of the main metallic constituents and carbon build-up of the sorbents and biomass char. Evident differences in the chemical composition of the sorbents should impact their gas cleaning properties as Al, Mg, and Ca all take part in the sorption of acidic contaminants. Char, on the other hand, is particularly important in the readsorption of compounds volatilized from biomass (Ca, K, Mg) as it has a highly developed surface, similar to activated carbons. Char is also known to show some degree of catalytic activity for the reduction of tars.

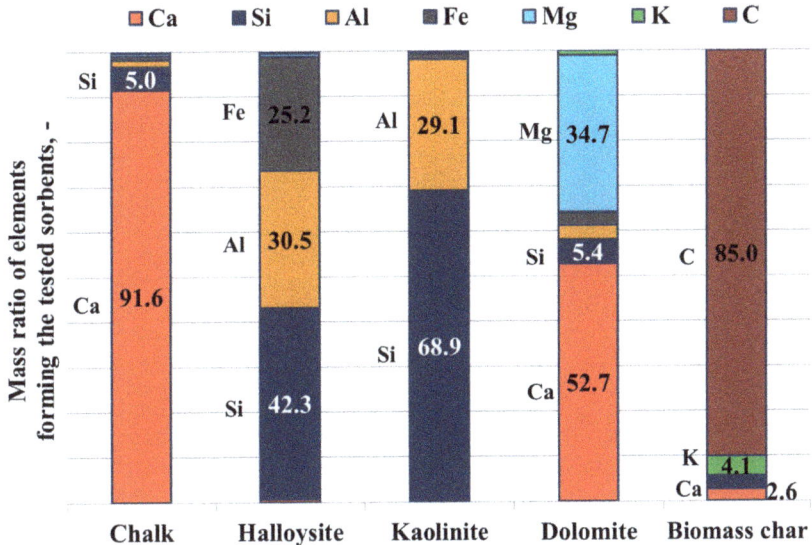

Figure 1. Comparison of the chemical composition of sorbents and biomass char. Content of main constituents (inorganic elements and carbon) that actively take part in the adsorption process.

In the study, all laboratory analyses of sorbents, biomass, and gasification products were done at the Institute for Chemical Processing of Coal.

3. Experimental Installation

The experimental set-up utilized for this research was localized in the Center for Clean Coal Technologies in Zabrze, Poland. The installation was based on a pilot GazEla reactor. The gasifier is a fixed-bed, mixed-flow reactor, where fuel is fed from the top and an air/steam mixture is introduced into three characteristic zones of the bed. The innovative part of the reactor is a method for the recovery of hot syngas directly from the gasification zone. For the gasification of biomass, the reactor is characterized by 60 kW$_{th}$ input and a cold gas efficiency (CGE) of ca. 67%. More results regarding the gasification of various fuels and a detailed description of its operation can be found in the literature [1,2].

The pilot installation was also devised for conducting research on different configurations of gas cleaning methods and, thus, the reactor was simultaneously connected to both dry and wet gas cleaning installations. As the main goal of this research was to develop a method for HT filtration and sorption of syngas, results presented here were obtained from one configuration of the dry gas cleaning route. Downstream of the GazEla reactor, raw syngas, at 450–550 °C, is directed toward the inlet of a ceramic filter. Before the filter, sorbents were injected into the syngas.

Because the sorbent plays here a double role, its particle size distribution was set such that it produces a well-structured filter cake of uniform porosity while creating a uniform, stable aerosol of the sorbent at every point of the filter. Thus, the largest particle size of sorbent was set such that, in the largest cross-section of the filter, the sorbent should not settle from the gas, but rather be entrained with the flow of syngas until its separation on the filter cake.

The filter is conventionally separated into two zones by a horizontal plate that supports the filter candles. Gas enters the dirty zone. When it travels through ceramic candle filters, it is dedusted and cleaned. For the pilot installation, the filter is composed of 10 ceramic filters (1 m long, 60 mm in outer diameter) divided into two cleaning sections. After passing through the candles, the dedusted gas enters a clean zone of the filter and passes to a next cleaning unit. During this research, the dedusted gas subsequently underwent cooling down from ca. 450 °C down to 40 °C for condensation of water and organic matter. After cooling, cold syngas contains a lot of water–tar mist which needs to be separated. In the gas cleaning set-up, this process can be done with demisting pads, as well as a coalescing filter, depending on the level of purity that is demanded by the final application of the syngas. In the research, cleaned syngas was combusted in a dual-fuel piston engine.

Figure 2 presented below represents a schematic diagram of the pilot gasification installation utilized during the research.

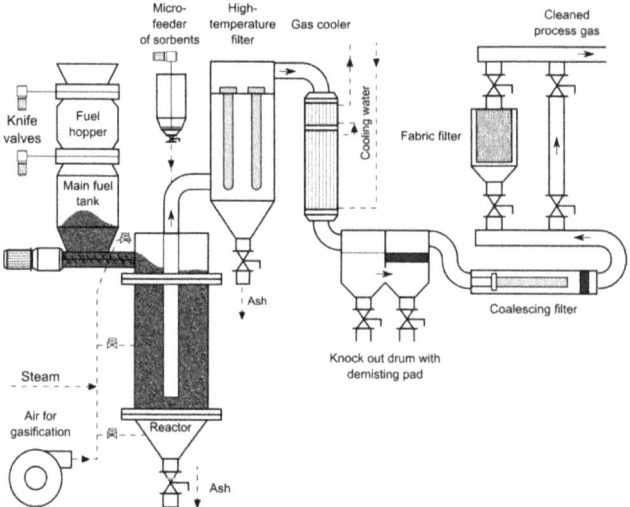

Figure 2. Schematic diagram of the pilot gasification installation with the fixed-bed GazEla reactor and the dry gas cleaning unit.

4. Measurements and Methods

4.1. Generation of Syngas

During all sorption tests, the reactor was operated on the same feedstock and its parameters were set to the same limits, so that the stream of syngas, its composition, the temperature, and the amount of impurities would be comparable. Start-up was performed after initial preheating of the reactor lining to ca. 350–450 °C. At this point, the process bed was formed and ignited by the addition of a hot air/steam mixture. The stream of the gasification agent was controlled and steadily increased up to the full capacity of the reactor. Until the bed reached its final temperature profile, and as long as the temperature of syngas at the reactor's outlet was not stable, the gas was directed to a burner without any gas cleaning. When the reactor reached stable operation (ca. 2 h), syngas was fed into the gas cleaning installation. After the installation reached stabilization (another 2 h), the sampling for syngas cleaning efficiencies was started.

4.2. Filtration Process and Filter Regeneration: Measurement and Method for Analysis of Data

The ceramic filter was continuously monitored with a range of temperature and pressure readings. The two most important parameters for its stable operation are readings of gas temperature at its inlet and outlet. It is known that the origin of syngas greatly influences the range of operational temperature for HT filtration (operational window). If a certain temperature limit is exceeded, filtration may lead to a complete failure of the filter as a result of a continuous increase of differential pressure (dP) on the filter. In such a case, the filter cake often undergoes some degree of sintering or obtains some degree of viscosity, both of which result in greater resistance for pulse-back cleaning and render the regeneration inefficient. Above the upper-temperature limit, the filter pores can become plugged as a result of condensation of mineral matter volatilized from fuel and chemical activity of tar, which leads to their polymerization and char formation. The lower limit, on the other hand, protects from the condensation of high-molecular-weight tars within filter pores. For this reactor and conventional biomasses, the operational window of HT filtration should be kept in the range of 350–500 °C.

The most important control parameter for ceramic filters is its differential pressure (dP) across the filters. The measured dP summarizes both the dP generated by filter media and the filter cake collected on it. As the amount of solids collected in the filter cake or its compressibility increases, so does its flow resistance and, thus, dP of the filter increases. At a given set point, the filter is subjected to pulse-back cleaning to recover a desirable low dP. To keep operational costs of gasification installation as low as possible, filters should be run at the smallest possible dP. In the long term, stable and reliable operation of HT filters at low dP is obtained by performing the filtration mostly as depth filtration within the volume of a filter cake. In this way, the filer medium does not become irreversibly blocked by dust particles and the efficiency of collection is kept high. For this unit, such results are attainable when it is operated at a dP of 1–2 kPa. Above these values, a build-up of excessive filter cake can lead to failure of the filtration process, e.g., as a result of filter cake bridging.

For a more precise comparison of different tests and standardization of analyzed data, a convention was set here, where the registered pressure drop across the filter is divided by the actual flow rate of gas through the filter.

During the research, a standardized procedure for the preparation of the filter candles and their pulse-back regeneration was set. Filters before each test were thoroughly cleaned through pressure pulses when the installation was offline (20 °C with a full reverse flow). After that, at process temperatures (ca. 450 °C), candles were pre-coated by a controlled layer of the tested sorbent. During a test, filtration was started when the reactor reached stable operational conditions and maintained them for a minimum of 2 h, whereby the temperature of gas exceeded 450 °C and its stream was constant. From the beginning of filtration, gas was injected with a sorbent, and, throughout the test, the stream of the sorbent was kept constant. Regeneration cycles were controlled automatically by system control and data acquisition (SCADA) and for the normal, continuous operation started at dP = 1 kPa.

4.3. Methods for Measurement of Contaminants Present in Syngas

Another aspect of the development of a new syngas cleaning method is to establish a methodology for the determination of its performance parameters. There exist standardized analytical procedures and sampling methods for the analysis of syngas composition and the amount of contaminants contained in it. For organic matter in the gas, a well-established standard set in the "Tar Protocol" was applied for years. The protocol proposes sorption in isopropyl alcohol or on solid sorbents as the best methods for the collection of organics from syngas. However, even though this method is applied globally for a long time now, the variety of results obtained from similar reactors and process conditions show that there still is a gap in the analytical procedures. The measurement conditions that occur in raw syngas increase the measurement error dramatically. To try and address this issue, many different approaches were proposed. One of the most resilient ideas for determination of the efficiency of syngas cleaning is an analysis of its combustion products through online measurement of the flue gas composition. This indirect method uses hot-sampling and FTIR analyzers, which give a very good range of measured compounds and very low detection levels for all contaminants in their oxidized forms. The drawback here is that post-combustion analysis gives information only about the overall efficiency of a gas cleaning system and it is impossible to point out the efficiency of a given apparatus.

To obtain more direct and reliable data, a different approach was adopted for this study. The first step of the method was a basic, on-line measurement of syngas composition using IR analyzers.

This robust measurement provides on-line information regarding changes in the gasification process and quality of generated syngas. Tar, water, and particle matter contents were measured by taking samples of the gas in parallel, before and after the ceramic filter. The remainder of the analysis, i.e., removal of AAEMs, halides, and H_2S, was determined from the analysis of liquid and solid products recovered from the gas cleaning units. Simultaneously, another analysis of halides and H_2S was performed through their absorption from syngas (impinger bottles with NaOH).

The liquid samples were analyzed for the presence of halides and H_2S.

4.3.1. Water, Tar, and Solid Particle Content

The tar content of syngas is inherently connected with the gasification process. Its content not only indicates the loss of conversion efficiency, but is also the most important issue that prevents larger market uptake and technical utilization of the gasification process.

The simultaneous filtration/sorption process is not devised to directly convert or reduce tars in syngas. However, there exist experimental results indicating that HT filters may induce a partial reduction of syngas tar content. In this research, an adapted method for the quantification of tars based on the Tar Protocol was used.

The sampling system consisted of a probe, two impinger bottles, and a tube filled with cotton wool. The probe was introduced axially into a syngas pipe. The end of the probe was connected to the first impinger bottle containing about 50 mL of isopropyl alcohol at ambient temperature. Here, most of the tar and dust was collected. Moreover, in this region, the syngas was also saturated with the solvent which prevented water from freezing further down the collection line. The first impinger bottle was connected with the rest via a Teflon tubing, which acts as an additional condenser. The second bottle, filled with 50 mL of isopropyl alcohol at -20 °C, collects the remaining water and low-boiling-point volatile organic carbon (VOC). Finally, the glass tube, which is filled with cotton wool, acts as a droplet collector. Producer gas is sampled through the system by a pump coupled with a flow meter and a regulator.

After sampling, both sorption solutions are combined and analyzed together for the content of volatile organic carbon (VOC), water, dust, and tars. In this research, no qualitative analysis of VOC was conducted.

The concentration of water in isopropyl alcohol solutions was determined by the Karl Fischer method with the use of a Mettler Toledo automatic titrator. Tars and solid particles were measured gravimetrically. The dust contained in isopropyl alcohol solutions was filtered off, washed with an additional portion of isopropyl alcohol, dried, and weighed. The mass of tar was measured after evaporation of the solvent under reduced pressure (0.1 bar, 80 °C) until a constant mass was reached.

4.3.2. Halides

The content of chlorine species present in syngas was measured with the use of two methods. The direct measurement from syngas was done through adsorption in NaOH performed with a set-up similar to the above-mentioned water/tar/particle measurement. The samples were analyzed with the use of ion chromatography. Due to Cl content in the gas, which is only marginally above the detection limit of the method, the measurement was done in parallel for raw gas after the reactor, as well as after the water condensation and demisting steps, to assure no Cl was present in syngas after the last stage of syngas cleaning. As an indirect measurement, the measurement of Cl content in the water condensate from the gas cooler was also done. A baseline for gasification of the biomass was detected for a scenario when the gas cleaning was performed without the use of any sorbents or pre-coats. Any result of chlorine ion content in the condensate lower than the baseline indicates that HCl was removed from syngas in the prior adsorption/filtration step.

For chromatography, a dual-channel reagent-free capillary Dionex ICS-5000 ion chromatograph was used. The set-up consisted of dual ion chromatography (IC) channels, each connected to an individual gradient pump, eluent generator cartridge, injection valve, column set, and detector, while an autosampler was also used. A Chromeleon® 6.7 (Dionex) Data Management system was used for instrument control and data handling.

For separation and analysis of cations, the IonPack CS-16 analytical column (250 mm × 3 mm) and IonPack CG-16 guard column were applied. An external standard method using a commercial six-cation solution was utilized for quantitative analysis. Furthermore, 40 mM methanesulfonic acid eluent of high purity was electrochemically produced by the eluent generator cartridge (EGC-MSA). The eluent flow was maintained at a rate of 0.340 mL/min. The column and detector compartments were thermostated at 30 °C and 20 °C, respectively, to obtain constant conditions. Chromatograms were recorded isocratically for 30 min.

4.3.3. AAEMs

The analysis was conducted by wavelength-dispersive X-ray fluorescence spectroscopy with the use of an ARL OPTIM'X spectrometer from Thermo Fisher Scientific with an X-ray Rh lamp of 200 W with a 75-μm Be slit. The sample for this method was prepared by ashing at 815 °C before pressing into tablets with a 20% supplement of wax.

4.3.4. Heavy Metals

The content of trace elements (heavy metals) in chars from biomass was determined with the use of the IChPW's internal procedure. Analyzed samples were firstly dissolved in concentrated acids (HNO_3 and HF) using a closed circuit, multistage microwave-assisted mineralization process. For this purpose, a 10-position microwave mineralizer Ethos 1 from Milestone was used. The maximum temperature of the mineralization was equal to 200 °C. Dissolved samples were analyzed using an inductively coupled plasma optical emission spectroscope (ICP-EOS) iCAP 6500 DUO from Thermo Fisher Scientific. For calibration of the EOS, single element standards from SCP Science were used.

5. Results and Discussion

5.1. Filtration Process

All filtration tests were carried out at filtration temperature equal to 450 ± 5 °C and a constant stream of sorbents. Three of the tested sorbents gave the desired filtration/regeneration patterns. Thus, the sorbents fulfilled their basic requirement of keeping the filtration process stable. Only dolomite (T4) led to a complete failure of the filter. With dolomite, the filter cake resisted pulse cleaning and, thus, a constant rise in dP was measured.

Figure 3 presented below depicts the results of filtration data for all four sorbents. For tests 1–3, the time from the start until the first pulse cleaning (formation of filter cake layer), ranged from 1 h 45 min to 2 h 40 min. These values correspond well with the data previously collected. Moreover, the time between subsequent pulses was consistent and repeatable (from 40 min to 1 h 15 min) as can be seen during the analysis of the registered pressure drop vs. time patterns. The repetitive regeneration cycles indicate a stable operation of the filter. For T1, another interesting characteristic feature can also be noted. After the fourth hour of operation, the filter cake collected adopted a self-cleaning (without pulses) characteristic. This phenomenon was noted before, when the filter cake adopted a loose structure and linear velocities of syngas were sufficiently small. Furthermore, a well-selected particle size of sorbent and ratio of sorbent to char in the filter cake positively influenced the observed self-cleaning property. This phenomenon is rarely registered for filtration of gasification gases; however, it is quite conventional for the HT filtration of flue gases.

During T4, the severity of pulse jet cleaning was increased gradually in the search for any signs of improvement. An increase in pulse times from the conventional 200 ms to 500 ms and in the pressure of the pulses from 6 bar to 8 bar did not give any signs of cake recovery. Conversely, the severity of pulses led to some mechanical degradation of the filter elements, and pit holes at the dirty side of the filters were noted during the later inspection.

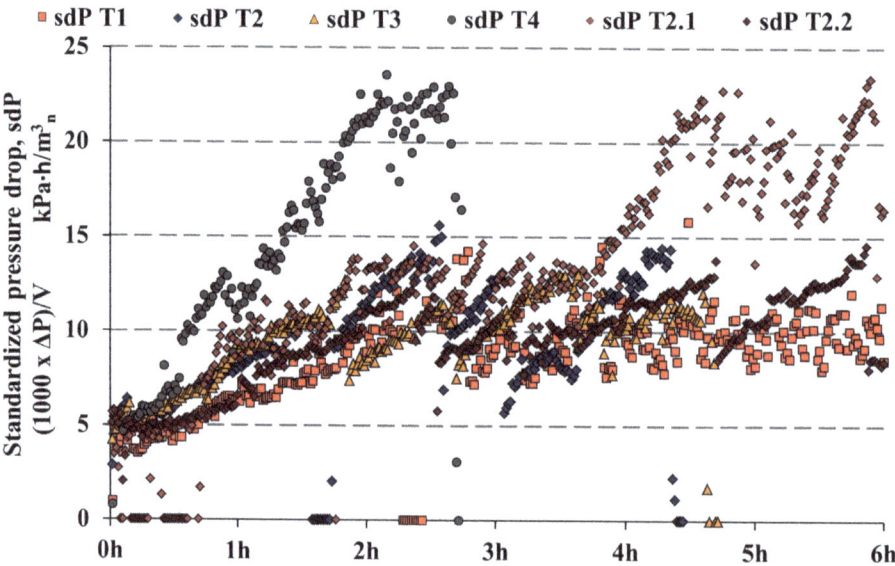

Figure 3. Change in pressure drop on the high-temperature (HT) filter during filtration/adsorption tests. Pressure drop (dP) presented in a standardized form that takes into account the actual flow rate of syngas.

Concerning previous experience with CaO-based sorbents, dolomite should not have brought such a quick and decisive filter failure as noted for T4. At 60%, it is composed of CaO and flows well, similarly to chalk; thus, it is surprising to discover this problem. In the past, operation of the filter was tested for CaO precoating and co-filtration runs where the ratio of char to chalk was higher than 1:0.2 and stable filtration patterns were recorded. Noteworthy, the optimum point for operation of the filter with a pressure drop of 1 kPa was found to be the ratio of 1:1–1:2.

From preliminary testing, halloysite was shown to be a very promising material that enables both good gas cleaning and filter regeneration properties. For this reason, it was chosen for further experiments where limits of its filtration-enhancing properties were determined; thus, two additional tests were done (T2.1 and T2.2) where the mass stream of halloysite was kept constant and gasification conditions were varied through the use of fuel with different moisture content. The changes in reactor operating conditions induced changes in the quality of produced syngas, which subsequently led to pronounced variations observed in the filtration process itself.

Analysis of the three halloysite experiments was primarily done with attention to temperatures of syngas at the reactor exit, at the filter inlet, and inside the dirty side of the filter. The data comparing the halloysite filtration tests are presented in Figures 4 and 5.

As is often common for pilot installations, here, the operation of the reactor and the gas cleaning unit was also stabilized by trace heating. For the GazEla installation, the trace heating is mounted starting from the reactor outlet, throughout and past the HT filter. Due to the thermal mass of the filter, it has the most stable characteristic of temperature changes. The heating is primarily used for start-ups, as well as to keep the operation of the cleaning unit safe when an intermittent, unstable operation of the reactor happens. In normal operation of the pilot GazEla reactor, 550 °C to 650 °C syngas is produced at its outlet. The raw syngas temperature depends on reactor power and quality of the gasified fuel. As a standard, the filter body is kept at a minimum temperature of 430 °C.

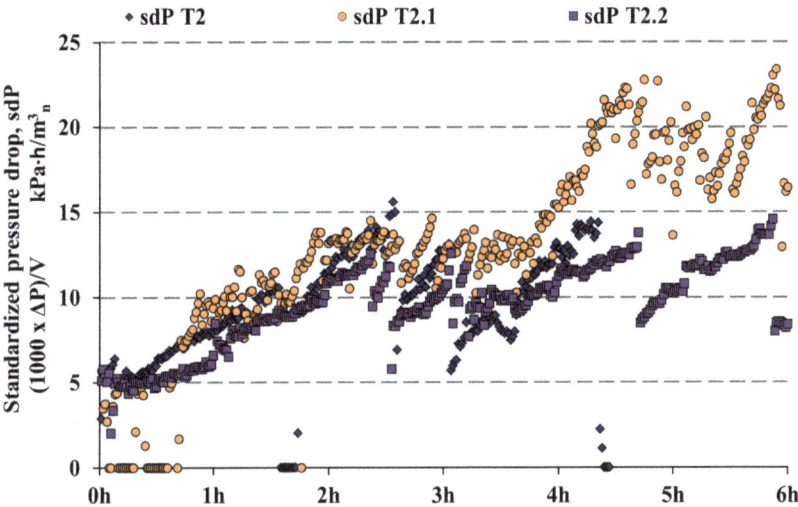

Figure 4. Change in pressure drop on the HT filter during filtration/adsorption tests: comparison of three experiments with halloysite.

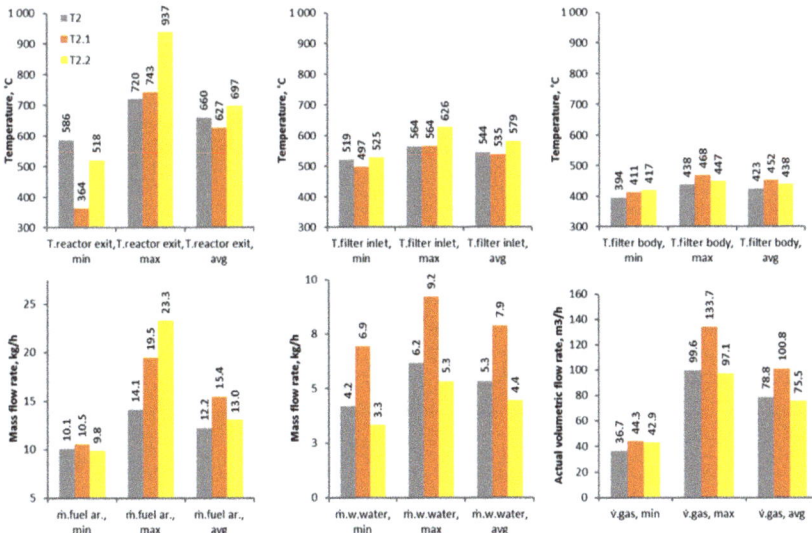

Figure 5. Changes in the main process parameters regarding the halloysite-assisted filtration of syngas.

For T2.1, the biomass was artificially wet to the point of 29 wt.%, while, for T2.2, the feedstock was dried to the point of 8.9 wt.%. The GazEla reactor was tested previously on biomasses of moisture content up to 40%; however, at this point, operation of the fixed bed changed dramatically, as did the tar content and characteristics of produced syngas. For the three experiments, the stream of fuel fed into the reactor (on a dry basis) was kept stable to simulate the conditions where the reactor power demand was kept constant and the quality of available fuel changed. This operational difference led to the variable stream of wastewaters condensed from syngas, as well as the stream of syngas calculated from the mass balance of the reactor. In a fixed bed, the instantaneous stream of fuel fed into the reactor cannot be measured directly with the precision known and needed for stable operation of FB gasifiers. Here, the changes in fuel stream were determined based on time-average mass balances and they were shown to be in good agreement with calculated mass and elemental balances.

In line with the above-mentioned findings, during T2.1, lower minimum and average temperatures of raw gas were noted at the reactor outlet. This stems from the higher demand for heat needed for evaporation of water contained in the feedstock. Thus, it may be counterintuitive that, during the same test, the temperature inside the filter body was on average 14–29 °C higher in comparison to the two other experiments. Such a characteristic was noted before for the GazEla reactor, where some residual oxygen was present in raw syngas. During the extended residence time in the HT filter body, and due to the high temperatures of the filters, the oxygen was consumed in exothermal reactions. Such phenomena are predominant for fixed-bed gasification of feedstocks with very high moisture content, and they are also often connected with the production of very high amounts of heavy tars. The reason for this is that the temperature of the pyrolysis zone in a fixed-bed reactor drops and shifts toward the production of higher-molecular-weight tars (lower thermal breakdown of the tars). The heavy tars are difficult to measure analytically as they are very reactive and easily convert to polyaromatic hydrocarbons (PAH). The PAHs in tar can then undergo spontaneous polymerization, as a result of the high availability of C–O–H bonds, e.g. in phenols, or they can polymerize/carbonize and form solid-state char. These changes in the chemistry of tars are also one of the main reasons why HT filters often lose their porosity and eventually increase their baseline dP. To support this notion, the authors hold experience gathered throughout the past five years of operation of the filter, whereby a situation never occurred where the filtration of post-combustion gases (flue gases) led to a complete failure of the filter elements (pressure drop exceeding 6kPa), despite the great range of tested operational parameters. In flue gases, the stable filtration behavior is adopted much more quickly, mostly during the first 1–2 h, and it is possible to cycle between operational baseline set points without problems. The filter also operates on flue gases at much higher temperatures without instability. Supporting this thesis seems also to be the experiences from entrained flow, membrane reactors, where completely tar-free syngas is produced, e.g., utilized in integrated gasification combined cycle power plants. There also, the syngas ceramic filters are commonly operated close to their upper temperature limits of 900–1000 °C with acceptably long maintenance cycles (>5000 h).

In Figure 5, it can be seen that, in T2.1, the filtration behavior adopted two separate patterns. In the first part of the tests, the pressure drop increased steadily, up to the set point of baseline dP (<2.5 h). At this time, regeneration pulses occurred every ca. 4–5 min, which was never observed for continuous stable runs, indicating that the filter cake kept sticking to the surface of the filters. To keep the unit in operation, the pressure drop set point was not changed until ca. the fourth hour of the test. From previous experience, a stable filtration pattern can be recovered for some cases where the thickness of the filter cake layer is deliberately, substantially increased. A thicker layer of filter cake tends to drop off the elements in larger flakes; however, this can also lead to patchy cleaning and irreversibly higher operational baseline of the filter. In this case, the method was successful, and, after the fifth hour, the time intervals between pulses started to increase. This can be observed by the dropping slopes of the dP curve between subsequent regeneration cycles. In stable filtration runs noted for both T2 and T2.2, the duration between regeneration pulses reached the conventional 1 h to 1 h 15 min.

Neither the temperature at the filter inlet nor the temperature in the filter body seemed to have any correlation with the moment when the filtration started to run stably. The moment when the filtration stabilized for T2.1 was in good relation to the time when the temperature profile of the pyrolysis zone of the reactor reached 500 °C. No noticeable change in the composition of syngas with respect to permanent gases (O_2, CO_2, CO, CH_4, H_2), the stream of fed fuel, or stream of recovered wastewater was registered. Thus, the three performed tests indicate the predominant role of tars in keeping the syngas filtration stable. Noteworthy for T2.1 was also the much higher actual stream of syngas filtered. The higher amount of water in raw syngas led to an increase in the filtration velocity, U_f (linear velocity of gas passing through a filter element, m/s). For HT applications, a commonly accepted value of maximum allowable U_f is 3 m/s. However, from experience, the filtration of syngases from fixed-bed reactors at $U_f > 1.5$ m/s cannot be kept with a pressure drop lower than 2kPa. Hence, to keep the dP low, the operational optimum was found for this unit to keep the U_f in the range from 1 m/s to 2 m/s. During T2.1, the U_f reached the maximum of 1.9 m/s.

5.2. Water, Tar, and Solid Particle Content

The water content of syngas at the outlet from the gasifier is directly connected to fuel composition, its moisture content, and the amount of steam used as a gasifying agent. Secondary reactions connected to hydrogen present in fuel further lead to the formation of steam, which can take part in tertiary reforming reactions. For the GazEla reactor, a fuel of moisture content 25 wt.%. was found to be optimal. The moisture content of the fuel and the amount of steam added for temperature control of the reactor's bed together have a high influence on the amount of tar produced. For the pilot-scale GazEla reactor, the content of organics in raw syngas can reach up to 50 g/Nm^3 for waste fuels such as sewage sludge. For the majority of conventional feedstocks, the value does not exceed 25 g/Nm^3. Table 4 presents concentrations of the basic contaminants in raw syngas (at reactor outlet) and after passing through the HT filter. For all tested points, one prevalent observation can be reported. Both water and tar contents were slightly reduced in the HT filter, even though the filter was operated at a moderate temperature of 450 °C and the filters were constructed from theoretically chemically non-active Al/Si material composites. For chalk, halloysite, kaolinite, and dolomite, the values of tar reduction were equal to 18.2%, 10.3%, 10.4%, and 16.9%, respectively. Thus, the CaO-based sorbent may give the best tar reforming characteristics while AlO–SiO materials take part in tar reforming to a lesser extent (halloysite, kaolinite). MgO is also known to participate in tar reforming in FB reactors; however, here, dolomite performed slightly worse than clean chalk. From current experience, it is hard to determine to what extent this effect is connected to the action of the sorbents themselves, because tar reforming on HT filters operated without sorbents (char alone) was also noted previously. It is noteworthy to say that most ceramic filters are built of mullite or other aluminosilicates and, thus, in their chemical composition, show similarities to halloysite and kaolinite. Thus, it was proposed that the reduction of tar content in the filter vessel may originate from a combination of basic thermal decomposition of high-molecular-weight tars or their polymerization and phase change to soot as a result of long residence time, which may be catalyzed by the presence of Ca, Mg, Al, and Si.

Table 4. Concentration of water, tar, and particle matter present in raw syngas (1) and after its filtration/sorption (2) in the HT filter.

	Unit	T1	T2	T3	T4
H$_2$O 1	g/Nm^3	188	128	160	145
H$_2$O 2	g/Nm^3	187	126	158	144
Tar 1	g/Nm^3	13.9861	12.7429	17.4362	24.6653
Tar 2	g/Nm^3	11.4424	11.4352	15.6267	20.4876
PM 1	g/Nm^3	4.5751	2.1822	1.4811	3.6773
PM 2	g/Nm^3	0.00	0.00	0.0162	0.00

In the analysis of particulate matter content of syngas, there is always great uncertainty connected with the determination of solids entrained from fixed-bed reactors. This subject is connected with technical limitations in the size of sampling probes, as well as the size distribution of particles entrained from a fixed bed of the reactor. For this reason, the particulate matter (PM) presented in Table 4 was only determined through the gas sampling of syngas downstream of the ceramic filter. The amount of PM in syngas upstream of HT filter was calculated through a mass balance of C and ash present in recovered filter cake. For T1, T2, and T4, no PM was measured downstream of the filter. Only for T3 was a small amount of PM registered at the outlet from the filter, indicating a breakthrough of dust which can happen if the layer of filter cake after pulse cleaning is too thin. Previously, another source of PM in syngas filtered on ceramic elements was also found. In HT dry gas cleaning systems, where the installation is subjected to large quantities of tars and does not operate continuously (the pilot system is a research installation), small amounts of polymerized carbon deposits can form at cold spots of the ceramic filter and pipelines where the temperature of syngas is not sufficiently high to keep condensation of high-molecular-mass tars from occurring. At such places, tars tend to condense and polymerize, thus creating deposits of very brittle carbon structures. However, this problem was never noted at installations where tars originate from FB reactors, because these tars are of much different composition (mostly light tars), and the concentrations of gravimetric tars rarely exceed 15 g/Nm^3.

Proximate and ultimate analysis of chars recovered from the ceramic filter unit was done to assess the ratio of char entrained from the reactor in relation to the stream of fed sorbent (Table 5). For T1 and T4, the ratio of sorbent to char in gas was similar as can be seen from the proximate and ultimate analysis of the filter cakes. Similarly, T2 and T3 had similar char-to-sorbent ratios. Filter cakes from T1 and T4 were much richer in C and showed much smaller amounts of ash. Regarding the filter cake obtained from the reference test, the sample was a mixture of biomass char and ash with high C, A, and Cl contents. It can also be seen that the ratio of char to the sorbent for T1–T4 was equal to 1:0.9; 1:4.2; 1:5.0, and 1:1.25, respectively.

Table 5. Proximate and ultimate analysis of filter cakes recovered from HT filtration/sorption of syngas (as received).

	Unit	T1	T2	T3	T4	Reference
H_2O	%	0.7	0.7	0.3	0.4	1.5
A	%	50.5	79.5	82.5	55.3	9.4
C	%	42.8	17.0	18.6	40.8	80.5
H	%	0.41	0.53	0.35	0.87	1.97
N	%	0.1	0.15	0.15	0.34	0.49
S	%	0.09	0.06	0.04	0.14	0.22
Cl	%	0.25	0.24	0.09	0.09	1.352

Subsequently, qualitative and quantitative analyses of AAEMs present in filter cake samples were performed, and Table 6 collates the results. Values presented in parentheses in each cell correspond to a given content of the ash species in the fresh sorbent. The last column of the table is fitted with data from the analysis of clean char from the gasification process of wood chips where no co-filtering agent or sorbent was used. Based on this comparison, the analysis of changes in ash chemistry was possible, giving grounds for raising conclusions as to which AAEMs are volatilized during gasification in the GazEla reactor and can be captured from syngas with the use of the filtration/sorption method.

It is visible that, for all test points, dilution effects of the sorbents with ashes from biomass took place. Thus, with the use of the mixing law, it was checked whether there were any high deviations from the mixing proportions of ashes obtained from the sorbents and biomass chars. Such findings can indicate the capture of species volatilized in a reactor, as well as volatilization from the filter cake, thus indicating lower efficiency for its removal.

To some extent, chars from different tests interact with each other and are extracted together from the filter vessel even after prior cleaning of the elements and their precoating with another sorbent. Thus, for T1 and T4, it can be seen that the sorbents used were rich in Ca and Ca–Mg, respectively. Sorbents applied in T2 and T3 had similar chemical compositions (rich in Al and Si); however, halloysite is also naturally rich in Fe. It can be seen here that the sample from T2 (halloysite) was partially polluted with chalk from T1, whereas samples from T4 (dolomite) contained some kaolinite from T3. It is interesting that halloysite, which contains high quantities of Fe, did not contaminate the sample from T3. This finding may be partially supported by the observed good filtration-enhancing properties and flowability of halloysite. Noteworthy also, even though the amount of ashes in biomass was small and, thus, the dominance of compounds originated from sorbents was very high (ratio of ashes

1:10–1:50) for all samples, the effect of ash enrichment in constituents characteristic for biomass ashes such as K and Na was high.

Table 6. Alkali and alkali earth metal content in chars forming filter cake on the surface of the ceramic filter (unit: wt.%).

	T1	T2	T3	T4	Reference
SiO_2	2.22 (2.79)	25.25 (35.75)	67.18 (65.68)	14.16 (2.79)	9.38
Al_2O_3	0.67 (0.85)	17.89 (25.77)	26.78 (27.77)	4.98 (1.61)	1.88
Fe_2O_3	4.41 (0.49)	14.75 (21.35)	1.22 (1.17)	1.82 (1.49)	1.18
CaO	52.7 (50.69)	18.17 (0.68)	1.63 (0.18)	34.77 (27.18)	14.22
MgO	0.36 (0.35)	0.57 (0.58)	0.21 (0.14)	23.24 (17.89)	2.73
P_2O_5	0.09 (0.02)	0.57 (0.77)	0.09 (0.03)	0.52 (0.3)	7.74
SO_3	1.22 (0.87)	0.32 (007)	0.08 (0.07)	1.28 (1.22)	4.35
Mn_3O_4	0.1 (0.02)	0.47 (0.56)	0.04 (0.02)	0.25 (0.17)	0.64
TiO_2	0.05 (0.04)	2.61 (3.78)	0.89 (0.92)	0.19 (0.09)	0.22
BaO	0.01 (0.01)	0.08 (0.11)	0.02 (0.01)	0.02 (0.01)	0.39
SrO	0.05 (0.05)	0.03 (0.02)	0.01 (0.01)	0.02 (0.01)	0.13
Na_2O	0.11 (0.06)	0.22 (0.22)	0.19 (0.15)	0.2 (0.08)	5.64
K_2O	1.04 (0.19)	1.75 (0.45)	0.98 (0.42)	1.7 (0.59)	22.46

For a more visual comparison of the results, Figure 6 was prepared where samples were compared through mass balance analysis, and the results are presented as ratios of species calculated from the mass balance related to the concentrations measured in samples. Thus, values above 1 indicate here that the content of the specie in the filter cake was lower than expected from the mixing law and may indicate lower removal efficiency of the element. On the other hand, values below 1 indicate that the content of an element in the filter cake was higher than predicted and, thus, may indicate its preferential removal from syngas. Results fitting into a range of ±25% are generally accepted to be in agreement with the mixing law. The streams of char and sorbents were all in good agreement, and the ratios of values calculated ranged from 0.92–0.97.

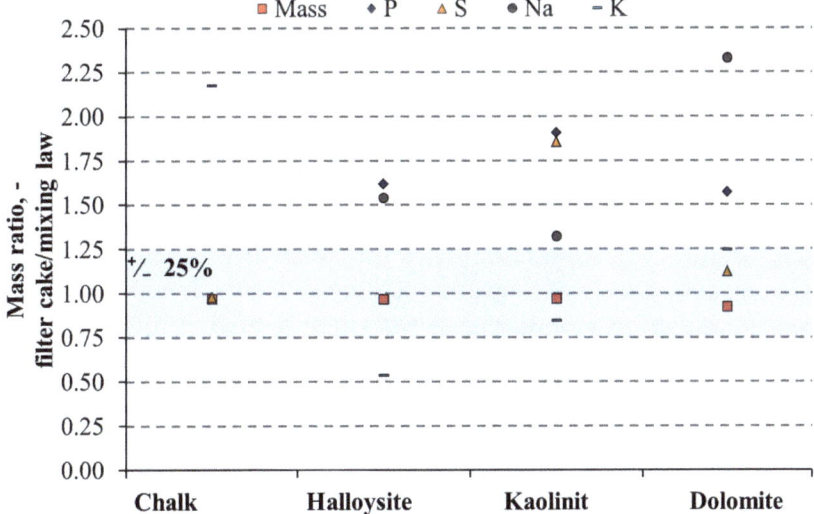

Figure 6. Ratio of the major elements in filter cake calculated from the mixing law of sorbents and char vs. values measured analytically in recovered filter cake.

The best correlation with the mixing law was found for dolomite, where only concentrations of P and Na in filter cake were well below half of the expected value, which may indicate lower retention in the filtration/sorption system. At the other end of the scale was chalk, which was found to be in good agreement only for S, while lower than expected contents of K (2.17), Na (5.17), and P (8.19) were noted (P and Na values are off the scale presented in Figure 6).

For halloysite, K was the only measured element which might indicate preferential adsorption (0.53) from syngas. For all sorbents, the scatter of the results indicates that it is very difficult to close mass balances of elements in fixed-bed gasification systems. Primarily for halloysite, S concentration in the filter cake was more than 20 times lower than the mass balance indicates (off the scale in Figure 6). On the other hand, for dolomite, the Na content in filter cake was equal to less than 4.5 times the prediction. For future work, it is, thus, advised to try applying other analytical procedures in the search for these elements in syngas; however, as already mentioned, ion chromatography of gas sampled through absorption in NaOH does not provide as clear results as hoped for.

The difficulty in closing mass balances of elements in gasification systems can be seen here. A sampling of char from syngas in the real installation is only possible for a test designed specifically to be done without the use of sorbents. Even though the biomass source was kept constant during the research, its exact composition, as well as its gasification conditions, varied. Taking into account that components of biomass ash volatilize in a manner related to their ash composition and process parameters, the char bed does not remain constant nor does its ability to readsorb AAEMs from raw syngas.

5.3. Heavy Metals

The starting assumption for the determination of collection efficiency of heavy metals is that, in process conditions of the filter, the species that contain them are attached to other solids present in the gas. When subjected to filtering, they should be recovered either directly through filtration (mechanical separation) or indirectly by adsorbing onto the surface of filter cake particles, as they already have a well-developed surface and the HT filter conditions are more favorable for adsorption than those that occur in the reactor. Table 7 collates the obtained results regarding the amount of heavy metals present in fresh sorbents, as well as in filter cakes recovered after the filtration/sorption process. Values presented in parentheses in each cell correspond to the given species content of the fresh sorbent.

Table 7. Concentration of heavy metals in samples of feedstocks and products from gasification and filtration/sorption processes (unit: mg/kg on a dry basis).

	T1	T2	T3	T4	Reference
As	1.71 (<1.7)	3.94 (<1.7)	1.71 (2.98)	2.23 (<1.7)	-
Cd	1.38 (<0.27)	2.01 (2.06)	0.475 (1.08)	0.808 (0.799)	-
Co	3.84 (1.08)	149 (280)	3.39 (5.4)	2.87 (2.28)	-
Cr	150 (5.83)	577 (790)	111 (55.5)	157 (15.4)	-
Cu	21.3 (3.16)	50 (70.9)	20.6 (20.8)	13.9 (8.28)	-
Mn	581 (95.6)	2483 (3749)	243 (66.9)	842 (1.002)	-
Mo	23 (1.32)	10.4 (4.98)	9.37 (3.91)	20.6 (1.54)	-
Ni	103 (4.46)	388 (558)	54.9 (15.5)	64.2 (11.8)	-
Pb	34.6 (10.1)	91.2 (26.5)	73.2 (81.3)	58.3 (16.4)	-
Sb	7.13 (<2.0)	7.27 (13.1)	1.76 (3.5)	3.72 (<2)	-
V	7.23 (7.15)	155 (340)	24.6 (50.7)	13.4 (15.3)	-
Zn	732 (43.8)	1014 (288)	414 (16.3)	539 (84.3)	-

In the case of five elements, a 10-fold increase in concentration was registered in comparison to the result of the fresh sorbent. These elements were Cr, Mn, Mo, Ni, and Zn. In the case of Zn and Mo, an increase in concentration in all samples of filter cakes was noted. For Zn, the highest gain was noted in T3 (25-fold for kaolinite), while the lowest was noted in T4 (six-fold for dolomite). Previous studies on the volatilization of metal compounds during the thermochemical treatment of ashes concur with this finding [12–14]. For Mo, higher recovery was noted in T1 (17-fold for chalk), while the lowest was noted for both T2 and T3 (two-fold for halloysite and kaolinite). For Cr, a higher increase in concentration was noted in the case of T1 (25-fold for chalk), while the lowest was noted for T3 (two-fold kaolinite). For Mn, T4 gave the highest increase in concentration (840-fold for dolomite), while the lowest was noted for T3 (3.6-fold for kaolinite). The increase in the amount on Mn in the T4 filter cake was substantial, and the determination of its origin needs further investigation. For Ni, the highest rise in concentration was noted for T1 (23-fold for chalk), whilst the lowest was noted for T3 (3.5-fold for kaolinite).

Thus, all sorbents showed the concentration of heavy metals in their relevant filter cakes higher than expected from mass balances. During the research, there was no method for the determination of any residual heavy metals in syngas downstream of the ceramic filter that could support any definitive conclusions on any preferential removal of heavy metals.

5.4. Halides and H_2S

Generally, it is known that K almost exclusively volatilizes with correlation to Cl. Thus, from the results presented above, it was expected that the Cl content in syngas should be low. For baseline conditions, the amount of HCl measured in syngas was close to the detection limit of the method. In raw syngas, the mean value of HCl present was equal to 5 ppm$_v$. To assess its removal in the gas cleaning system, the analysis of Cl content in side-products was necessary. No F or Br content in the gas was detected. Table 8 presented below depicts all results collected regarding Cl content in condensates from gas cooling. It can be seen here that, in syngas after treatment with kaolinite and dolomite, Cl concentration was so low that it reached the lower detection limit of the method.

Table 8. Concentration of chlorine in water–tar condensate recovered from the gas cooler.

	Unit	T1	T2	T3	T4
Cl	mg/kg	17.8	14.2	<0.07	<0.07

For filter cakes, the results of Cl content are presented in Table 5. The results indicate that biomass char contained 16.9-fold higher Cl concentration than the biomass before thermochemical treatment, which supports the theory of Cl readsorption on the bed of char or its preference to remain in the concentrated solid phase composed of ash and char.

During balancing of Cl present in the system, it was noted that only up to 30 wt.% can be accounted for with all the measured inputs and outputs. The remaining stream of Cl presumably stayed in reactor bottom ash, as known from the literature [3,16]. However, due to technical limitations of the lab-scale conditions and the insufficient amount of bottom ash generated from clean biomass like wood chips, it was not possible to adequately sample and analyze this stream.

Performed analyses allowed determining the amount of HCl present in raw syngas, whereas no HCl could be measured after the last step of gas cleaning. To summarize its removal, Figure 7 was prepared. The figure relates the actual streams of HCl trapped in filter cake to Cl content in condensate from syngas cooling. For T1 and T2, 90% of Cl present in syngas was recovered in the filtration/adsorption step. This value should be compared with Cl content measured for the reference case, where filtration was carried out without any sorbent and Cl was preferentially recovered in the condensate from syngas cooling.

Finally, during ion chromatography of the liquid samples, the detection of H_2S was also expected. However, for all tested samples, no S ions were detected. This result indicates that the applied analytical procedure needs to be further developed in the future to give better precision in the determination of syngas composition.

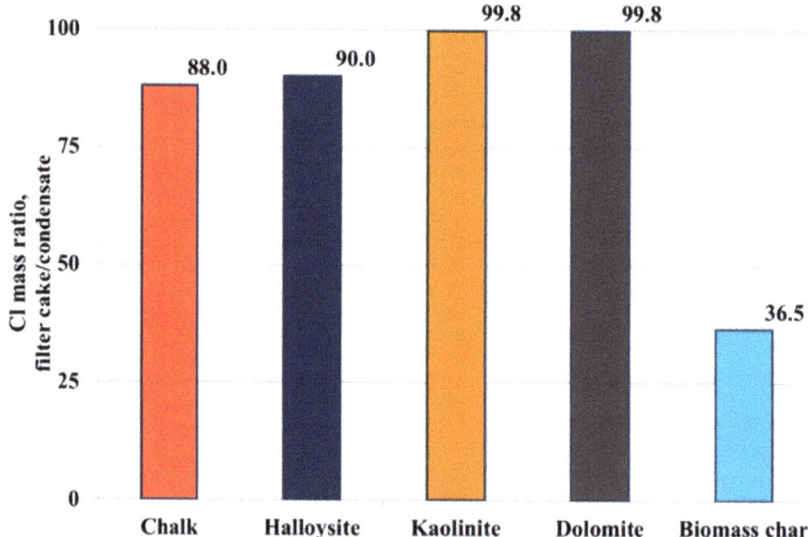

Figure 7. Ratio of Cl recovered from syngas in filter cake and condensate from syngas cooling.

6. Conclusions

To summarize, during the research, the use of four sorbents of distinctive chemical and physical properties was tested for HT conditioning of syngas, namely, chalk (CaO), dolomite (CaO–MgO), halloysite (AlO–SiO), and kaolinite (AlO–SiO). The two representative aluminosilicates differed from each other in their tertiary structure.

The research showed that co-filtration of dolomite with solids from the gasification process may inversely affect the regeneration of HT filters but lead to quick and irreversible filter failure. Even though the sorbent itself is easily flowable, is easy to handle, effectively pre-coats the filters, and gives good sorption parameters, it is also the first material on the list of tested sorbents which cannot be reported to have filtration-enhancing properties until further research on merits of its application is available.

For the rest of the tested sorbents, design criteria regarding both pulse-back regeneration of the filter and sorption of syngas were satisfied. Collected data allowed concluding that the concept for one-vessel multicomponent gas cleaning is possible and can be successful.

All tested materials adsorbed 90%–99% of Cl present in raw syngas, thus reducing its content to levels below 5 ppm$_v$, which, for most energetic applications, is considered as acceptable.

HT filtration and sorption provides a means for the reduction of high-molecular-weight tar content in the range of 10% to 18%, depending on the sorbent used. From preliminary results, it was found that CaO- and CaO–MgO-based materials have higher tar reforming capabilities.

In pilot-scale gasification process conditions, no direct sign of preferential removal of AAEMs on the filter cake-containing sorbents was found. Char in the fixed bed acts as the first stage adsorber and causes retention of compounds volatilized from ash, leading to the high variability of the obtained results.

After filtration on precoated ceramic filters, no presence of PM could be measured in the pilot-scale gasification installation. The only source of solids determined in gas after filtration originated from a temporary breakthrough of particles after pulse-back cleaning or marginal condensation and polymerization of heavy tars.

The application of high-efficiency, HT filter for dedusting of syngas was proven to be the preferred technology for the first step of syngas cleaning. Dust-free syngas allows for its direct cooling and condensation of organic matter, as well as cleaning in oil scrubbers or its further upgrading without problems related to fouling of the installation. In syngas cleaning, however, it is vital to remove tars as quickly as possible and accordingly to their condensation temperature. The fouling of heat exchangers in dust-free syngas conditions follows the path of condensation of heavy tars on contact with "cold tubes" of the heat exchanger and further polymerization of the liquefied tars. A solution to this problem was found in all applications where syngas cooling is either more rapid and goes down to temperatures below the water dew point or does not cool syngas down to levels where heavy tars can condense.

Furthermore, detection of H_2S was expected during ion chromatography of the condensed phase of syngas, as well as from the dedicated sampling of syngas through NaOH absorption. However, for all tested samples, no S ions were measured. This result indicates that the applied analytical procedure needs to be further developed in the future to give better precision in the determination of syngas composition.

Finally, from the analysis of obtained data, halloysite was shown to give the best overall performance in syngas cleaning through sorption-enhanced HT filtration. It had superior filtration and pulse-back cleaning properties, and it remained flowable in all apparatuses used for its storage, handling, and feeding. For very high filtration velocity (1.9 m/s) and syngas of high water and heavy tar content, it allowed the filtration to remain stable even though the set point of dP had to be increased to 2 kPa. In terms of gas cleaning properties, CaO remains the best solution because of its higher tar reforming properties and the comparable ability for the removal of Cl.

7. Prospects for Future Work

- Optimization of sorbent-to-char ratio.
- Determination of maximum removal efficiency for H_2S/HCl.
- Validation tests confirming the unsatisfactory filtration properties of dolomite.

Author Contributions: Conceptualization, M.S.; Methodology, M.S. and T.I.; Validation, M.S. and A.S.; Formal Analysis, M.S.; Investigation, M.S. and T.I.; Resources, M.S.; Data Curation, M.S.; Writing-Original Draft Preparation, M.S.; Writing-Review & Editing, M.S.; Visualization, M.S.; Supervision, T.I.; Project Administration, M.S. and T.I.; Funding Acquisition, A.S.", All authors have read and agreed to the published version of the manuscript.

Funding: This research was funded by the Polish Ministry of Science and Higher-Education (*Ministerstwo Nauki i Szkolnictwa Wyższego*) for the research project entitled *Improvement of biomass and waste gasification technology in fixed and fluidized bed reactors (GazEla, IZPS, IZOP)* grant number 11.18.018/11.19.018

Conflicts of Interest: The authors declare no conflict of interest.

References

1. Sobolewski, A.; Iluk, T.; Szul, M. SRF gasification in GazEla pilot fixed bed gas generator for CHP units. *J. Power Technol.* **2017**, *97*, 158–162.
2. Kotowicz, J.; Sobolewski, A.; Iluk, T. Energetic analysis of a system integrated with biomass gasification. *Energy* **2013**, *52*, 265–278. [CrossRef]
3. Abdoulmoumine, N.; Adhikari, S.; Kulkarni, A.; Chattanathan, S. A review on biomass gasification syngas cleanup. *Appl. Energy* **2015**, *155*, 294–307. [CrossRef]
4. Nakamura, S.; Kitano, S.; Yoshikawa, K. Biomass gasification process with the tar removal technologies utilizing bio-oil scrubber and char bed. *Appl. Energy* **2016**, *170*, 186–192. [CrossRef]
5. Shen, Y. Chars as carbonaceous adsorbents/catalysts for tar elimination during biomass pyrolysis or gasification. *Renew. Sustain. Energy Rev.* **2015**, *43*, 281–295. [CrossRef]

6. Obernberger, I.; Thek, G.; Brunner, T.; Nowak, P.; Mandl, C.; Kerschbaum, M.; Borjabad, E.; Mediavilla, I.; Peña, D.; Carrasco, J. Next Generation Fuel Flexible Residential Biomass Heating Based on an Extreme Air Staging Technology with Ultra-low Emissions. In Proceedings of the European Biomass Conference and Exhibition Proceedings, 26th EUBCE-Copenhagen, Copenhagen, Denmark, 14–18 May 2018; pp. 7–15.
7. Crittenden, B.; Thomas, W.J. *Adsorption Technology and Design*; Elsevier: Amsterdam, The Netherland, 1998; ISBN 978-0-08-048997-1.
8. Werle, S.; Ziółkowski, Ł.; Bisorca, D.; Pogrzeba, M.; Krzyżak, J.; Milandru, A. Fixed-Bed Gasification Process—The Case of the Heavy Metal Contaminated Energy Crops. *Chem. Eng. Trans.* **2019**, *61*, 1392–1398.
9. Stec, M.; Czaplicki, A.; Tomaszewicz, G.; Słowik, K. Effect of CO_2 addition on lignite gasification in a CFB reactor: A pilot-scale study. *Korean J. Chem. Eng.* **2017**, *1*, 129–136. [CrossRef]
10. Schmid, J.C.; Fuchs, J.; Benedikt, F.; Mauerhofer, A.M.; Müller, S.; Hofbauer, H.; Stocker, H.; Kieberger, N.; Bürgler, T. Sorption Enhanced Reforming with the Novel Dual Fluidized Bed Test Plant at TU Wien. In Proceedings of the European Biomass Conference and Exhibition Proceedings, 25th EUBCE-Stockholm, Stockholm, Sweden, 12–15 June 2017; pp. 421–428.
11. Qin, K.; Lin, W.; Jensen, P.A.; Jensen, A.D. High-temperature entrained flow gasification of biomass. *Fuel* **2011**, *93*, 589–600. [CrossRef]
12. Froment, K.; Defoort, F.; Bertrand, C.; Seiler, J.M.; Berjonneau, J.; Poirier, J. Thermodynamic equilibrium calculations of the volatilization and condensation of inorganics during wood gasification. *Fuel* **2013**, *107*, 269–281. [CrossRef]
13. Okuno, T.; Sonoyama, N.; Hayashi, J.; Li, C.-Z.; Sathe, C.; Chiba, T. Primary Release of Alkali and Alkaline Earth Metallic Species during the Pyrolysis of Pulverized Biomass. *Energy Fuels* **2005**, *19*, 2164–2171. [CrossRef]
14. Hirohata, O.; Wakabayashi, T.; Tasaka, K.; Fushimi, C.; Furusawa, T.; Kuchonthara, P.; Tsutsumi, A. Release Behavior of Tar and Alkali and Alkaline Earth Metals during Biomass Steam Gasification. *Energy Fuels* **2008**, *22*, 4235–4239. [CrossRef]
15. Simell, P.A.; Hakala, N.A.K.; Haario, H.E.; Krause, A.O.I. Catalytic Decomposition of Gasification Gas Tar with Benzene as the Model Compound. *Ind. Eng. Chem. Res.* **1997**, *36*, 42–51. [CrossRef]
16. Sonoyama, N.; Okuno, T.; Mašek, O.; Hosokai, S.; Li, C.-Z.; Hayashi, J. Interparticle Desorption and Re-adsorption of Alkali and Alkaline Earth Metallic Species within a Bed of Pyrolyzing Char from Pulverized Woody Biomass. *Energy Fuels* **2006**, *20*, 1294–1297. [CrossRef]

© 2020 by the authors. Licensee MDPI, Basel, Switzerland. This article is an open access article distributed under the terms and conditions of the Creative Commons Attribution (CC BY) license (http://creativecommons.org/licenses/by/4.0/).

Article

Feedstock-Dependent Phosphate Recovery in a Pilot-Scale Hydrothermal Liquefaction Bio-Crude Production

Ekaterina Ovsyannikova *, Andrea Kruse and Gero C. Becker *

Department of Conversion Technologies of Biobased Resources, Institute of Agricultural Engineering, University of Hohenheim, 70599 Stuttgart, Germany; andrea_kruse@uni-hohenheim.de
* Correspondence: e.ovsyannikova@uni-hohenheim.de (E.O.); gero.becker@uni-hohenheim.de (G.C.B.); Tel.: +49-711-459-23413 (E.O.); +49-711-459-24785 (G.C.B.)

Received: 3 December 2019; Accepted: 11 January 2020; Published: 13 January 2020

Abstract: Microalgae (*Spirulina*) and primary sewage sludge are considerable feedstocks for future fuel-producing biorefinery. These feedstocks have either a high fuel production potential (algae) or a particularly high appearance as waste (sludge). Both feedstocks bring high loads of nutrients (P, N) that must be addressed in sound biorefinery concepts that primarily target specific hydrocarbons, such as liquid fuels. Hydrothermal liquefaction (HTL), which produces bio-crude oil that is ready for catalytic upgrading (e.g., for jet fuel), is a useful starting point for such an approach. As technology advances from small-scale batches to pilot-scale continuous operations, the aspect of nutrient recovery must be reconsidered. This research presents a full analysis of relevant nutrient flows between the product phases of HTL for the two aforementioned feedstocks on the basis of pilot-scale data. From a partial experimentally derived mass balance, initial strategies for recovering the most relevant nutrients (P, N) were developed and proofed in laboratory-scale. The experimental and theoretical data from the pilot and laboratory scales are combined to present the proof of concept and provide the first mass balances of an HTL-based biorefinery modular operation for producing fertilizer (struvite) as a value-added product.

Keywords: struvite; HTL; biorefinery; renewable fuel; HyFlexFuel

1. Introduction

Hydrothermal liquefaction (HTL) of biomass presents a promising procedure [1,2] for overcoming dependency on fossil fuels and advancing toward sustainable decarburization of the transportation sector. Hydrothermal liquefaction enables the conversion of wet biomass or waste materials into bio-crude oil by using hot-compressed water (287–375 °C) [3,4] that refines downstream to liquid fuel [5]. It is of special interest in jet fuel. In biorefinery, sustainability has an important role in achieving the viability of a large-scale bio-crude facility. Nutrient recovery presents an attractive option for improving sustainability and adding value to the production chain. After HTL, the macro (P, K, N) and micro (S, Mg, Ca, Fe) nutrients that might be recovered for biomass production systems are distributed between its products, which include the HTL oil (also known as bio-crude), liquid, and solid phases. This distribution of nutrients among HTL phases depends on not only process conditions but also the nature and composition of the biomass feedstocks [6]. Furthermore, it presupposes variations in nutrient recovery strategies among various biomass feedstocks.

Significant research on HTL, and by association, nutrient recovery, has been performed in small-scale batch systems and mainly utilized algae as feedstock [7–9]. The next step toward the commercialization of biofuel production, which involves an HTL pilot-scale plant that operates in a continuous mode, has received considerably less attention [10]. Only a few published reports relate to nutrient recovery following bio-crude production via continuous HTL. One study by Edmundson et al. [11] has demonstrated that soluble phosphate, which was recovered through acid extraction from the HTL solid phase that originated from continuous HTL of algae feedstock (total volume of system ~1.6 L), could be recycled for algae production. However, replacing nitrogen in a growth medium by using the HTL liquid phase can have a negative effect on the growth rate of algae [11]. McGinn's study [12] has considered the HTL liquid phase that results from continuous HTL of algae (total volume of system ~0.45 L). He has noted that the complete decoupling of phosphate (via struvite $MgNH_4PO_4 \cdot 6H_2O$ precipitation) and nitrogen as ammonia (through ammonia stripping) from the HTL liquid phase suggests a flexible method of recycling them for algae cultivation. To the author's knowledge, no study has previously reported on the feedstock-related fate of nutrients in a pilot-scale HTL unit or discussed nutrient recovery in a large-scale HTL scenario in relation to the distribution of nutrients. This study provides such data, which were gathered in the framework of the HyFlexFuel project. It assesses two campaigns of a continuous HTL pilot plant. Each campaign used a different feedstock—primary sewage sludge (PSS) and the microalgae *Spirulina* (SPR)—both of which are highly relevant to nutrient recovery.

Interest in converting sewage sludge to bio-crude via HTL has increased in the past few years [13–15] because of its availability in large volumes as the main byproduct of wastewater treatment plants, its high water content, and its embedded energy potential. In addition, it is an attractive secondary phosphate source [16]. Previous studies have found the phosphate from sewage sludge primarily in the HTL solid phase as a result of a substantial content of metal ions, such as calcium, iron, and aluminum, and the low solubility of their phosphate salts [6]. While some studies have focused on the immobilization of heavy metals from sewage sludge in the HTL solid phase [17,18], no publication has been found that targets the study of phosphate recovery from HTL solid residue of sewage sludge. Phosphate recovery from the HTL solid phase may be possible in the form of struvite ($MgNH_4PO_4 \cdot 6H_2O$), as previously demonstrated for the related process of hydrothermal carbonization (HTC). Becker et al. [19] have successfully performed phosphate and nitrogen recovery based on HTC of digested sewage sludge via precipitation of struvite from a mixture of the HTC liquid phase and acid leachate from hydrochar. Other studies have also revealed the potential of struvite production on the basis of hydrothermal treatment of various kinds of biomass [20–22]. Bauer et al. [23] have evaluated a possible approach to HTL liquid phase management by way of struvite precipitation for a wide range of biomass that includes pre-digested and digested sludge. Moreover, they have suggested that struvite precipitation from the HTL liquid phase may be economically appealing.

Microalgae are another promising feedstock for HTL. Their advantages include high annual biomass productivity, an ability to grow in poor-quality water, high water, and energy content. Phosphate is significant for various aspects of cellular metabolism of microalgae [24] and is essential for their growth. Prior studies have demonstrated that the HTL of algae biomass results predominantly in phosphate as well as nitrogen and other nutrients, such as K, Na, S, and Mg, in the HTL liquid phase [6,25–27]. The values of nutrients that are recovered in the HTL liquid phase depend heavily on the process parameter and initial algae composition (e.g., marine or freshwater species). Several studies that have been driven by the development of scalable algae-based biofuel production have already addressed nutrient recovery [28]. However, most research has focused on the feasibility of reusing the HTL liquid phase for algae cultivation to reduce the consumption of fertilizers for their growth [29]. Leng et al. [30] and Gu et al. [31] have provided a detailed overview of studies that concern the recycling of nutrients for algae cultivation. Generally, the HTL liquid phase has to be heavily diluted to avoid the inhibitory effect of organic compounds, such as phenols, furans, aromatic hydrocarbons, and nitrogenous compounds [26,32], on algal growth rate. Alba et al. [33] have indicated that the lack of essential nutrients (besides N and P) in the HTL liquid phase, which is enhanced by dilution, can lead to a reduction in the growth rate. Only a few studies have applied an integrated approach and precipitated the nutrients from the HTL liquid phase in the form of high-value products, such as struvite. As noted above, McGinn et al. [12] have decoupled phosphate in the form of struvite from the HTL liquid phase to minimize the inhibition of algae growth through the presence of organic compounds. In addition, Shanmugam et al. [34] have used phosphate recovery via struvite precipitation as a pretreatment for anaerobic digestion of the HTL liquid phase and discovered that the biogas production from the struvite-recovered HTL liquid phase was 3.5x higher than that from the non-struvite-recovered HTL liquid phase.

Struvite precipitation is a well-known technology in wastewater treatment for the recovery of phosphate in the form of slow-release fertilizer [35]. This technique can be coupled with other processes. Some examples of coupling of struvite precipitation with HTL/HTC have been explained above, and another example can be found in reference [36]. The major benefit of struvite crystallization is the production of slow-release fertilizer that is established in the market, transportable, and suitable for long-term storage. Moreover, struvite production does not require severe process conditions; it is precipitated through pH adjustments to approximately pH 9, and the addition of a magnesium source, an ammonium source, or both, if necessary [37]. The production of struvite from HTL byproducts (HTL solid and HTL liquid phases) and its commercialization present an advantage in terms of the economic performance of fuel production and sustainability.

The aim of this study is to examine the feedstock-related application of HTL byproducts for nutrient extraction and evaluate the potential of phosphate recovery in the form of struvite in pilot-scale HTL bio-crude production. The specific research questions of this investigation are as follows: (1) To determine the relationship between feedstock and nutrient occurrence in HTL products at the pilot-scale; (2) to outline perspectives on the use of HTL byproduct streams to produce marketable fertilizer; and (3) to verify the possibility of efficiently recovering phosphate in the form of struvite from HTL byproducts by means of a laboratory-scale study.

2. Materials and Methods

2.1. Hydrothermal Liquefaction and its Products

The HTL products that were used in this study were produced at Aarhus University in Denmark by a continuous pilot-scale HTL reactor with a feed capacity of up to 100 L·h^{-1}, and a total volume of the system of ~20 L. The detailed HTL reactor system description and the procedure can be found in reference [38]. In brief, it includes a feed introduction system, heat exchanger, trim heater, reactor, an oscillation system, a take-off system, and a product collection zone. Primary sewage sludge was collected in February 2018 from the primary treatment of wastewater at a plant in Viborg, while the *SPR* was purchased from Inner Mongolia Rejuve Biotech Co. Ltd. The initial dry matter content of feedstock slurries amounted to 4 wt% and 16.4 wt% for the PSS and *SPR*, respectively. Hydrothermal liquefaction was performed under sub-critical conditions at 220 bars and 350 °C. The flow rate of the slurries was 60 L·h^{-1}. The total duration of runs was 5 h and 6 h for the PSS and *SPR*, respectively. The HTL gaseous stream was separated from other HTL product streams in hydro-cyclone. The HTL oil phase that was produced was gravimetrically separated from the HTL liquid and HTL solid byproducts, which were subsequently separated by filtration [38]. The collected HTL oil, liquid, and solid phases were distributed to the HyFlexFuel partners for further investigation. The pre-dried feedstock and HTL solid phase that were obtained from Aarhus were dried at 105 °C and stored at room temperature for future handling. The HTL liquid phase that was obtained was immediately analyzed for elemental concentration and subsequently stored in a freezer at −24 °C prior to processing. The HTL oil phase that was obtained was stored at 4 °C. Figure 1 presents the simplified sample flow diagram and sample ID.

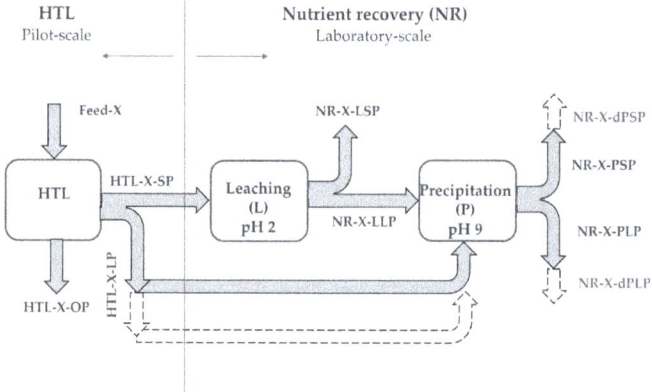

Figure 1. Simplified sample flow diagram and sample ID. SP solid phase; LP liquid phase; OP oil phase. X corresponds to primary sewage sludge (PSS) or *Spirulina* (SPR) and d to direct.

2.2. Phosphate Recovery.

2.2.1. Solubility Experiments

Solubility experiments were performed to evaluate the potential of phosphate release from the HTL solid phase if applied directly as a fertilizer or as a phosphate-containing source for subsequent extraction.

To identify the phosphate form that was presented in the HTL solid phase, the study applied semi-sequential fractionation extraction following the standards, measurements, and testing (SMT) harmonized protocol [39,40]. This procedure included 3 separate extraction proceedings to obtain 5 phosphorus fractions: Inorganic (IP) and organic (OP) phosphorus fraction; apatite inorganic (AP: Ca-bound) and nonapatite inorganic (NAIP: Associated to Al, Fe oxides, and hydroxides) phosphorus fraction; and, finally, total phosphorus (TP) as an overall characteristic. In addition, independent solubility behaviors of phosphate and other nutrients from the HTL solid phase in a range of acid-basic environments were studied. 10 mL of a leaching agent was added to 1 g of the dry HTL solid phase. The resulting slurry was shaken overnight at room temperature. After phase separation by centrifugation or filtration with 13 μm filter paper, the equilibrium pH of leachate was measured with a pH electrode HACH HQ40d. Two experimental setups were conducted. The 1st examined various leaching agents (1M HCl, 1M H_2SO_4, 1M NaOH, 1M citric acid). In the 2nd, sulfuric acid was used as a leaching agent, and their concentrations were varied (0.01M, 0.1M, 0.5M, 1M, 2M). The extraction capacity for phosphorus and other elements was calculated from the amount of the element that was extracted in leachate divided by the sample mass:

$$\text{Extraction capacity} = \frac{c_{\text{leachate}} \times V_{\text{leaching agent}}}{m_{\text{HTL-X-SP}}} \quad (1)$$

where c_{leachate} is the concentration of the element in leachate [mg·L^{-1}], and $V_{\text{leaching agent}}$ is the volume of the leaching agent [L].

To estimate the percentage of crop-available phosphate in the HTL solid phase, the study applied the calcium-acetate-lactate (CAL) extraction method (pH 4) [41] that was established in Germany for soil testing. Extraction was performed according to VDLUFA standard methods [42].

The elemental concentration in all extracts and leachates was measured by ICP-OES (Agilent 715). Analyses by SMT fractionation, all leaching tests, and the CAL extractions were conducted twice.

2.2.2. Struvite Production

To verify the possibility of phosphate recovery via struvite precipitation, the following procedure was conducted:

(1) 5 g of the dry HTL solid phase was suspended in 50 mL Milli-Q water to ensure pumpability at future large scales, and to reproduce the average output moisture condition from the HTL pilot plant. The slurry was subsequently leached with 1 molar sulfuric acid at pH 2 for 2 h (SI Analytics, TitroLine 7000).
(2) 5 mL of phosphate-rich leachate that was separated from the solid residual via filtration was mixed with the HTL liquid phase, which was high in ammonium ions, in a 1-to-6 ratio to guarantee an oversaturation state for struvite crystallization [37].
(3) After the addition of a previously estimated amount of $MgCl_2 \cdot 6H_2O$ to test the underdose and overdose scenarios of Mg^{2+}, the mixture was stirred for 2 h at constant pH 9 that was adjusted by 1M NaOH (SI Analytics, TitroLine 7000).
(4) Finally, the solid that precipitated was filtered with 0.13 μm filter paper and dried at 35 °C for future analysis. The HTL liquid phase from the *SPR* was also directly subjected to the struvite precipitation through supplementation of the magnesium source and maintained constant pH 9 for 2 h. Each trial was performed in duplicate.

The release rate and recovery rate for identifying the phosphate recovery performance were calculated as follows (sample ID presented in Figure 1):

$$\text{Release rate} = \frac{m_{HTL-X-SP} - m_{NR-X-LSP}}{m_{HTL-X-SP}} \times 100, \, [\%] \quad (2)$$

$$\text{Recovery rate} = \frac{(m_{NR-X-LLP} + m_{HTL-X-LP}) - m_{NR-X-PLP}}{(m_{NR-X-LLP} + m_{HTL-X-LP})} \times 100, \, [\%], \quad (3)$$

where m is the mass of phosphorus in the corresponding stream.

2.3. Analyses

2.3.1. Solid and Oil Phase

The inorganic elemental compositions of the feedstocks, the HTL solid and organic phases and the solid phases that were obtained during nutrient recovery were determined by ICP-OES (Agilent 715) after acid microwave-assisted digestion. The feedstock, HTL solid phase, and solid phase after leaching and precipitation (Figure 1) were digested by means of the INVERSE AQUA REGIA method in accordance with the procedure that has been described in reference [43]. The Feed-X, HTL-X-SP, and HTL-X-LSP were not completely dissolved during digestion since recalcitrant minerals, such as silica, were not affected; nevertheless, they are not of interest in the current study. Hence, in spite of the recognition of the applied procedure, the total element concentration may differ from that which was obtained due to the limitations of the digestion method. For digestion of the oil phase, the first step was to add 10 mL of 65% HNO_3 and 1 mL of 37% HCl to 100 mg of the organic phase. The solution was then digested in a microwave (CEM Discover) for 5 min at 220 °C. Subsequently, the completely dissolved samples were diluted to 25 mL with Milli-Q water. All digested solutions were correspondingly diluted with 1% HNO_3 to avoid matrix effects and to adjust to the calibration range of ICP-OES. The nitrogen content of the HTL oil and solid-phases was identified by an elemental analyzer (HekaTech, Euro EA). The ash content of the HTL solid samples was determined according to DIN 51719. The analyses were performed twice.

To specify the minerals that are associated with HTL solid phases and precipitates, the samples were characterized by powder x-ray diffraction (XRD). The diffraction patterns were recorded on a PANanalytical X'Pert Pro X-Ray diffractometer with a monochromatic CuKα radiation source. The raw data were processed with Xpert-II software, and the mineral phases were identified with the Inorganic Crystal Structure Database (ICSD). The morphology of the precipitates was characterized by scanning electron microscopy (SEM; GeminiSEM 500 from Zeiss).

2.3.2. Liquid Phase

The HTL liquid phase, as well as liquid streams from the nutrient recovery (Figure 1), underwent analysis for pH, elemental concentration, and ammonium nitrogen (NH_4-N). NH_4-N was determined with the Hach-Lange cuvette test LCK 304. The inorganic element content was measured with ICP-OES (Agilent 715) by injection diluted 1-to-10 with 1% HNO_3 HTL liquid samples. The nitrogen content was determined by an elemental analyzer (HekaTech, Euro EA), and the pH was measured by HACH HQ40d. The analyses were conducted twice.

3. Results and Discussion

3.1. Nutrient Distribution in Feedstocks and Between HTL Products

Table 1 illustrates the difference between PSS and SPR in terms of composition. The ash content, which reflected the inorganic matter of SPR (5.8 wt%), was similar to those reported by other studies [22,38,44,45] and significantly lower than the ash content of PSS (19.3 wt%). Its value in the literature ranged from 7.5 wt% [14,15] to over 30 wt% [46,47]. This variation was linked to the strong dependencies of the composition of PSS on the sewage sludge source and treatment techniques of wastewater. The nutrient content in the SPR mainly reflected the elemental concentration of the cultivation medium that was used [48]. Thus, the analytes of potassium (1.7 wt%), phosphorus (1.1 wt%), sodium (0.4 wt%), and magnesium (0.3 wt%) constituted the main inorganics of SPR, which (with the exception of magnesium) was in accordance with previous studies [26]. By contrast, calcium (4.0 wt%), phosphorus (2.1 wt%), iron (0.6 wt%), and aluminum (0.4 wt%) constituted the major inorganic composition of PSS [49]. The higher protein content in SPR [50] resulted in higher nitrogen and sulfur contents compared to those in PSS.

Table 1. Nutrient distribution in feedstocks and between hydrothermal liquefaction (HTL) products (mean value (MV) ± standard deviation (SD) of two replicates).

Sample ID	Ash	P	N	K	S	Na	Mg	Ca	Fe	Al
-	wt%	$mg·g^{-1}$	wt%	$mg·g^{-1}$	$mg·g^{-1}$	$mg·g^{-1}$	$mg·g^{-1}$	$mg·g^{-1}$	$mg·g^{-1}$	$mg·g^{-1}$
Feed-PSS [a]	19.3	21.2	2.9	2.2	3.7	2.7	1.4	40.2	6.4	3.6
±	0.6	2.5	0.0	0.4	0.1	0.6	0.2	4.7	0.4	0.2
Feed-SPR [a]	5.8	11.1	11.1	17.1	8.0	4.5	3.2	1.1	0.7	0.1
±	0.0	1.1	0.0	1.7	0.7	0.5	0.3	0.1	0.1	0.0
HTL-PSS-OP [b]	28.4 [c]	28.5	2.0	0.7	2.9	0.3	3.1	48.2	7.6	4.2
±		3.9	0.1	0.1	0.4	0.0	0.4	11.8	0.6	0.7
HTL-SPR-OP [b]	6.6 [c]	<0.1	6.5	<0.1	8.3	<0.1	<0.1	<0.1	0.9	<0.1
±			0.1		0.7				0.0	
HTL-PSS-SP [a]	79.6	99.2	0.9	2.6	2.0	1.0	14.5	220.4	14.7	18.4
±	0.2	1.8	0.0	0.1	0.1	0.0	0.1	15.4	3.4	0.0
HTL-SPR-SP [a]	32.0	58.2	5.3	15.9	7.8	11.4	38.4	22.6	7.2	1.8
±	0.2	0.8	0.0	0.3	0.4	0.2	0.8	0.6	0.1	0.0
	wt%	$mg·L^{-1}$	wt%	$mg·L^{-1}$	$mg·L^{-1}$	$mg·L^{-1}$	$mg·L^{-1}$	$mg·L^{-1}$	$mg·L^{-1}$	$mg·L^{-1}$
HTL-PSS-LP	NA [d]	15.8	0.1	97.6	54.3	157.8	13.9	3.2	<0.3	<1.5
±		1.4	0.0	5.4	0.4	8.5	1.4	0.1		
HTL-SPR-LP	NA [d]	1082.4	1.1	2243.7	403.4	541.0	0.2	4.5	0.9	<1.5
±		58.0	0.2	176.6	22.5	53.7	0.1	0.5	0.1	

[a] on dry basis; [b] analyzed as received; [c] from [38]; [d] not analyzed.

Differences in the composition and structure of feedstocks resulted in disparate distributions of nutrients in the HTL products (Table 1). The inorganics that were found in the oil phase can have an impact on upgrading to fuel [51–53]. For example, phosphor- and iron-containing species could settle on the upgrading catalyst, thereby blocking access to the catalyst interior and thus deactivating it. The pollution of the HTL oil phase with nitrogen also appears to inhibit the upgrading step [54]. The low level of sulfur in the HTL oil phase could have a negative effect on the activity of conventional sulfided upgrading catalysts, e.g., CoMo catalyst [52,55]. In the *SPR* oil phase, iron was present in abundant concentrations (0.1 wt%). These results concur with previous studies of HTL of microalgae [44,56,57]. Iron-associated proteins that were present in microalgae biomass [58] appeared to be stable to decomposition during HTL and can be recovered in bio-crude as iron-porphyrins [51]. The demineralization regarding iron was difficult [59] and required more detailed investigation. The PSS oil phase was contaminated primarily by the elements of calcium (4.5 wt%) and phosphorus (2.8 wt%). The presence of inorganic elements in HTL-PSS-OP may be related to solid particles [60]. To investigate this hypothesis, the PSS oil phase was dissolved in ethanol, and the measurement of the particle size distribution was conducted with dynamic image analysis after destroying agglomerates in an ultrasonic bath. It was detected that a noticeable number of particles with a mean diameter of 38 μm were suspended in the PSS oil phase (see Figure S1). In contrast, by the same method, a number of solid particles in HTL-*SPR*-OP were negligible. The transport of solid particles in the oil phase is not yet known. Their content could be reduced by, for example, an in-line filtration system during HTL, which was not in operation during the present study [38,61].

A proportion of biomass phosphorus (as phosphate) was recovered in both the HTL solid and liquid phases. The HTL solid phase was the main reserve of phosphate from the PSS, as in Marrone's study [15], while the phosphate from the *SPR* was found in both the HTL solid and liquid phases (Table 1). Moreover, the content of phosphorus in HTL-*SPR*-LP was in the range of previous studies [26,32]. The distribution of phosphate can relate to the presence of other dissolved multivalent metal ions, such as Ca^{2+}, Mg^{2+}, Al^{3+}, and $Fe^{2+/3+}$ [6,62], in the reaction solution that tended to precipitate in the form of phosphate salts under a subcritical condition [63–65]. Their high concentration can shift the phosphate from the liquid phase to the solid phase. Therefore, the considerable concentration of metal ions (primarily Ca^{2+}) in PSS compared to *SPR* could be responsible for the recovery of phosphate in the HTL solid phase. While the multivalent ions demonstrated a tendency to precipitate under the studied hydrothermal conditions—which was reflected in the fact that calcium, magnesium, iron, and aluminum were mainly recovered in the HTL solid phase—the monovalent ions (K^+, Na^+) remained dissolved in the HTL liquid phase. A similar tendency has been described in references [66,67]. The nitrogen that was recovered in the HTL liquid phase, which resulted from the degradation of proteins under hydrothermal conditions [15,26], was clearly higher for HTL-*SPR*-LP than for HTL-PSS-LP. The different levels of nitrogen (as ammonium) in the HTL liquid phase between PSS and *SPR* resulted in different pH levels [68]. The pH of the HTL liquid phase from the processing of *SPR* became basic (pH 8.5), and that of PSS became neutral (pH 7.0), which in turn induced alteration of the mineral solubility.

Figure 2 presents the estimated elemental balance for the experimental HTL pilot plant runs of PSS and *SPR*. By taking into account the mean yield of the oil phase (24.5 wt% and 32.9 wt% on a dry basis [38]), the dry matter content of feedstock slurries, the flow rate of slurries, and the total duration of runs, the elemental balance was calculated from the average elemental content in the HTL products that were obtained after the reactor runs (Table 1). The density of slurries was calculated with consideration to the percentage of solids in the slurry and by a solid density of 500 kg·m^{-3} [38]. It is important to note that the yields of gas, liquid, and solid HTL byproducts were not directly measured at this stage of the development of the HTL pilot plant. The yields for the gas and solid phases, which were determined on the basis of the dry matter of feedstock, were taken from previous studies at the batch system as average values. The yields for the HTL solid and gas phases were assumed to be 10 wt% and 14 wt% [47,69] for PSS and 5 wt% and 20 wt% [9,44,47,50] for *SPR*. The amount of the HTL liquid phase after the run was calculated as the difference between the mass of feedstock slurry and the mass of oil, solid, and gas phases that were produced in each run. The percentage of the respective elements for the HTL runs of PSS and *SPR* was calculated from the division of the mean amount of the element in the corresponding phase by the mean amount of this element in the biomass feedstock. The SD of the percentage was calculated from the SD of the elemental content in the biomass feedstock and the HTL product (Table 1); the SD of the yield of the HTL product was not taken into account due to lack of information.

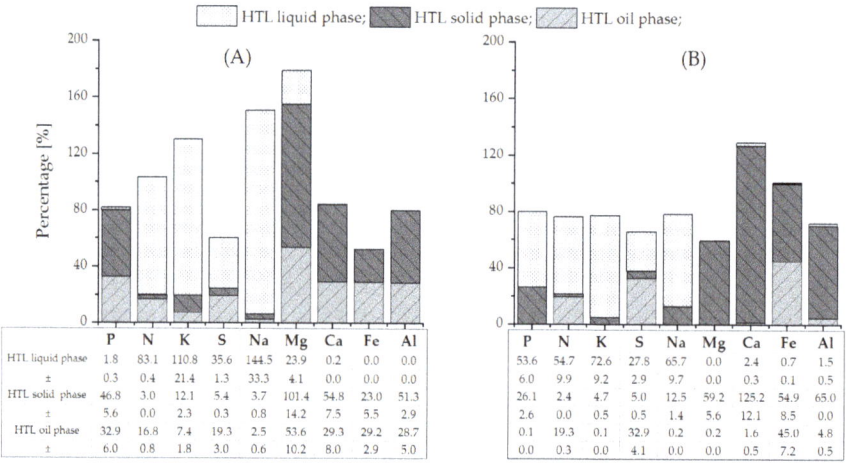

Figure 2. Estimated elemental balance for HTL runs of PSS (**A**) and *SPR* (**B**) (MV ± SD; error bars are not shown in the figure).

Such assumptions might lead to an over- or underestimation of the elemental mass balance. In general, a lower balance closure was observed for the HTL run of PSS than for the HTL run of *SPR*. The mass balance for HTL of PSS (Figure 2) indicated an overestimation of good soluble elements, such as potassium and sodium, and underestimation of poor soluble elements, such as phosphorus, calcium, iron, and aluminum. The causes for this over- or underestimation are unclear but might originate from an overestimation of the amount of the HTL liquid phase or an overestimation of the elemental content by, for instance, evaporation of some HTL liquid phase during sampling (for K and Na) or from fouling and deposition in the reactor (for P, Ca, Fe, and Al). The especially high overestimation of magnesium in PSS is questionable and might be due to contamination of the pilot plant from previous runs. The calculated mass balance for the HTL run of *SPR* (Figure 2) is more or less closed. Only the balance for calcium exceeded 100%, which may be linked to the dissolution of Ca deposits in the reactor. In addition, the sampling of a multi-phase system is always challenging since the ratio of the phases might not be correct in each sample. Despite the over- and underestimation, several general tendencies regarding mass balance can be identified. The preliminary mass balance can assist in selecting a promising strategy for nutrient extraction.

The phosphorus balance was closed at 82% and 80% for the HTL run of PSS and *SPR*. The balance suggests that approximately half of the phosphorus from the processed *SPR* was extracted in the HTL liquid phase, while half of the phosphorus from PSS in the HTL solid phase. The smaller fraction of phosphorus from *SPR* was recovered in the HTL solid phase, and from PSS in the bio-crude. The phosphorus distribution determines the possibility of using the HTL solid and liquid phases as an indirect or direct source of phosphates in the case of HTL of PSS and *SPR*, respectively. Nonetheless, there appears to be a strong tendency to recover phosphates with the PSS bio-crude. As mentioned, a substantial load of particles with a mean diameter of 38 µm were impurifying the crude oil. Therefore, the phosphate remains in mineral form and can be filtered from the crude oil. The data for nitrogen also provided an acceptable balance. Approximately four-fifths of the recovered nitrogen was in the HTL liquid phase for the HTL run of PSS, and three-fourths for the HTL run of *SPR*. Thus, the availability of nitrogen for recovery from the liquid phase was increased by HTL. Magnesium, calcium, iron, and aluminum from PSS exhibited the same tendency as the phosphorus distribution. Meanwhile, these metals from *SPR*, with the exception of iron, were recovered primarily in the HTL solid phase. The occurrence of iron in the form of recalcitrant porphyrin in bio-crude has already been discussed. The majority of the potassium and sodium that were initially present in PSS and *SPR* was found in the HTL liquid phase, which was consistent with previous studies on batch systems [66].

3.2. Strategy of Phosphate Recovery

The mass balance that has been presented in Section 3.1 demonstrates that the HTL solid byproduct could be an appropriate point for phosphate recycling and reuse. In view of the aforementioned disadvantages, this work does not further consider the direct use of the HTL liquid. The HTL solid phase from PSS and *SPR* with the agronomic relevant composition (N-P_2O_5-K_2O) of 0.9-22.7-0.4 and 5.3-13.3-1.9 (Table 1), respectively, may represent a material with properties that are relevant to the application as a fertilizer. For this purpose, the phosphate must be released from the HTL solid phase to become available for the crops. The release behavior of phosphate is related to its form.

Figure 3 presents the respective content of the various forms of phosphate in the HTL solid phases from PSS and *SPR*, as well as the percentage of crop-available phosphate. The phosphate was mainly present in IP form; in both cases, the OP amounted to less than 1%. Moreover, IP consisted mainly of AP (97 wt% and 65 wt% for HTL-PSS-SP and HTL-*SPR*-SP, respectively). The phosphate form distribution was in line with those in previous studies of phosphate behavior under hydrothermal conditions [70–73]. The higher percentage of AP in the HTL solid phase from PSS could relate to the higher level of calcium in PSS (Table 1) compared to in *SPR* [73]. The Ca-bound phosphate was the most prevalent phosphate form in HTL-PSS-SP, which was consistent with XRD analysis. Such analysis concluded that calcium phosphate with inclusions of sodium and magnesium (($Ca_{3.892}Na_{0.087}Mg_{0.021}$)($Ca_{5.491}Na_{0.121}Mg_{0.028}$)($PO_4$)$_{5.1}$) was the most abundant phosphate crystalline mineral in the HTL solid residual from PSS (see Figure S2). This finding was consistent with findings of minerals by other researchers after hydrothermal treatment of biomass that is rich in metals [43,74,75]. In turn, the recorded diffraction patterns of HTL-*SPR*-SP corresponded to a dittmarite mineral ($NH_4MgPO_4 \cdot H_2O$) (see Figure S2). In contrast, Roberts et al. [64] have detected hydroxyapatite after HTL of microalgae biomass. The differences may be explained by the lower calcium content in *SPR* and the presence of calcium phosphate in HTL-*SPR*-SP in an amorphous form [74] that cannot be detected by XRD. The CAL-P extraction illustrated that these minerals might not be used directly as fertilizer. The crop-available form of phosphate was only 1 wt% in HTL-PSS-SP and 31 wt% in HTL-*SPR*-SP. Furthermore, previous studies [17,18] have identified enrichment of the HTL solid phase with heavy metals. Thus, a concentration of phosphate in high-value fertilizer, such as struvite was required. Struvite production through the approach of precipitating phosphate that was released from the HTL solid phase [19] does not seem difficult to implement at the industrial scale and is investigated further in the next section.

3.3. Phosphate Recovery

3.3.1. Release of Phosphate from the HTL Solid Phase

Section 3.2 has identified the HTL solid phase as a phosphate-containing source. The release of phosphate from mineral phases by wet chemical extraction (also known as leaching) is the most common technique, as it offers high efficiency and low energy demand. On the other hand, the consumption of the leaching agent and the co-dissolution of inorganic species other than phosphate salts can negatively affect the extraction. The leaching study was carried out with the aim of identifying optimal conditions for the production of phosphate-rich leachate with high purity for struvite precipitation.

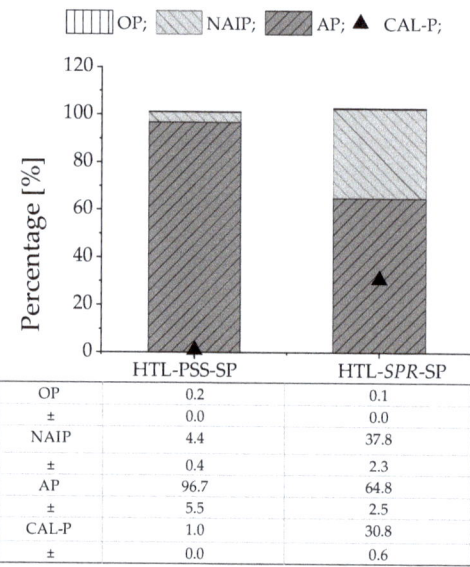

	HTL-PSS-SP	HTL-SPR-SP
OP	0.2	0.1
±	0.0	0.0
NAIP	4.4	37.8
±	0.4	2.3
AP	96.7	64.8
±	5.5	2.5
CAL-P	1.0	30.8
±	0.0	0.6

Figure 3. Different forms of phosphate as well as the crop-available form of phosphate in HTL solid phase (MV ± SD of two replicates; error bars are not shown in the figure). The percentage was calculated as follows: (AP(IP, OP, or CAL-P)/TP) × 100%.

Figure 4a indicates that sulfuric and hydrochloric acids provided a considerable phosphate extraction capacity for the HTL solid phase from both PSS and SPR (equilibrium pH of leachate <2). The application of NaOH as a leaching agent was limited. The environmentally beneficial citric acid demonstrated a lower efficiency than that of mineral acids for phosphate recovery from HTL-PSS-SP. The reason for this lower efficiency could be the precipitation of secondary nonapatite phosphate in the case of citric acid. Together with calcium phosphates, other acid-soluble compounds that present in HTL-PSS-SP, such as $CaCO_3$ and Mg-, Fe-, and Al-containing compounds, might be decomposed as well. The released from calcium phosphate PO_4^{3-} ion can instantly bind with available aluminum or iron ions that have a high affinity for phosphate and precipitates, such as secondary Al- and Fe-phosphate salts. According to a study of phosphate solubility by Stumm and Morgan [76], when decreasing the pH, the Ca-phosphate dissolved first, with the Al-phosphate and Fe-phosphate following, respectively. Almost complete acidic phosphate dissolution can be expected at pH < 2 [77,78]. Thus, the precipitation of secondary Al- and Fe-phosphate salts may explain why weaker citric acid (equilibrium pH of leachate >2) exhibited a poorer extraction performance compared to sulfuric and hydrochloric acid for HTL-PSS-SP. This suggestion was also supported by the higher extraction capacities of aluminum and iron ions with H_2SO_4 than with citric acid. For sulfuric acid as a leaching agent, the extraction capacities of aluminum and iron were 5 mg·g^{-1} and 20 mg·g^{-1}, respectively. For citric acid, the capacities were 2 mg·g^{-1} and 10 mg·g^{-1}, respectively. In contrast, the extraction efficiency of citric acid for HTL-SPR-SP was comparable to those of sulfuric and hydrochloric acids. This finding can be linked to the higher solubility of the phosphate forms in HTL-SPR-SP compared to the phosphate forms in HTL-PSS-SP (Figure 3). Furthermore, the contents of aluminum and iron in the HTL solid phase from SPR were lower (0.2 wt% and 0.7 wt%, respectively) than from PSS (1.8 wt% and 1.5 wt%, respectively). The molar relation between metal ions and phosphate in the HTL solid phase from SPR (Al/P ~0.03 and Fe/P ~0.07) was lower than that from PSS (Al/P ~0.21 and Fe/P ~0.08), which can result in less co-precipitation of secondary Al- and Fe-phosphate salts [77]. While citric acid provided a high extraction capacity for HTL-SPR-SP, its application may be limited by its negative effect on subsequent precipitation [79]. Figure 4a conveys that sulfuric acid is more selective to phosphate release compared to other leaching agents. Significantly less calcium was found in the extract with the application of sulfuric acid, which could relate to the simultaneous co-precipitation of calcium sulfate (gypsum) [80]. This co-precipitation can be seen specifically in the decreasing calcium ion concentrations with increasing concentrations of sulfuric acid in Figure 4b. Large-scale use of sulfuric acid is beneficial from an economic perspective because it offers a low cost as a byproduct of the desulfurization of natural gas and petroleum. Consequently, sulfuric acid was selected as the leaching agent for the following study.

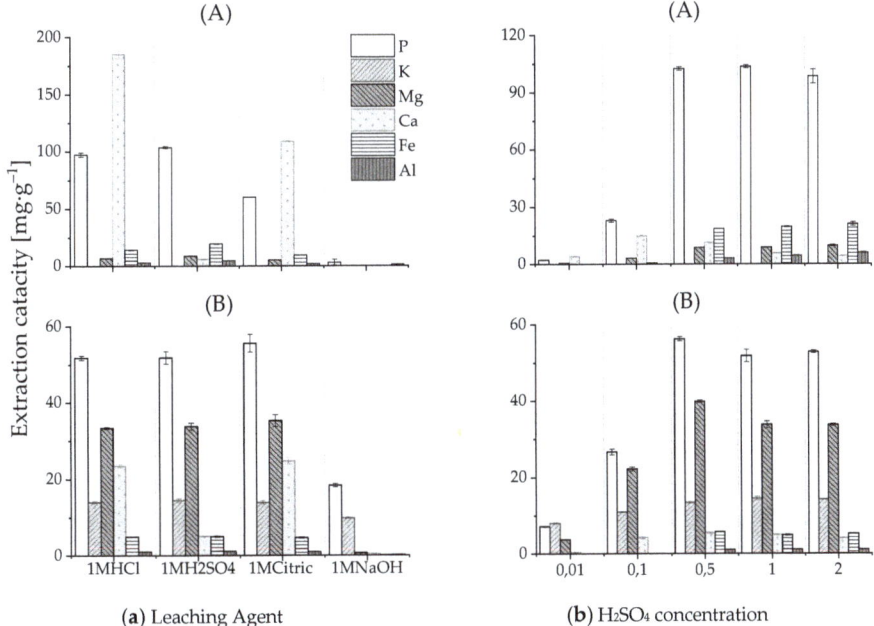

Figure 4. Effect of the leaching agent (a) and H_2SO_4 concentration (b) on the extraction capacity of elements P, K, Mg, Ca, Fe, and Al for the HTL solid phase from PSS (A) and *SPR* (B).

Phosphate was extracted at multiple sulfuric acid concentrations. Figure 4b indicates that an acid concentration of 0.1M provided incomplete extraction (22% in the case of HTL-PSS-SP and 47% in the case of HTL-*SPR*-SP). The maximum extraction capacity from the HTL solid phase was achieved at 0.5M. Higher acid concentrations, which imply an increase of H^+ per g of the HTL solid phase, did not result in a significant improvement in extraction results. The estimation of acid consumption for the phosphate release was essential for the technical feasibility of phosphate recovery technologies. The amount of acid that is required depends on the chemical composition of the HTL solid phase. Calcium phosphate is its main constituent (Figure 3) and reacts essentially with acid; thus, it is the main acid consumer. The literature has reported an average acid consumption for solid residuals that are rich in calcium phosphate was around 3 mol H^+ pro mol P [77,80,81]. If one assumes an average consumption of sulfuric acid of 3 mol H^+ pro mol P to dissolve the phosphates in the HTL solid phase at a given liquid-to-solid ratio (10:1), then sulfuric acid with concentrations of 0.5M and 0.3M should be sufficient to release phosphate from HTL-PSS-SP and HTL-*SPR*-SP, respectively. These calculated concentrations are positively reflected in the experimental data in Figure 4b.

In the context of sustainability, the potential to recycle the remaining acid-insoluble solid residue should also be considered. This acid-treated residual contains low concentrations of phosphorus. The concentrations of other major elements are also altered. The byproduct of leaching of the HTL solid phase from PSS could be used, for example, for the production of activated carbon [21] or as pozzolan in concrete [82]. Of course, the high sulfur content due to gypsum precipitation in the leaching step (with sulfuric acid) warrants attention. Further work with acid-washed HTL residues is required to improve the current understanding of this material.

3.3.2. Phosphate Precipitation in the Form of Struvite

This section examines phosphate separation in the form of struvite ($MgNH_4PO_4 \cdot 6H_2O$) by mixing phosphate-rich leachate with the HTL liquid phase that contains an ammonium ion. Ammonium nitrogen (NH_4-N) amounted to 320 mg·L^{-1} in HTL-PSS-LP and to 6800 mg·L^{-1} in HTL-*SPR*-LP. The measured concentration can be lower than the real one because of the possible loss of some ammonium during the thawing of the liquid samples. Since half of the phosphate from *SPR* (Figure 2) remains after HTL in the liquid phase, the possibility of direct struvite crystallization from the HTL liquid phase was examined. Table 2 presents the nutrient distribution in process streams that circulated in nutrient recovery (Figure 1), while Table 3 illustrates the performance of the nutrient recovery.

Table 2. Nutrient distribution in the streams circulated by phosphate precipitation.

Sample ID	P	N	K	Mg	Ca	NH_4-N
-	mg·g^{-1}	wt%	mg·g^{-1}	mg·g^{-1}	mg·g^{-1}	mg·g^{-1}
NR-PSS-LSP ±	47.5 3.0	NA [a]	2.1 0.1	7.9 0.2	109.8 5.8	-
NR-*SPR*-LSP ±	7.2 0.6	NA [a]	2.5 0.1	2.7 0.4	25.5 0.1	-
NR-PSS-PSP ±	97.9 8.2	3.9 0.0	1.2 0.2	75.2 6.0	15.7 2.2	-
NR-*SPR*-PSP ±	90.0 8.6	5.3 0.0	2.2 0.2	68.8 6.8	7.7 0.5	-
NR-*SPR*-dPSP ±	95.6 4.9	5.2 0.0	1.4 0.2	87.1 11.5	0.2 0.0	-
-	mg·L^{-1}	wt%	mg·L^{-1}	mg·L^{-1}	mg·L^{-1}	mg·L^{-1}
NR-PSS-LLP ±	3890.1 146.9	NA [a]	5.0 0.0	466.3 21.7	1066.3 7.5	NA [a]
NR-*SPR*-LLP ±	4309.1 33.4	NA [a]	1087.0 25.8	2232.2 2.5	440.6 5.1	NA [a]
NR-PSS-PLP ±	5.2 0.2	NA [a]	51.4 1.3	846.0 30.3	119.9 10.8	53.5 7.5
NR-*SPR*-PLP ±	449.4 44.2	NA [a]	1100.6 119.7	0.1 0.0	9.4 0.6	4080.0 510
NR-*SPR*-dPLP ±	1.0 0.0	NA [a]	1063 75.7	292.2 57.1	3.8 1.3	44,650 525

[a] not analyzed.

The release rate of phosphate from the HTL solid phase from PSS was lower than that from HTL solid phase from *SPR*. The previous section has suggested that approximately 3 mol H$^+$ pro mol P was required to dissolve the phosphate from the HTL solid phase from PSS. To set and maintain pH 2, 2.7 mL H_2SO_4 pro g of the HTL solid phase from PSS was piped into the system. Assuming complete dissociation of H_2SO_4 corresponded to approximately 1.6 mol H$^+$ pro mol P at the given liquid-to-solid ratio. There does not seem to be sufficient H$^+$ to provide complete dissolution of primary and secondary phosphate. In the case of *SPR*, the H_2SO_4 guided into the system was equivalent to approximately 2 mol H$^+$ pro mol P. It was sufficient to result in the dissolution of a major part of the phosphate and conforms to the previously specified considerations.

Table 3. Performances of phosphate recovery.

X	Unit	PSS	SPR	dSP
Leaching				
HTL-X-SP/H$_2$O	-	1:10	1:10	-
1M H$_2$SO$_4$/HTL-X-SP	mL·g^{-1}	2.7 ± 0.3	2.2 ± 0.1	-
Extraction capacity of P	mg$_P$·g$^{-1}_{SP}$	49.4 ± 2.3	52.2 ± 0.1	-
Release rate P	%	49.9 ± 2.2	90.2 ± 0.8	-
Precipitation				
Leachate/HTL-X-LP	-	1:6	1:6	-
NH$_4^+$:PO$_4^{3-}$	mol·mol^{-1}	1.1	8.4	13.9
Mg^{2+}:PO$_4^{3-}$	mol·mol^{-1}	2.6	0.4	1.5
Mg^{2+}:Ca^{2+}	mol·mol^{-1}	12.3	11.8	467
Recovery rate P	%	99.0 ± 0.0	66.5 ± 2.9	99.9 ± 0.0
Recovery rate Mg	%	22.6 ± 3.0	99.9 ± 3.0	67.9 ± 4.3
NH$_4$-N recovery	%	79.4 ± 2.8	19.4 ± 9.3	8.0 ± 4.4
Masse precipitate/P in initial mix solution	g·g$^{-1}_P$	11.1 ± 0.2	5.8 ± 0.1	9.3 ± 0.0

The recovery rate of phosphorus from the mix solution was approximately 99% and 66% for PSS and *SPR*, respectively, and approximately 99.9% for direct precipitation from the HTL liquid phase of *SPR*. The pH, molar ratio of the participating ions (PO$_4^{3-}$, Mg^{2+}, NH$_4^+$), and presence of foreign ions (e.g., Ca^{2+}) are among the major parameters that affect the struvite crystallization [83]. The formation of struvite occurs with the creation of supersaturation (index of the deviation of a dissolved salt from its equilibrium), which is the driving force of crystallization. Supersaturation may be achieved by increasing any or all concentrations of ammonium, magnesium, phosphate, and pH in the solution. In general, pH 9 is optimal for struvite precipitation [84]. To attain oversaturation in the mixed solution (spontaneous formation of struvite), the phosphate-rich leachate was mixed with the HTL liquid phase that was high in ammonium ions in a 1-to-6 volume ratio. The results were the molar ratios of 1.1 and 8.4 for PSS and *SPR*, respectively. The molar ratio of NH$_4^+$:PO$_4^{3-}$ in the HTL process water of *SPR* was 13.9. The excess of ammonium is beneficial for struvite crystallization [85,86] and could positively affect the purity of the precipitate, as supported by the fact that the nitrogen content in the precipitate from *SPR* (5.3% and 5.2%) was higher than from PSS (3.9%) and within range of the theoretical value of 5.7%. In contrast, the overdose of ammonium resulted in a low recovery rate of ammonium ions (only 19% and 8% for *SPR* compared to 79% for PSS). The magnesium-to-phosphate molar ratio was also key during struvite crystallization. The magnesium content in leachate and process water was low relative to the phosphate content. Thus, the Mg^{2+}:PO$_4^{3-}$ ratio had to be adjusted by the addition of a magnesium source, which provided the Mg^{2+} that was required for oversaturation and offsets the negative effect of the calcium ion, which competed with the magnesium ion for the phosphate ion [83]. In the case of PSS, the Mg^{2+}:PO$_4^{3-}$ ratio was adjusted to approximately 2. The underdose and overdose scenarios were compared for *SPR* as feedstock with a low concentration of Ca^{2+}. In the case of an underdose of magnesium, the recovery rate was approximately 66%, which indicated that the magnesium ion was a limiting factor for struvite precipitation. This finding was in line with reference [86]. Thus, to improve the performance of phosphate recovery for *SPR*, a higher magnesium ion dose was necessary. However, it is notable that the optimal design of phosphate recovery entails a compromise between the performances and chemical consumption.

The percentage of each element (Table 2) in the precipitate was compared with the theoretical value for pure struvite as a reference (12.6 wt% P; 5.7 wt% N; 9.9 wt% Mg). It can be concluded that the precipitate correlated well with struvite. This suggestion was further confirmed by the XRD analysis. The XRD patterns of the precipitate matched the reference struvite (see Figure S3). Figure 5 presents the SEM image of the precipitate that was obtained. The SEM images revealed coarse, irregularly shaped crystals of various sizes. The most commonly observed crystals in the precipitate from PSS had an average length of 10 μm. The crystals were larger in the case of *SPR*. An excess of ammonium ions may account for the larger size, as already illustrated by other research [83,87]. In addition to the struvite crystals, other solid precipitates were found (marked with yellow arrow). This co-precipitates might be amorphous calcium phosphate. The smaller amount of Ca^{2+} in the case of *SPR* can result in fewer impurities in the form of calcium phosphate, as evident in the SEM images.

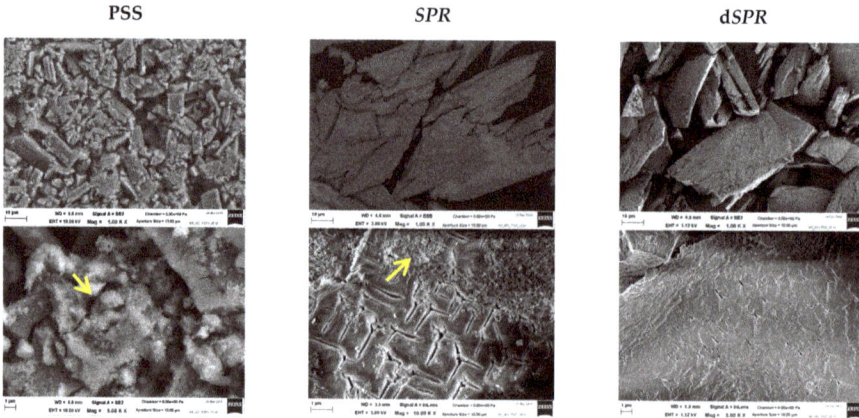

Figure 5. Scanning electron microscopy images of the precipitates that were obtained by mixing phosphate-rich leachate from the HTL solid phase with the HTL liquid phase (PSS and *SPR*) and by direct precipitation from the HTL liquid phase (d*SPR*). Solid precipitates different from struvite are marked with a yellow arrow.

It has been indicated that acid dissolving of phosphate followed by precipitation of struvite is an effective approach for HTL-based phosphorus recovery. To scale up and optimize the performance of this approach in terms of the quantity of struvite that it generates, the characteristics (size and purity) of the precipitate, and the consumption of chemicals, it is necessary to gain additional insight into struvite formation, which requires a detailed study.

3.3.3. Overall Consideration of Process and Mass Flow of Macronutrients

The phosphate recovery performance in the laboratory-scale study and the elemental balance that was calculated for the HTL pilot plant were used to calculate the mass flow diagram of macronutrients during the HTL coupling with nutrient recovery (see Figure 1). The calculated mass balance can help to reduce the process development time and identify future research potential.

The mass flow for PSS (Figure 6a) implies a relatively low recovery rate of phosphate from unprocessed PSS in the form of struvite. It can firstly be linked to the recovery of a considerable amount of phosphate in the HTL oil phase and, secondly, to the non-optimal efficiency of the leaching step. The increase of acid consumption (Section 3.3.2) to approximately 3 mol H^+ pro mol P might lead to the complete dissolution of the phosphate and an improvement in leaching efficiency. Moreover, the release of more phosphate may result in an increase in the HTL liquid phase and the Mg source consumption to cover the corresponding ion ratio as well as in the NaOH consumption for adjusting the pH. For example, an increase of 1M H_2SO_4 up to 1 L·h^{-1} (corresponds to 3 mol H^+ pro mol P) can require an increased HTL liquid phase to provide $NH_4^+:PO_4^{3-}$ of 1.6 up to 55 L·h^{-1}. This amount may conflict with the amount of the liquid phase that originates from HTL (calculated at approximately 57 L·h^{-1}). The increase in chemical consumption that [88] has been identified as a major part of struvite production costs could result in heightened operating costs. The improvement of nutrient performance necessitates a trade-off between the amount and quality of struvite and the consumption of ammonium and magnesium sources as well as NaOH.

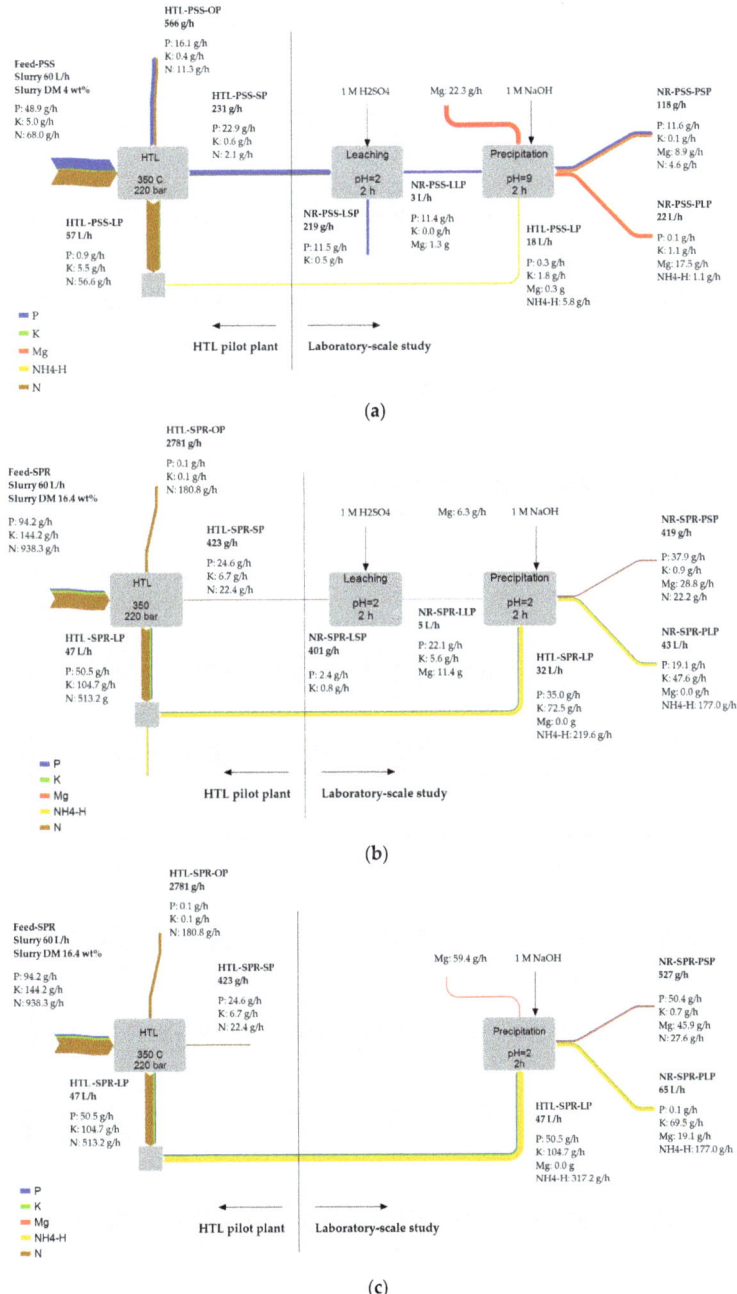

Figure 6. Overall process considerations and mass flow of macronutrients for HTL of PSS (**a**) and SPR (**b**,**c**). The sample ID can be found in Figure 1.

Figure 6b,c presents macronutrients flow from *SPR* in the precipitation struvite from the mix solution of leachate and the HTL liquid phase and in the direct precipitation from the HTL liquid phase.

Approximately 40% and 54% of the phosphate from the unprocessed *SPR* was recovered in the struvite in Figure 6b,c, respectively. The HTL liquid phase from *SPR* is the ammonium-rich stream, and an excess of ammonium ions presents in both cases, which can cause a low recovery rate of ammonium ions. In the first case, the amount of ammonium that is required for struvite precipitation can be regulated and optimized by adjusting the mix ratio of leachate and the HTL liquid phase. Meanwhile, in the case of direct precipitation, it is more difficult to control and adjust the optimal $NH_4^+:PO_4^{3-}$. Further research should consider how to approach the ammonium-rich post-precipitation liquid phase. Possible strategies include stripping ammonia and recovering it in the form of the fertilizer ammonium sulfate [89] or using activated carbon as a sorbent for ammonium separation. In such a case, the struvite crystallization could be beneficial as a pre-treatment technique.

In summary, the recovery of phosphate from the HTL residual stream was successfully performed. The process variations (H_2SO_4, NaOH, and Mg dosages as well as the amount of the HTL liquid phase) and the composition of the initial solution are the main challenges in developing an efficient and cost-effective process design for the struvite precipitation.

4. Conclusions

Biorefineries that are based on HTL and utilize feedstocks with high nutrient loads can add value to the production chain of liquid biofuels relatively easily through the addition of struvite-producing units. For high-protein and low-ash feedstock, such as *SPR*, a one-step approach to directly precipitate struvite from the HTL liquid phase recovers most of the phosphate. Still, ammonium is present in such substantial loads that additional treatment becomes mandatory. Since ammonia stripping is usually a disadvantage of struvite precipitation, such effect might be forced to the extent that ammonia is recovered as, for example, ammonia sulfate during the production of struvite. The sewage sludge that is processed by HTL provides a solid residue that is suitable for recovering phosphate as struvite by means of acid leaching and the addition of released phosphate to the ammonium-rich HTL liquid phase. The larger problem in recovering phosphate is to recover the fine particulate solid residue that is dispersed in the bio-crude oil. Further development of possible in-line filtration during HTL might resolve this problem and enable, along with a future optimization study of precipitation and leaching step, a way to higher phosphate recovery rates for the HTL solid residue, as illustrated by the mass flow.

Supplementary Materials: The following are available online at http://www.mdpi.com/1996-1073/13/2/379/s1, As supplementary materials are available: Figure S1. The particle-size distribution in HTL oil phase from PSS; Figure S2. XRD analysis of HTL solid phase from PSS and *SPR*; and Figure S3. XRD analysis of precipitate from PSS and *SPR*.

Author Contributions: Conceptualization: E.O., G.C.B.; analysis and investigation: E.O.; data analyses: E.O., G.C.B.; writing and preparation of manuscript: E.O., G.C.B., A.K.; funding acquisition: G.C.B. All authors have read and agreed to the published version of the manuscript.

Funding: This research was funded by the European Union's Horizon 2020 research and innovation program under grant agreement No 764734 (HyFlexFuel-Hydrothermal liquefaction: Enhanced performance and feedstock flexibility for efficient biofuel production).

Acknowledgments: We gratefully thank the Department of Engineering at Aarhus University (Patrick Biller and Konstantinos Anastasakis) for providing samples for investigation. We are grateful to Michael Zimmermann (Karlsruhe Institute of Technology (KIT)) for XRD and SEM analyses.

Conflicts of Interest: The authors declare no conflict of interest.

References

1. Jong, S.D.; Hoefnagels, R.; Faaij, A.; Slade, R.; Mawhood, R.; Junginger, M. The feasibility of short-term production strategies for renewable jet fuels—A comprehensive techno-economic comparison. *Biofuels Bioprod. Bioref.* **2015**, *9*, 778–800. [CrossRef]
2. Biller, P.; Roth, A. Hydrothermal Liquefaction: A Promising Pathway Towards Renewable Jet Fuel. In *Biokerosene: Status and Prospects*; Kaltschmitt, M., Neuling, U., Eds.; Springer: Berlin/Heidelberg, Germany, 2018; pp. 607–635. ISBN 978-3-662-53065-8.
3. Peterson, A.A.; Vogel, F.; Lachance, R.P.; Fröling, M.; Michael, J.; Antal, J.R.; Tester, J.W. Thermochemical biofuel production in hydrothermal media: A review of sub- and supercritical water technologies. *Energy Environ. Sci.* **2008**, *1*, 32–65. [CrossRef]
4. Savage, P.E.; Levine, R.B.; Huelsman, C.M. Hydrothermal Processing of Biomass. In *Thermochemical Conversion of Biomass to Liquid Fuels and Chemicals*; Crocker, M., Ed.; RSC Publishing: Cambridge, UK, 2010; Chapter 8; pp. 192–221. ISBN 978-1-84973-035-8.
5. Castello, D.; Haider, M.S.; Rosendahl, L. Catalytic upgrading of hydrothermal liquefaction biocrudes: Different challenges for different feedstocks. *Renew. Energy* **2019**, *141*, 420–430. [CrossRef]
6. Ekpo, U.; Ross, A.B.; Camargo-Valero, M.A.; Williams, P.T. A comparison of product yields and inorganic content in process streams following thermal hydrolysis and hydrothermal processing of microalgae, manure and digestate. *Bioresour. Technol.* **2016**, *200*, 951–960. [CrossRef] [PubMed]
7. Elliott, D.C.; Biller, P.; Ross, A.B.; Schmidt, A.J.; Jones, S.B. Hydrothermal liquefaction of biomass: Developments from batch to continuous process. *Bioresour. Technol.* **2015**, *178*, 147–156. [CrossRef] [PubMed]
8. Elliott, D.C. Review of recent reports on process technology for thermochemical conversion of whole algae to liquid fuels. *Algal Res.* **2016**, *13*, 255–263. [CrossRef]
9. López Barreiro, D.; Prins, W.; Ronsse, F.; Brilman, W. Hydrothermal liquefaction (HTL) of microalgae for biofuel production: State of the art review and future prospects. *Biomass Bioenergy* **2013**, *53*, 113–127. [CrossRef]
10. Castello, D.; Pedersen, T.; Rosendahl, L. Continuous Hydrothermal Liquefaction of Biomass: A Critical Review. *Energies* **2018**, *11*, 3165. [CrossRef]
11. Edmundson, S.; Huesemann, M.; Kruk, R.; Lemmon, T.; Billing, J.; Schmidt, A.; Anderson, D. Phosphorus and nitrogen recycle following algal bio-crude production via continuous hydrothermal liquefaction. *Algal Res.* **2017**, *26*, 415–421. [CrossRef]
12. McGinn, P.J.; Park, K.C.; Robertson, G.; Scoles, L.; Ma, W.; Singh, D. Strategies for recovery and recycling of nutrients from municipal sewage treatment effluent and hydrothermal liquefaction wastewaters for the growth of the microalga Scenedesmus sp. AMDD. *Algal Res.* **2019**, *38*, 101418. [CrossRef]
13. Suzuki, A.; Nakamura, T.; Yokoyama, S.-Y.; Ogi, T.; Koguchi, K. Conversion of sewage sludge to heavy oil by direct thermochemical liquefaction. *J. Chem. Eng. Jpn. JCEJ* **1988**, *21*, 288–293. [CrossRef]
14. Vardon, D.R.; Sharma, B.K.; Scott, J.; Yu, G.; Wang, Z.; Schideman, L.; Zhang, Y.; Strathmann, T.J. Chemical properties of biocrude oil from the hydrothermal liquefaction of Spirulina algae, swine manure, and digested anaerobic sludge. *Bioresour. Technol.* **2011**, *102*, 8295–8303. [CrossRef] [PubMed]
15. Marrone, P.A.; Elliott, D.C.; Billing, J.M.; Hallen, R.T.; Hart, T.R.; Kadota, P.; Moeller, J.C.; Randel, M.A.; Schmidt, A.J. Bench-Scale Evaluation of Hydrothermal Processing Technology for Conversion of Wastewater Solids to Fuels. *Water Environ. Res.* **2018**, *90*, 329–342. [CrossRef] [PubMed]
16. Peccia, J.; Westerhoff, P. We Should Expect More out of Our Sewage Sludge. *Environ. Sci. Technol.* **2015**, *49*, 8271–8276. [CrossRef] [PubMed]
17. Huang, H.-J.; Yuan, X.-Z. The migration and transformation behaviors of heavy metals during the hydrothermal treatment of sewage sludge. *Bioresour. Technol.* **2016**, *200*, 991–998. [CrossRef] [PubMed]
18. Shao, J.; Yuan, X.; Leng, L.; Huang, H.; Jiang, L.; Wang, H.; Chen, X.; Zeng, G. The comparison of the migration and transformation behavior of heavy metals during pyrolysis and liquefaction of municipal sewage sludge, paper mill sludge, and slaughterhouse sludge. *Bioresour. Technol.* **2015**, *198*, 16–22. [CrossRef]
19. Becker, G.C.; Wüst, D.; Köhler, H.; Lautenbach, A.; Kruse, A. Novel approach of phosphate-reclamation as struvite from sewage sludge by utilising hydrothermal carbonization. *J. Environ. Manag.* **2019**, *238*, 119–125. [CrossRef]

20. Yu, Y.; Lei, Z.; Yuan, T.; Jiang, Y.; Chen, N.; Feng, C.; Shimizu, K.; Zhang, Z. Simultaneous phosphorus and nitrogen recovery from anaerobically digested sludge using a hybrid system coupling hydrothermal pretreatment with MAP precipitation. *Bioresour. Technol.* **2017**, *243*, 634–640. [CrossRef]
21. Zhao, X.; Becker, G.C.; Faweya, N.; Rodriguez Correa, C.; Yang, S.; Xie, X.; Kruse, A. Fertilizer and activated carbon production by hydrothermal carbonization of digestate. *Biomass Convers. Biorefin.* **2018**, *8*, 423–436. [CrossRef]
22. Zhao, X.; Stökle, K.; Becker, G.C.; Zimmermann, M.; Kruse, A. Hydrothermal carbonization of Spirulina platensis and Chlorella vulgaris combined with protein isolation and struvite production. *Bioresour. Technol. Rep.* **2019**, *6*, 159–167. [CrossRef]
23. Bauer, S.; Cheng, F.; Colosi, L. Evaluating the Impacts of ACP Management on the Energy Performance of Hydrothermal Liquefaction via Nutrient Recovery. *Energies* **2019**, *12*, 729. [CrossRef]
24. Dyhrman, S.T. Nutrients and Their Acquisition: Phosphorus Physiology in Microalgae. In *The Physiology of Microalgae*; Beardall, J., Raven, J.A., Borowitzka, M.A., Eds.; Springer: Cham, Switzerland, 2016; pp. 155–183. ISBN 978-3-319-24945-2.
25. Valdez, P.J.; Nelson, M.C.; Wang, H.Y.; Lin, X.N.; Savage, P.E. Hydrothermal liquefaction of Nannochloropsis sp.: Systematic study of process variables and analysis of the product fractions. *Biomass Bioenergy* **2012**, *46*, 317–331. [CrossRef]
26. Jena, U.; Vaidyanathan, N.; Chinnasamy, S.; Das, K.C. Evaluation of microalgae cultivation using recovered aqueous co-product from thermochemical liquefaction of algal biomass. *Bioresour. Technol.* **2011**, *102*, 3380–3387. [CrossRef] [PubMed]
27. Bagnoud-Velásquez, M.; Schmid-Staiger, U.; Peng, G.; Vogel, F.; Ludwig, C. First developments towards closing the nutrient cycle in a biofuel production process. *Algal Res.* **2015**, *8*, 76–82. [CrossRef]
28. Barbera, E.; Bertucco, A.; Kumar, S. Nutrients recovery and recycling in algae processing for biofuels production. *Renew. Sustain. Energy Rev.* **2018**, *90*, 28–42. [CrossRef]
29. López Barreiro, D.; Bauer, M.; Hornung, U.; Posten, C.; Kruse, A.; Prins, W. Cultivation of microalgae with recovered nutrients after hydrothermal liquefaction. *Algal Res.* **2015**, *9*, 99–106. [CrossRef]
30. Leng, L.; Li, J.; Wen, Z.; Zhou, W. Use of microalgae to recycle nutrients in aqueous phase derived from hydrothermal liquefaction process. *Bioresour. Technol.* **2018**, *256*, 529–542. [CrossRef]
31. Gu, Y.; Zhang, X.; Deal, B.; Han, L. Biological systems for treatment and valorization of wastewater generated from hydrothermal liquefaction of biomass and systems thinking: A review. *Bioresour. Technol.* **2019**, *278*, 329–345. [CrossRef]
32. Biller, P.; Ross, A.B.; Skill, S.C.; Lea-Langton, A.; Balasundaram, B.; Hall, C.; Riley, R.; Llewellyn, C.A. Nutrient recycling of aqueous phase for microalgae cultivation from the hydrothermal liquefaction process. *Algal Res.* **2012**, *1*, 70–76. [CrossRef]
33. Garcia Alba, L.; Torri, C.; Fabbri, D.; Kersten, S.R.A.; Brilman, D.W.F. Microalgae growth on the aqueous phase from Hydrothermal Liquefaction of the same microalgae. *Chem. Eng. J.* **2013**, *228*, 214–223. [CrossRef]
34. Shanmugam, S.R.; Adhikari, S.; Shakya, R. Nutrient removal and energy production from aqueous phase of bio-oil generated via hydrothermal liquefaction of algae. *Bioresour. Technol.* **2017**, *230*, 43–48. [CrossRef] [PubMed]
35. Peng, L.; Dai, H.; Wu, Y.; Peng, Y.; Lu, X. A comprehensive review of phosphorus recovery from wastewater by crystallization processes. *Chemosphere* **2018**, *197*, 768–781. [CrossRef] [PubMed]
36. Zhang, T.; He, X.; Deng, Y.; Tsang, D.C.W.; Jiang, R.; Becker, G.C.; Kruse, A. Phosphorus recovered from digestate by hydrothermal processes with struvite crystallization and its potential as a fertilizer. *Sci. Total Environ.* **2019**, *698*, 134240. [CrossRef]
37. Le Corre, K.S.; Valsami-Jones, E.; Hobbs, P.; Parsons, S.A. Phosphorus Recovery from Wastewater by Struvite Crystallization: A Review. *Crit. Rev. Environ. Sci. Technol.* **2009**, *39*, 433–477. [CrossRef]
38. Anastasakis, K.; Biller, P.; Madsen, R.; Glasius, M.; Johannsen, I. Continuous Hydrothermal Liquefaction of Biomass in a Novel Pilot Plant with Heat Recovery and Hydraulic Oscillation. *Energies* **2018**, *11*, 2695. [CrossRef]
39. Ruban, V.; López-Sánchez, J.F.; Pardo, P.; Rauret, G.; Muntau, H.; Quevauviller, P. Harmonized protocol and certified reference material for the determination of extractable contents of phosphorus in freshwater sediments—A synthesis of recent works. *Fresenius J. Anal. Chem.* **2001**, *370*, 224–228. [CrossRef] [PubMed]

40. González Medeiros, J.J.; Pérez Cid, B.; Fernández Gómez, E. Analytical phosphorus fractionation in sewage sludge and sediment samples. *Anal. Bioanal. Chem.* **2005**, *381*, 873–878. [CrossRef]
41. Schüller, H. Die CAL-Methode, eine neue Methode zur Bestimmung des pflanzenverfügbaren Phosphates in Böden. *Z. Pflanzenernaehr. Bodenk.* **1969**, *123*, 48–63. [CrossRef]
42. Thun, R.; Hoffmann, G. *Die Untersuchung von Böden, 4., neubearb. u. erw. Aufl*; VDLUFA-Verl.: Darmstadt, Germany, 2012; ISBN 9783941273139.
43. Ovsyannikova, E.; Arauzo, P.J.; Becker, G.C.; Kruse, A. Experimental and thermodynamic studies of phosphate behavior during the hydrothermal carbonization of sewage sludge. *Sci. Total Environ.* **2019**, *692*, 147–156. [CrossRef]
44. Jena, U.; Das, K.C.; Kastner, J.R. Effect of operating conditions of thermochemical liquefaction on biocrude production from Spirulina platensis. *Bioresour. Technol.* **2011**, *102*, 6221–6229. [CrossRef]
45. Haider, M.; Castello, D.; Michalski, K.; Pedersen, T.; Rosendahl, L. Catalytic Hydrotreatment of Microalgae Biocrude from Continuous Hydrothermal Liquefaction: Heteroatom Removal and Their Distribution in Distillation Cuts. *Energies* **2018**, *11*, 3360. [CrossRef]
46. Lemoine, F.; Maupin, I.; Lemée, L.; Lavoie, J.-M.; Lemberton, J.-L.; Pouilloux, Y.; Pinard, L. Alternative fuel production by catalytic hydroliquefaction of solid municipal wastes, primary sludges and microalgae. *Bioresour. Technol.* **2013**, *142*, 1–8. [CrossRef] [PubMed]
47. Madsen, R.B.; Glasius, M. How Do Hydrothermal Liquefaction Conditions and Feedstock Type Influence Product Distribution and Elemental Composition? *Ind. Eng. Chem. Res.* **2019**. [CrossRef]
48. Juneja, A.; Ceballos, R.; Murthy, G. Effects of Environmental Factors and Nutrient Availability on the Biochemical Composition of Algae for Biofuels Production: A Review. *Energies* **2013**, *6*, 4607–4638. [CrossRef]
49. Tabatabai, M.; Frankenberger, W. Chemical composition of sewage sludges in Iowa. *Res. Bull.* **1979**, *36*, 1.
50. Biller, P.; Ross, A.B. Potential yields and properties of oil from the hydrothermal liquefaction of microalgae with different biochemical content. *Bioresour. Technol.* **2011**, *102*, 215–225. [CrossRef]
51. Jarvis, J.M.; Sudasinghe, N.M.; Albrecht, K.O.; Schmidt, A.J.; Hallen, R.T.; Anderson, D.B.; Billing, J.M.; Schaub, T.M. Impact of iron porphyrin complexes when hydroprocessing algal HTL biocrude. *Fuel* **2016**, *182*, 411–418. [CrossRef]
52. Jensen, C.U. PIUS—Hydrofaction(TM) Platform with Integrated Upgrading Step. Ph.d.-serien for Det Ingeniør- og Naturvidenskabelige Fakultet. Ph.D. Thesis, Aalborg Universitet, Aalborg, Denmark, 2018.
53. Speight, J.G. *Handbook of Petroleum Product Analysis*, 2nd ed.; Wiley: Hoboken, NJ, USA, 2015; ISBN 9781118986370.
54. Jensen, C.U.; Rosendahl, L.A.; Olofsson, G. Impact of nitrogenous alkaline agent on continuous HTL of lignocellulosic biomass and biocrude upgrading. *Fuel Process. Technol.* **2017**, *159*, 376–385. [CrossRef]
55. Kubička, D.; Horáček, J. Deactivation of HDS catalysts in deoxygenation of vegetable oils. *Appl. Catal. A Gen.* **2011**, *394*, 9–17. [CrossRef]
56. Jiang, J.; Savage, P.E. Metals and Other Elements in Biocrude from Fast and Isothermal Hydrothermal Liquefaction of Microalgae. *Energy Fuels* **2018**, *32*, 4118–4126. [CrossRef]
57. Jiang, J.; Savage, P.E. Influence of process conditions and interventions on metals content in biocrude from hydrothermal liquefaction of microalgae. *Algal Res.* **2017**, *26*, 131–134. [CrossRef]
58. Marchetti, A.; Maldonado, M.T. Iron. In *The Physiology of Microalgae*; Beardall, J., Raven, J.A., Borowitzka, M.A., Eds.; Springer: Cham, Switzerland, 2016; pp. 233–279. ISBN 978-3-319-24945-2.
59. Jiang, J.; Savage, P.E. Using Solvents to Reduce the Metal Content in Crude Bio-oil from Hydrothermal Liquefaction of Microalgae. *Ind. Eng. Chem. Res.* **2019**. [CrossRef]
60. Agblevor, F.A.; Besler, S. Inorganic Compounds in Biomass Feedstocks. 1. Effect on the Quality of Fast Pyrolysis Oils. *Energy Fuels* **1996**, *10*, 293–298. [CrossRef]
61. Elliott, D.C.; Hart, T.R.; Schmidt, A.J.; Neuenschwander, G.G.; Rotness, L.J.; Olarte, M.V.; Zacher, A.H.; Albrecht, K.O.; Hallen, R.T.; Holladay, J.E. Process development for hydrothermal liquefaction of algae feedstocks in a continuous-flow reactor. *Algal Res.* **2013**, *2*, 445–454. [CrossRef]
62. Hable, R.D.; Alimoradi, S.; Sturm, B.S.M.; Stagg-Williams, S.M. Simultaneous solid and biocrude product transformations from the hydrothermal treatment of high pH-induced flocculated algae at varying Ca concentrations. *Algal Res.* **2019**, *40*, 101501. [CrossRef]
63. Leusbrock, I.; Metz, S.J.; Rexwinkel, G.; Versteeg, G.F. The solubilities of phosphate and sulfate salts in supercritical water. *J. Supercrit. Fluids* **2010**, *54*, 1–8. [CrossRef]

64. Roberts, G.W.; Sturm, B.S.M.; Hamdeh, U.; Stanton, G.E.; Rocha, A.; Kinsella, T.L.; Fortier, M.-O.P.; Sazdar, S.; Detamore, M.S.; Stagg-Williams, S.M. Promoting catalysis and high-value product streams by in situ hydroxyapatite crystallization during hydrothermal liquefaction of microalgae cultivated with reclaimed nutrients. *Green Chem.* **2015**, *17*, 2560–2569. [CrossRef]
65. Riman, R.E.; Suchanek, W.L.; Byrappa, K.; Chen, C.-W.; Shuk, P.; Oakes, C.S. Solution synthesis of hydroxyapatite designer particulates. *Solid State Ionics* **2002**, *151*, 393–402. [CrossRef]
66. Christensen, P.S.; Peng, G.; Vogel, F.; Iversen, B.B. Hydrothermal Liquefaction of the Microalgae Phaeodactylum tricornutum: Impact of Reaction Conditions on Product and Elemental Distribution. *Energy Fuels* **2014**, *28*, 5792–5803. [CrossRef]
67. Anastasakis, K.; Ross, A.B. Hydrothermal liquefaction of the brown macro-alga Laminaria Saccharina: Effect of reaction conditions on product distribution and composition. *Bioresour. Technol.* **2011**, *102*, 4876–4883. [CrossRef]
68. Kruse, A.; Koch, F.; Stelzl, K.; Wüst, D.; Zeller, M. Fate of Nitrogen during Hydrothermal Carbonization. *Energy Fuels* **2016**, *30*, 8037–8042. [CrossRef]
69. Biller, P.; Johannsen, I.; Dos Passos, J.S.; Ottosen, L.D.M. Primary sewage sludge filtration using biomass filter aids and subsequent hydrothermal co-liquefaction. *Water Res.* **2018**, *130*, 58–68. [CrossRef] [PubMed]
70. Huang, R.; Tang, Y. Speciation Dynamics of Phosphorus during (Hydro)Thermal Treatments of Sewage Sludge. *Environ. Sci. Technol.* **2015**, *49*, 14466–14474. [CrossRef] [PubMed]
71. Huang, R.; Fang, C.; Lu, X.; Jiang, R.; Tang, Y. Transformation of Phosphorus during (Hydro)thermal Treatments of Solid Biowastes: Reaction Mechanisms and Implications for P Reclamation and Recycling. *Environ. Sci. Technol.* **2017**, *51*, 10284–10298. [CrossRef] [PubMed]
72. Feng, Y.; Ma, K.; Yu, T.; Bai, S.; Pei, D.; Bai, T.; Zhang, Q.; Yin, L.; Hu, Y.; Chen, D. Phosphorus Transformation in Hydrothermal Pretreatment and Steam Gasification of Sewage Sludge. *Energy Fuels* **2018**, *32*, 8545–8551. [CrossRef]
73. Xu, Y.; Yang, F.; Zhang, L.; Wang, X.; Sun, Y.; Liu, Q.; Qian, G. Migration and transformation of phosphorus in municipal sludge by the hydrothermal treatment and its directional adjustment. *Waste Manag.* **2018**, *81*, 196–201. [CrossRef]
74. Huang, R.; Fang, C.; Zhang, B.; Tang, Y. Transformations of Phosphorus Speciation during (Hydro)thermal Treatments of Animal Manures. *Environ. Sci. Technol.* **2018**, *52*, 3016–3026. [CrossRef]
75. Yakaboylu, O.; Harinck, J.; Gerton Smit, K.G.; Jong, W. Supercritical water gasification of manure: A thermodynamic equilibrium modeling approach. *Biomass Bioenergy* **2013**, *59*, 253–263. [CrossRef]
76. Stumm, W.; Morgan, J.J. *Aquatic Chemistry. Chemical Equilibria and Rates in Natural Waters*, 3rd ed.; John Wiley & Sons Inc.: New York, NY, USA; Chichester, UK; Brisbane, Australia; Toronto, ON, Canada; Singapore, 1996; ISBN 9780471511854.
77. Petzet, S.; Peplinski, B.; Cornel, P. On wet chemical phosphorus recovery from sewage sludge ash by acidic or alkaline leaching and an optimized combination of both. *Water Res.* **2012**, *46*, 3769–3780. [CrossRef]
78. Heilmann, S.M.; Molde, J.S.; Timler, J.G.; Wood, B.M.; Mikula, A.L.; Vozhdayev, G.V.; Colosky, E.C.; Spokas, K.A.; Valentas, K.J. Phosphorus reclamation through hydrothermal carbonization of animal manures. *Environ. Sci. Technol.* **2014**, *48*, 10323–10329. [CrossRef]
79. Zhang, Q.; Zhao, S.; Ye, X.; Xiao, W. Effects of organic substances on struvite crystallization and recovery. *Desalin. Water Treat.* **2016**, *57*, 10924–10933. [CrossRef]
80. Donatello, S.; Tong, D.; Cheeseman, C.R. Production of technical grade phosphoric acid from incinerator sewage sludge ash (ISSA). *Waste Manag.* **2010**, *30*, 1634–1642. [CrossRef] [PubMed]
81. Franz, M. Phosphate fertilizer from sewage sludge ash (SSA). *Waste Manag.* **2008**, *28*, 1809–1818. [CrossRef] [PubMed]
82. Donatello, S. Characteristics of Incinerated Sewage Sludge Ashes: Potential for Phosphate Extraction and Re-Use as a Pozzolanic Material in Construction Products. Ph.D. Thesis, Imperial College London, London, UK, 2009.
83. Li, B.; Huang, H.M.; Boiarkina, I.; Yu, W.; Huang, Y.F.; Wang, G.Q.; Young, B.R. Phosphorus recovery through struvite crystallisation: Recent developments in the understanding of operational factors. *J. Environ. Manag.* **2019**, *248*, 109254. [CrossRef] [PubMed]
84. Liu, Y.; Kumar, S.; Kwag, J.-H.; Ra, C. Magnesium ammonium phosphate formation, recovery and its application as valuable resources: A review. *J. Chem. Technol. Biotechnol.* **2013**, *88*, 181–189. [CrossRef]

85. Pastor, L.; Mangin, D.; Barat, R.; Seco, A. A pilot-scale study of struvite precipitation in a stirred tank reactor: Conditions influencing the process. *Bioresour. Technol.* **2008**, *99*, 6285–6291. [CrossRef]
86. Stratful, I.; Scrimshaw, M.D.; Lester, J.N. Conditions influencing the precipitation of magnesium ammonium phosphate. *Water Res.* **2001**, *35*, 4191–4199. [CrossRef]
87. Li, B.; Boiarkina, I.; Yu, W.; Huang, H.M.; Munir, T.; Wang, G.Q.; Young, B.R. Phosphorous recovery through struvite crystallization: Challenges for future design. *Sci. Total Environ.* **2019**, *648*, 1244–1256. [CrossRef]
88. Jaffer, Y.; Clark, T.A.; Pearce, P.; Parsons, S.A. Potential phosphorus recovery by struvite formation. *Water Res.* **2002**, *36*, 1834–1842. [CrossRef]
89. Ghyselbrecht, K.; Monballiu, A.; Somers, M.H.; Sigurnjak, I.; Meers, E.; Appels, L.; Meesschaert, B. Stripping and scrubbing of ammonium using common fractionating columns to prove ammonium inhibition during anaerobic digestion. *Int. J. Energy Environ. Eng.* **2018**, *9*, 447–455. [CrossRef]

 © 2020 by the authors. Licensee MDPI, Basel, Switzerland. This article is an open access article distributed under the terms and conditions of the Creative Commons Attribution (CC BY) license (http://creativecommons.org/licenses/by/4.0/).

Article

Synthesis and Application of Heterogeneous Catalysts Based on Heteropolyacids for 5-Hydroxymethylfurfural Production from Glucose

Jéssica Siqueira Mancilha Nogueira [1], João Paulo Alves Silva [1], Solange I. Mussatto [2] and Livia Melo Carneiro [1,*]

1. Department of Chemical Engineering, Engineering School of Lorena, University of São Paulo, 12602-810 Lorena/SP, Brazil; jessicasmn@usp.br (J.S.M.N.); jpalves80@usp.br (J.P.A.S.)
2. Novo Nordisk Foundation Center for Biosustainability, Technical University of Denmark, 2800 Kongens Lyngby, Denmark; smussatto@biosustain.dtu.dk
* Correspondence: liviacarneiro@usp.br; Tel.: +55-12-31595162

Received: 22 December 2019; Accepted: 26 January 2020; Published: 4 February 2020

Abstract: This study aimed to evaluate the synthesis and application of heterogeneous catalysts based on heteropolyacids for 5-hydroxymethylfurfural (HMF) production from glucose. Initially, assays were carried out in order to establish the most favorable catalyst synthesis conditions. For such purpose, calcination temperature (300 or 500 °C), type of support (Nb_2O_5 or Al_2O_3), and active phase ($H_3PW_{12}O_{40}$—HPW or $H_3PMo_{12}O_{40}$—HPMo) were tested and combined based on Taguchi's L_8 orthogonal array. As a result, HPW-Nb_2O_5 calcined at 300 °C was selected as it presented optimal HMF production performance (9.5% yield). Subsequently, the reaction conditions capable of maximizing HMF production from glucose using the selected catalyst were established. In these experiments, different temperatures (160 or 200 °C), acetone-to-water ratios (1:1 or 3:1 v/v), glucose concentrations (50 or 100 g/L), and catalyst concentrations (1 or 5% w/v) were evaluated according to a Taguchi's L_{16} experimental design. The conditions that resulted in the highest HMF yield (40.8%) consisted of using 50 g/L of glucose at 160 °C, 1:1 (v/v) acetone-to-water ratio, and catalyst concentration of 5% (w/v). Recycling tests revealed that the catalyst can be used in four runs, which results in the same HMF yield (approx. 40%).

Keywords: 5-hydroxymethylfurfural; glucose; heteropolyacid catalysts

1. Introduction

In an attempt to reduce global dependence on fossil resources, which are associated with important negative environmental impacts, new technologies have been developed that aim to use renewable feedstock (lignocellulosic raw materials) for the production of fuels and chemicals [1]. Lignocellulosic biomass is an interesting raw material for such application, since it is widely available in the form of agricultural, agro-industrial, and forest residues, inexpensive, and rich in sugars that can be used for producing numerous compounds of industrial interest. Among the compounds that can be produced from lignocellulosic materials, furans such as furfural and 5-hydroxymethylfurfural (HMF) are molecules of enormous interest, since they have numerous applications in the chemical industry [2].

HMF is a building block platform chemical that can be used to produce various other compounds, including 2,5-dimethylfuran (DMF) and liquid fuels, as well as high added-value products such as polyesters, dialdehydes, ethers, among others [3,4]. Due to its wide-ranging applications, it has been considered as one of the 10 highest value platform molecules by the United States Department of Energy [5,6]. Recently, there has been enormous interest in new processes aiming to obtain HMF in order to supply the booming market and provide more sustainable production alternatives. Leading companies in renewable technologies, e.g., Avantium and AVA Biochem, have sought processes in order to produce HMF from lignocellulosic biomass in a pilot scale aiming at the production of bioplastics and other compounds [4,7]. Production of 2,5-furan-dicarboxylic acid (FDCA) is particularly highlighted due to its application as precursor monomer of the bioplastic polyethylene furanoate (PEF), which is a potential replacement for the conventional polymer polyethylene terephthalate (PET) [5,6].

HMF can be obtained by dehydrating hexose sugars, such as glucose or fructose. However, glucose costs less when compared to fructose and can be found in greater amounts in the form of cellulose in lignocellulosic materials, therefore being more attractive to be used in large-scale HMF production. The process to convert glucose into HMF is influenced by several variables, such as the temperature, type of catalyst, reaction time, and reaction medium composition. Regarding the reaction medium, a variety of solvents has been evaluated for such a purpose, including aqueous, organic, and biphasic systems (water mixtures and organic solvents), as well as ionic liquids [8–11]. Glucose dehydration reactions tend to be more selective in the presence of aprotic solvents, e.g., dimethylsulfoxide (DMSO), tetrahydrofuran, acetone, and n-butanol. Aqueous media have resulted in low yields, i.e., close to 20% [7,12], as they favor the formation of undesirable products (humin and furfural) and HMF rehydration reactions, which lead to the production of levulinic and formic acids [5,13]. Ionic liquids have provided conversion yields of over 30% [14,15]; however, final product separation is more difficult when using these solvents, in addition to being quite costly and toxic [3]. Taking all these considerations into account, biphasic systems (water/solvent) have been considered the most interesting alternative to HMF production.

The catalyst is also a significant variable that affects glucose conversion into HMF. When compared to homogeneous catalysts, the use of heterogeneous catalysts has shown better selectivity and lower costs, especially because they ease product separation, as well as catalyst recovery and reuse [7,16]. Several heterogeneous acid catalysts such as zeolites, metal oxides, silica, aluminosilicates, alumina, sulfated and tungsten zirconia, and superacid catalysts, have been studied for such a process. However, catalysts based on heteropolyacids have been slightly explored for HMF production, although these catalysts have shown promising results in reactions such as esterification [17], transesterification [18], hydrodesulfurization [19], glycerol dehydration [20], benzaldehyde acetylation [21], and isomerization [22]. Some studies showed HMF yields over 30% using $H_3PW_{12}O_{40}$ —HPW as homogeneous catalyst combined with boric acid in liquid ionic media [23] or using HPW as heterogeneous catalyst by the reaction with a liquid ionic [24] or $AgNO_3$ [25] in a biphasic system.

Thus, this study aimed to define the conditions to prepare heterogeneous catalysts based on heteropolyacids to be used in HMF production from glucose. Initially, assays were carried out to establish the conditions for catalyst synthesis. Different conditions were tested, such as calcination temperature (300 or 500 °C), type of support (Nb_2O_5 or Al_2O_3), and active phase ($H_3PW_{12}O_{40}$—HPW or $H_3PMo_{12}O_{40}$—HPMo). The catalyst that presented optimal performance to convert glucose into HMF was selected, and then the reaction conditions capable of maximizing HMF production using it were established. Finally, the possibility of catalyst recycling was also investigated.

2. Materials and Methods

2.1. Catalyst Preparation

Heteropolyacid catalysts were synthesized in duplicate according to Taguchi's L_8 orthogonal array presented in Table 1, through which different calcination temperatures (300 or 500 °C), 30% (w/w) active phases ($H_3PW_{12}O_{40}$—HPW or $H_3PMo_{12}O_{40}$—HPMo), and supports (Nb_2O_5 or Al_2O_3) were combined, resulting in 16 different catalysts. The choice of the supports and active phases were made so that their combination resulted in catalysts with Lewis and Brønsted acid sites, which was derived from the support and active phase, respectively. The supports Nb_2O_5 (HY-340) and Al_2O_3 were supplied by Companhia Brasileira de Metalurgia e Mineração (CBMM) and Alcoa, respectively.

The catalysts were prepared according to the incipient wetness impregnation method which uses an amount of solvent which is lower than or equal to that required to fill in the support pores. For such a purpose, the active phase (HPW or HPMo) was dissolved in 70% ethanol solution at ambient temperature and mixed with the support (Nb_2O_5 or Al_2O_3) in three successive steps until reaching a final metal concentration of 30%. The catalytic solid was then dried at 100 °C for 2 h and subsequently calcined at 300 or 500 °C for 3 h, according to the conditions shown in Table 1. The assays were carried out in duplicate.

2.2. HMF Production

The catalytic performance of different heteropolyacid catalysts prepared according to the experimental design given in Table 1 was evaluated regarding catalytic performance in order to convert glucose into HMF. The catalytic reactions carried out using 100 mL of reaction medium at 160 °C using 1% *w/v* of catalyst, 100 g/L of glucose, 1:1 *w/w* acetone-to-water ratio, and 300 rpm over 30 min. The catalyst that achieved optimal HMF production performance was then used in subsequent experiments with the aim of optimizing reaction conditions.

For optimization experiments, different reaction conditions, such as temperature (160 or 200 °C), acetone-to-water ratio (1:1 or 3:1 *v/v*), glucose concentration (50 or 100 g/L), and catalyst concentration (1% or 5% *w/v*) were used based on Taguchi's L_{16} experimental design. The column "E" had no factor associated to estimate the experimental design error. All catalytic tests were performed in pressurized stainless steel reactors (Parr series 4566). At the end of the runs, the catalyst was recovered from the reaction mixture by centrifugation at $2000 \times g$ for 20 min and calcined at 300 °C for 3 h.

Catalyst stability was evaluated by reusing it in successive batch runs performed under optimized process conditions.

2.3. Analytical Methods

The crystalline structure of the catalyst, support, and active phase was evaluated by X-Ray Powder Diffraction (XRD) using a PANalytical Model Empyrean X-ray diffractometer with CuKα 2θ radiation (λ = 1.5418 Å) at 40 kV and 30 mA, angle ranging between 10° and 90°, at step size of 0.02° and countdown time of 50 s per run. The catalyst surface morphology and structure were analyzed by SEM/EDX using Hitachi TM 3000 and Swifted 3000 equipment.

The textural properties of catalysts were investigated by specific surface area analysis using Quantachrome NOVA 2200e. For performing the analysis, a sample of 0.2 g was added to a glass cell and heated at 200 °C for 2 h under vacuum in order to remove impurities adsorbed on the catalyst surface. Specific surface area and pore volume were calculated by the Brunauer–Emmett–Teller (BET) the Barrett–Joyner–Halenda (BJH) methods, respectively.

The catalyst surface acidity was determined by acid-base titrations. In this analysis, 0.1 g of catalyst was suspended in 20 mL of 0.1 M NaOH. The suspension was stirred for 3 h at room temperature and then titrated with 0.1 M HCl in the presence of phenolphthalein. The catalyst surface acidity was expressed in mmol H^+/g catalyst. The active phase thermal stability was evaluated by thermogravimetry (TGA) and derivative thermogravimetry (DTG) analysis using a Shimadzu TGA 50 equipment with 50 mL/min of nitrogen flow, heating rate of 10 °C/min in a temperature range of 30 to 1000 °C.

Glucose concentration was determined by high performance liquid chromatography (HPLC) using an Agilent Technologies 1260 Infinity chromatograph equipped with an isocratic pump, a refractive index detector and a Bio-Rad Aminex HPX-87H column (300 × 7.8 mm). Operational conditions were temperature of 45 °C, 0.005 mol/L using sulfuric acid as eluent at a flow rate of 0.6 mL/min and sample volume of 0.02 mL. HMF concentration was also determined by HPLC, but using a UV detector (at 276 nm), a Waters Spherisorb C18 5 µm column (100 × 4.6 mm) at room temperature, 1:8 *v/v* acetonitrile-to-water ratio using 1% of acetic acid as eluent, flow rate of 0.8 mL/min, and sample volume of 0.02 mL. Glucose conversion (X_{Glu}) and HMF yield (Y_{HMF}) were calculated according to Equations (1) and (2).

$$X_{Glu}(\%) = \frac{[Glucose]_{initial} - [Glucose]_{final}}{[Glucose]_{initial}} \cdot 100, \quad (1)$$

$$Y_{HMF}(\%) = \frac{[HMF]_{produced}}{[Glucose]_{initial}} \cdot 100. \quad (2)$$

3. Results and Discussion

3.1. Catalyst Synthesis

Table 1 summarizes the conditions used to prepare the catalysts as well as their performance to convert glucose into HMF in terms of HMF yield (Y_{HMF}) and glucose conversion (X_{Glu}). As it can be seen, HMF yield ranged from 0.7% to 9.5% and glucose conversion from 65.3% to 93.6% according to the conditions established to prepare the catalyst. These results show that preparation conditions had an important influence on the effectiveness of glucose conversion into HMF. In some cases, the produced catalyst allowed achieving high glucose conversion rates, i.e., about 90% (assays 1, 3, 5, and 7). However, the highest HMF yield, about 9.5%, was achieved in assay 2, in which glucose conversion was only 75%. A high glucose conversion without a proportional HMF yield, as observed in other cases, suggests the formation of reaction by-products. In fact, the formation of humin, levulinic acid, furfural, and formic acid, which have been reported in literature as by-products of a glucose dehydration reaction to HMF, is often associated with reaction conditions, mainly to the use of solvents and high temperatures [5,26].

Table 1. Taguchi's L_8 orthogonal array to evaluate the effect of calcination temperature, support and active phase on heterogeneous catalysts preparation on 5-Hydroxymethylfurfural yield (Y_{HMF}) and glucose conversion rate (X_{Glu}).

Catalyst	Experimental Conditions (Factors and Interactions)							Response Variables	
	A Temp. (°C)	B Support	AB	C Active phase	AC	BC	ABC	Y_{HMF} [1](%)	X_{Glu} [2](%)
1	300	Nb	1	HPMo	1	1	1	3.0 / 2.8	92.3 / 93.6
2	300	Nb	1	HPW	2	2	2	9.5 / 9.4	75.0 / 75.7
3	300	Al	2	HPMo	1	2	2	4.4 / 3.7	90.3 / 85.3
4	300	Al	2	HPW	2	1	1	0.8 / 0.8	69.8 / 67.2
5	500	Nb	2	HPMo	2	1	2	6.9 / 7.0	91.8 / 90.7
6	500	Nb	2	HPW	1	2	1	0.7 / 0.8	65.3 / 66.1
7	500	Al	1	HPMo	2	2	1	7.3 / 7.7	91.0 / 87.6
8	500	Al	1	HPW	1	1	2	0.7 / 0.7	70.6 / 68.9

[1] Y_{HMF}: 5-Hydroxymethylfurfural yield (%). [2] X_{Glu}: Glucose conversion (%). [1,2] All results are in duplicate.

The statistical significance of main effects and their interactions on response variables was verified by the analysis of variance (ANOVA). As shown in Table 2, the variation percentages explained by HMF yield and glucose conversion achieved a high coefficient of determination (R^2 = 99.8% and 98.6%, respectively). These results reveal that the variations observed for response variables (HMF yield and glucose conversion) can be effectively explained by catalyst preparation conditions.

With respect to HMF yield, calcination temperature, support, active phase, and interaction effects (AB, AC, BC, and ABC) were significant at confidence level of 95%; moreover, glucose conversion, calcination temperature, support, active phase, and AB and AC interactions were also significant at 95% confidence level.

Table 2. Analysis of variance of the main effects and their interactions on 5-hydroxymethylfurfural (HMF) yield (Y_{HMF}) and glucose conversion (X_{Glu}) based on Taguchi's L_8 orthogonal array.

Source of Variation	Response Variables	
	p-Value for Y_{HMF}	p-Value for X_{Glu}
(A) Temperature	0.0155 *	0.0413 *
(B) Support	<0.0001 *	0.0237 *
(D) Active phase	<0.0001 *	0.0001 *
AB	<0.0001 *	0.0041 *
AD	<0.0001 *	0.0472 *
BD	<0.0001 *	0.2526
ABD	<0.0001 *	0.0626
R^2	0.9978	0.9862

* Significant at 95% confidence level: p test < 0.05.

The interaction effects of support and temperature for different active phases on HMF yield and glucose conversion are shown in Figure 1. Note that the highest HMF yields (Figure 1b) were obtained from reactions in which the catalyst was prepared by using HPW active phase supported on Nb_2O_5 and calcined at 300 °C. For the HPMo active phase, the type of support (Al or Nb) had no influence on final HMF yield, on the other hand, for this active phase, the catalysts calcined at 500 °C achieved higher HMF (7.0%) yields if compared to those calcined at 300 °C (3.5%). For the HPW active phase calcined at 500 °C, the type of support (Al or Nb) exerted no influence on the final HMF concentration. However, as for catalyst preparation using the same active phase calcined at 300 °C, the catalyst supported in Nb was more effective at HMF production and achieved yields of over 9%. These results suggest that the catalyst produced using HPW as active phase can achieve greater HMF production; however, such a result can only be reached when using Nb_2O_5 as support at 300 °C of calcination temperature. The highest HMF yield reached with Nb_2O_5 is probably associated with the presence of Lewis and Brönsted acid sites in this support, while Al_2O_3 only has Lewis sites on the surface [27].

For glucose conversion (Figure 1b), the highest conversions (above 85%) were obtained by using catalyst with HPMo as active phase, regardless of the type of support or calcination temperature conditions. Regarding the HPW active phase, the catalyst supported in Nb_2O_5 and calcined at 300 °C resulted in higher glucose conversion (75%) than the one calcined at 500 °C (65%); while for the catalyst supported in Al_2O_3, calcination temperature had no influence on glucose conversion (70%). Despite the high glucose conversion for catalysts prepared with HPMo active phase, HMF production was low, which suggests that reactions using catalyst with HPMo as active phase result in higher by-product formation and lower HMF selectivity. Thereby, the HPW/Nb_2O_5-300 °C catalyst was selected as the most desirable catalyst for converting glucose into HMF.

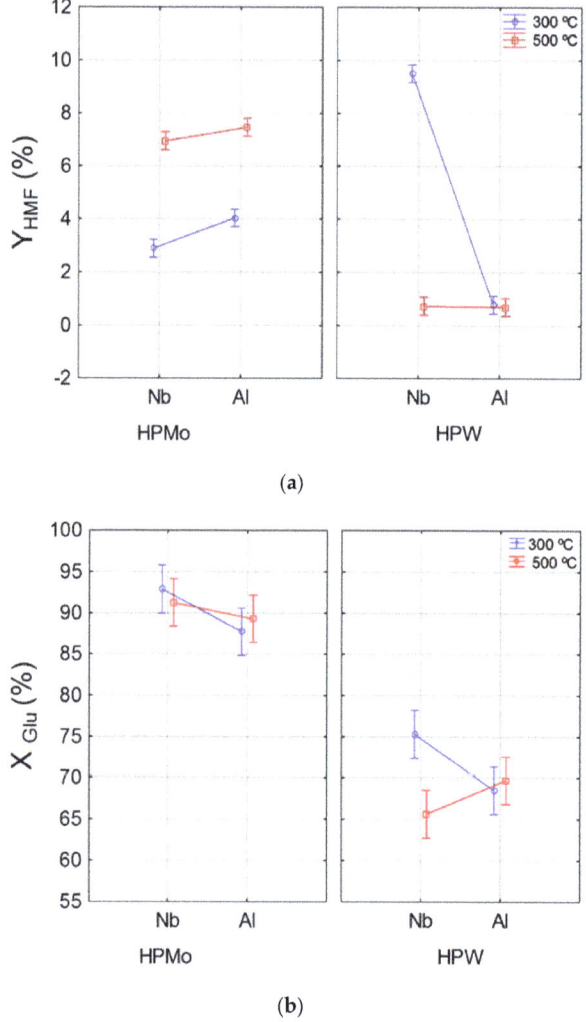

Figure 1. Interaction effect between temperature and support for different active phases (HPW and HPMo) on HMF yield (Y_{HMF}) (**a**) and glucose conversion (X_{Glu}) (**b**).

3.2. Catalyst Characterization

The active phase thermal stability of the selected catalyst was evaluated by thermogravimetric analysis (TGA) with the aim of determining whether the Keggin structure would remain at 300 °C or not, since it is the calcination temperature of the catalyst that achieved optimal performance. TGA profile shown in Figure 2 revealed that the main weight loss (approx. 7%) occurred until 208 °C. The DTG (derivative thermogravimetry) curve revealed four main stages of weight loss. In the first stage, corresponding to peaks of 75 and 120 °C, loss of water physically adsorbed in the material is observed. In the second stage, corresponding to a peak of 208 °C, crystallization water loss of the solid structure was observed, thus forming an anhydrous acid (Equation (3)). In the third stage, corresponding to a peak of 305 °C, acidic proton loss and anhydride structure formation was observed (Equation (4)). Finally, in the fourth stage, corresponding to a peak of 545 °C, the Keggin structure starts decomposing (Equation (5)). According to Kozhevnikov et al. [28] and Alsalme et al. [29], this stage occurs at approximately 600 °C. Therefore, it can be concluded that the HPW Keggin structure remains after calcination at 300 °C. However, the Keggin structure of the HPW catalyst calcined at 500 °C is close to its decomposition temperature, which explains the low yield observed for this catalyst.

$$H_3[PW_{12}O_{40}].nH_2O \rightarrow H_3[PW_{12}O_{40}] + nH_2O, \quad (3)$$

$$H_3[PW_{12}O_{40}] \rightarrow [PW_{12}O_{38.5}] + 1.5H_2O, \quad (4)$$

$$[PW_{12}O_{38.5}] \rightarrow 0.5P_2O_5 + 12WO_3. \quad (5)$$

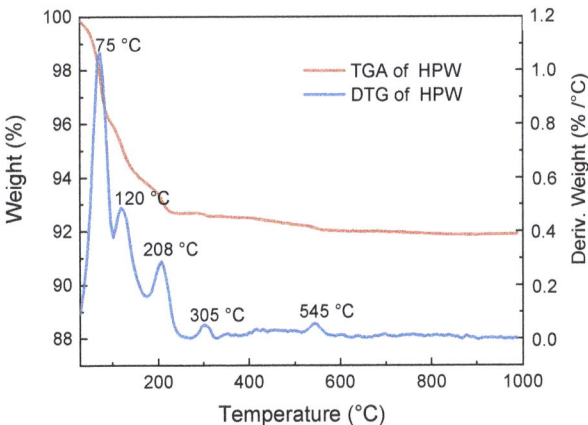

Figure 2. Thermogravimetric analysis (TGA) and derivative thermogravimetry (DTG) analysis of HPW samples.

The crystalline structures of the support (Nb_2O_5), active phase (HPW), and catalyst (HPW/Nb_2O_5) calcined at 300 °C were evaluated by X-ray powder diffraction (XRD) (Figure 3). The XRD pattern of the HPW active phase showed typical diffraction peaks of a HPW Keggin structure at 10.3°, 20.7°, 23.1°, 25.4°, and 29.5° [30,31]. Such result indicates that the Keggin structure has not been decomposed into WO_3 after calcination at 300 °C, which could occur for the active phase calcined at 500 °C. This result is in agreement with the thermogravimetry analysis results. The XRD pattern of the support (Nb_2O_5) revealed an amorphous structure, with broad and diffuse diffraction peaks, which is characteristic of Nb_2O_5 calcined below 500 °C [32,33]. Finally, the XRD pattern of the HPW/Nb_2O_5-300 °C catalyst was similar to that of the support with an amorphous characteristic, and exhibited no diffraction peak connected with HPW. This result indicates that HPW was well dispersed on the support surface, with no active phase agglomerations.

The HPW/Nb_2O_5/300 °C catalyst morphology and structure, and HPW dispersion over the niobium pentoxide surface was investigated by scanning electron microscopy coupled with energy dispersive X-ray spectroscopy (SEM/EDX). As it can be seen in Figure 4, it is composed of non-uniform crystals with an irregular surface and particles of various sizes. It was not possible to identify a predominant crystalline structure in this catalyst, which was expected, since Nb_2O_5 calcined at 300 °C exhibited an amorphous characteristic. According to the X-ray emission mapping of niobium and tungsten (Figure 4c,d), it is noted that the heteropolyacid was highly dispersed on the support surface, with no active phase agglomeration, which is in agreement with the XRD pattern obtained for the catalyst.

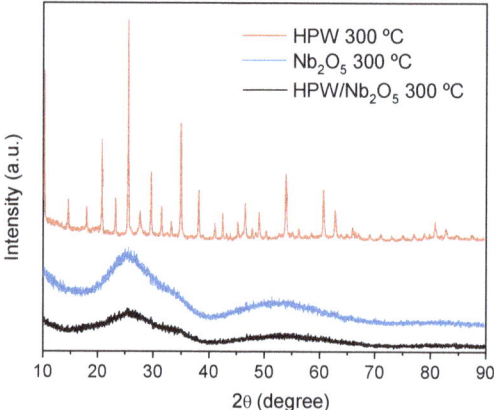

Figure 3. X-ray powder diffraction (XRD) patterns of the HPW active phase, Nb_2O_5 support, and HPW/Nb_2O_5 catalyst calcined at 300 °C.

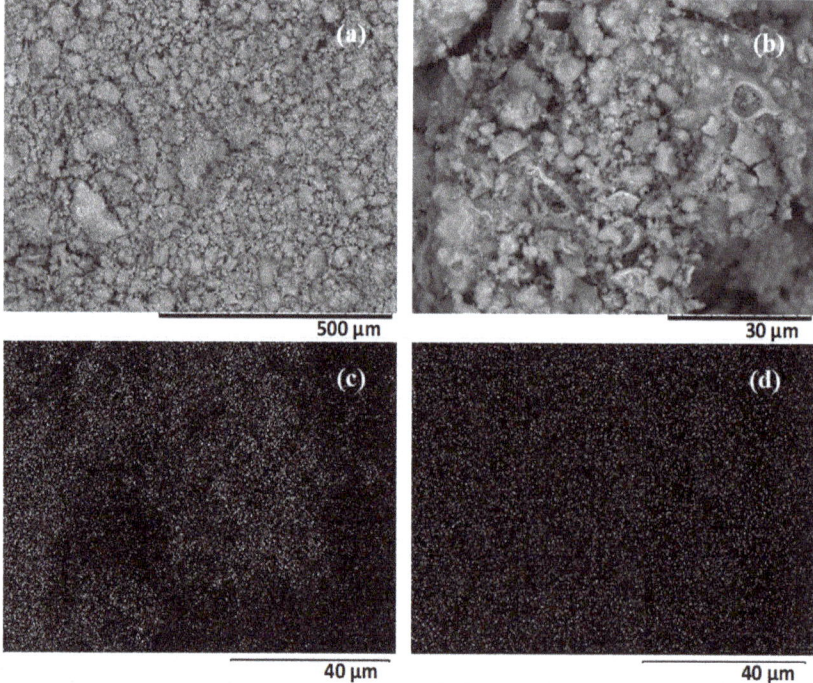

Figure 4. Scanning electron micrographs of HPW/Nb_2O_5-300 °C catalyst at (**a**) 200× magnification (scale bar: 500 µm), (**b**) 2000× magnification (scale bar: 30 µm), and X-ray emission mapping of niobium (**c**) and tungsten (**d**) obtained at 2000× magnification (scale bar: 40 µm).

Textural and acidity properties of the niobium support and selected catalyst (HPW/Nb_2O_5-300 °C) were also analyzed. As depicted in Table 3, the support showed values of surface area and volume of pores greater than those obtained for the catalyst HPW/Nb_2O_5/300 °C, since HPW impregnation onto the support leads to a massive reduction in these parameters. This occurs because the HPW Keggin structure is dispersed within the support pores (Nb_2O_5), causing a decrease in average pore volume and surface area. Decreased surface area of the catalyst $HPW/Nb_2O_5/300\ ^\circ C$ can be understood as an evidence of the chemical interaction between HPW and the support. Surface acidity is an important property which affects catalyst performance in the reaction and may be indicative of an impregnation success. $HPW/Nb_2O_5/300\ ^\circ C$ showed acidity of 106.98 μmol H^+/m^2, i.e., much higher than that of Nb_2O_5 used as support.

Table 3. Textural and acidity properties of Nb_2O_5 and HPW/Nb_2O_5 catalyst.

Sample	Surface Area (m^2/g)	Pore Volume (cm^3/g)	Surface Acidity (μmol H^+/m^2)
Nb_2O_5	130.91	0.14	0.31
HPW/Nb_2O_5-300 °C	36.08	0.031	106.98

3.3. Reaction Conditions Evaluation

In this section, Taguchi's L_{16} experimental design was used for optimizing reaction conditions so as to produce HMF from glucose using the selected catalyst $HPW/Nb_2O_5/300\ ^\circ C$. Different reaction conditions and responses are summarized in Table 4. As it can be seen, HMF yield obtained in these experiments ranged from 7.6% to 40.8%. As in the previous stage, high glucose conversion was obtained, i.e., ranging from 63.7% to 98.4%, which could also be associated with by-products formation. The highest HMF yield (40.8%) was obtained under the conditions of assay 2 (160 °C, 5% *w/v* of catalyst, 50 g/L of glucose, 1:1 *v/v* acetone-to-water, 300 rpm and reaction time of 30 min).

Table 4. Taguchi's L_{16} orthogonal array to evaluate the effect of temperature, acetone-to-water ratio, glucose concentration, and catalyst concentration on 5-Hydroxymethylfurfural production from glucose.

																Response Variables	
Exp.	A Temp. (°C)	B Acetone: Water (v/v)	AB	C C_{Glu} (g/L)	AC	BC	CE	D Cat. (%w/v)	AD	BD	CE	CD	BE	AE	E[1]	Y_{HMF}[2] (%)	X_{Glu}[3] (%)
1	160	1:1	1	50	1	1	1	1	1	1	1	1	1	1	1	15.4	71.3
2	160	1:1	1	50	1	1	1	5	2	2	2	2	2	2	2	40.8	93.3
3	160	1:1	1	100	2	2	2	1	1	1	1	2	2	2	2	8.8	63.7
4	160	1:1	1	100	2	2	2	5	2	2	2	1	1	1	1	30.1	86.1
5	160	3:1	2	50	1	2	2	1	1	2	2	1	1	2	2	17.5	80.9
6	160	3:1	2	50	1	2	2	5	2	1	1	2	2	1	1	40.5	93.3
7	160	3:1	2	100	2	1	1	1	1	2	2	2	2	1	1	7.6	63.9
8	160	3:1	2	100	2	1	1	5	2	1	1	1	1	2	2	32.6	88.0
9	200	1:1	2	50	2	1	2	1	2	1	2	1	2	1	2	11.5	97.4
10	200	1:1	2	50	2	1	2	5	1	2	1	2	1	2	1	14.2	97.9
11	200	1:1	2	100	1	2	1	1	2	1	2	2	1	2	1	8.5	96.0
12	200	1:1	2	100	1	2	1	5	1	2	1	1	2	1	2	9.8	98.4
13	200	3:1	1	50	2	2	1	1	2	2	1	1	2	2	1	20.8	95.9
14	200	3:1	1	50	2	2	1	5	1	1	2	2	1	1	2	14.0	96.4
15	200	3:1	1	100	1	1	2	1	2	2	1	2	1	1	2	18.8	73.0
16	200	3:1	1	100	1	1	2	5	1	1	2	1	2	2	1	17.0	85.5

[1] E: Column used for error estimative. [2] Y_{HMF}: HMF yield. [3] X_{Glu}: glucose conversion.

The statistical significance of the main effects and interactions on response variables was found by the analysis of variance (ANOVA) (Table 5). Percentages of variation in HMF yield and glucose conversion showed high correlation coefficients (R^2), i.e., over 95% confidence level. The statistical analysis showed that, for HMF yield, reaction temperature (A), acetone-to-water ratio (B), glucose (C) and catalyst concentration (D), and AC and AD interactions were significant at 95% confidence level. Regarding glucose conversion, reaction temperature (A), glucose (C) and catalyst concentration (D), and AB and AD interactions were significant at the same confidence level.

Table 5. Analysis of variance of response variables evaluated in the experimental design Taguchi L_{16}.

Source of Variation	Response Variables	
	p-Value for Y_{HMF}	p-Value for X_{Glu}
(A) $T_{reaction}$	0.000 *	0.002 *
(B) Acetone:Water	0.027 *	0.168
(C) $C_{glucose}$	0.007 *	0.008 *
(D) $C_{catalyst}$	0.000 *	0.002 *
AB	0.058	0.031 *
AC	0.030 *	0.889
AD	0.000 *	0.012 *
BC	0.448	0.063
BD	0.290	0.902
CD	0.874	0.185
R^2	0.984	0.960

* Significant at 95% confidence level: p test < 0.05.

Figure 5 shows the signal-to-noise ratio (SN ratio) diagrams of HMF yield (A) and glucose conversion (B). It is possible to determine optimal conditions for HMF production from glucose thereof. According to Figure 5A, the highest HMF yield was achieved when reaction temperature and glucose concentration were at the lowest level (temperature of 160 °C and 50 g/L of glucose), while catalyst concentration was at the highest level (5% *w/v*). Acetone-to-water ratio was insignificant for HMF yield in the range of values studied herein.

With respect to glucose conversion (Figure 5B), the highest conversion rates were achieved when reaction temperature, acetone-to-water ratio, and catalyst concentration were at the highest level (200 °C, 3:1 *v/v*, 5% *w/v*, respectively), while glucose concentration was insignificant within the studied range of values. It is interesting to note that glucose conversion was the only response variable that obtained a better result when at high reaction temperatures. This suggests that, for the studied range of values, high reaction temperatures may favor by-product formation.

According to results of the statistical analysis and considering that this paper aimed to obtain the highest HMF yield, optimal conditions for HMF production from glucose in the evaluated range of values was temperature of 160 °C, 1:1 (*v/v*) acetone-to-water ratio, glucose concentration of 50 g/L, and catalyst concentration of 5% *w/v*.

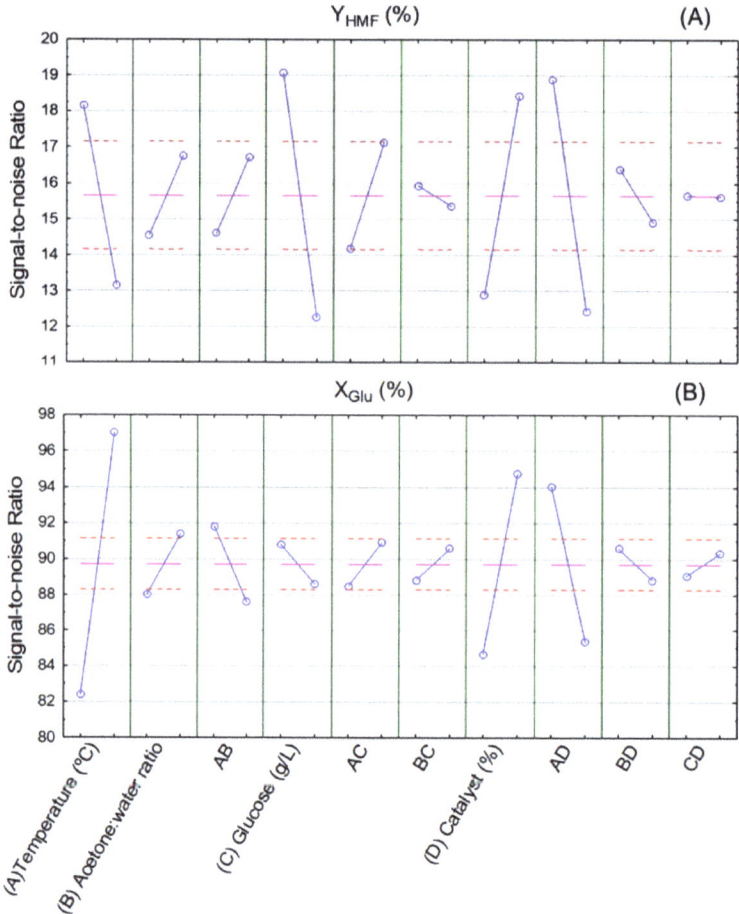

Figure 5. Signal-to-noise ratio (SN ratio) of (**A**) HMF yield (Y_{HMF} %) and (**B**) glucose conversion (X_{Glu} %) response variables.

The catalytic performance of HPW/Nb$_2$O$_5$ was compared with some recent reported studies in literature that produce HMF from glucose, as shown in Table 6. The HMF yield (40.8%) obtained in the present study using HPW/Nb$_2$O$_5$ as catalyst and water/acetone was higher than most studies reported in the literature which used heterogeneous catalyst [34–36], even when organic phase [37] or ionic liquid [11] was applied as solvent. Teimouri et al. [38], who used the same solvent as the present work (water/acetone), achieved lower HMF yield (34.6%), even using higher reaction temperature and time. Huang et al. [26], who used the same reaction temperature and time but different solvent (Water/γ-valerolactone), also obtained lower HMF yield compared to the present work. Shahangi et al. [9], Shen et al. [39], Zhang et al. [10], and Moreno-Recio et al. [40] achieved similar HMF yield obtained in this work, however all of them applied more drastic reaction conditions, as higher temperature and/or time. In general, even using acetone/water as a reaction medium, we achieved HMF yield equivalent to the highest values reported in the literature, but with milder reaction conditions.

Table 6. Comparison of the performance of HPW/Nb$_2$O$_5$-300 °C catalyst with some recent works reported in literature on HMF production from glucose.

Catalyst	Solvent	C$_{glu}$	T (°C)	T (min)	Y$_{HMF}$ (%)	Ref.
Nano-POM [a]	Water/acetone	4 g/L	190	240	34.6	[38]
/nano-ZrO$_2$/nano-γ-Al$_2$O$_3$						
Sn/γ-Al$_2$O$_3$	Water/DMSO (1:4 v/v)	4 wt.%	150	60	27.5	[35]
GO [b]-Fe$_2$O$_3$	DMSO	180 g/L	140	240	16	[37]
Al$_2$O$_3$ treated with 0.05 M NaOH	EMIMCl	10 wt.%	140	120	36	[11]
ACBL2 [c]	Water	11.25 g/L	160	480	15	[8]
TSA350 [d]	Water	36 g/L	120	360	19	[7]
Acid γ-Al$_2$O$_3$	Aq. solution of CaCl$_2$/MIBK (1:2.3 v/v)	30 g/L	175	15	23	[36]
Al-KCC-1 [e]	DMSO	125 g/L	170	120	39	[9]
S-TsC [f]	Water/γ-valerolactone (GVL) (0.3:4.7 v/v)	20 g/L	160	30	27.8	[26]
HMOR_20 [g]	Aq. solution of NaCl/MIBK (2.6:4 w/w)	3.33 wt.%	180	60	27	[34]
SO$_4^{2-}$/In$_2$O$_3$-ATP [h]	Water/GVL (1.9 v/v)	2 wt.%	180	60	40.2	[39]
Al-SPFR [i]	Water/GVL (1:10 v/v)	12.1 g/L	180	50	41.5	[10]
H-ZSM-5 zeolite	Aq. solution of NaCl/MIBK (1.5:3.5 v/v)	30 g/L	195	30	42	[40]
HPW/Nb$_2$O$_5$-300 °C	Water/acetone (1:1 v/v)	50 g/L	160	30	40.8	Present study

[a] POM: Polyoxometalates. [b] GO: Graphene oxide. [c] ACBL2; Activated carbon with acid treatment 18M H$_2$SO$_4$ and activated carbon with 15 wt.% of zinc. [d] TSA$_{350}$; Alumina-promoted sulfated tin oxide calcined at 350 °C. [e] Al-KCC-1; Aluminosilicate with Si/Al = 5. [f] S-TsC; Sulfonated tobacco stem-derived porous carbon. [g] HMOR_20; zeolite with SiO$_2$/AlO$_3$ = 20. [h] ATP; Attapulgite. [i] Al-SPFR; Al^{3+}-modified formaldehyde-p-hydroxybenzenesulfonic acid resin catalyst.

3.4. Catalyst Recycling Test

Finally, the HPW/Nb$_2$O$_5$-300 °C catalyst stability was evaluated through a recycling test using optimal reaction conditions for HMF production from glucose. Figure 6 shows that the catalyst was capable of maintaining the HMF yield constant at approximately 40% during four runs. This means that a simple calcination process was sufficient to eliminate by-products (humin) that could have covered some active sites of the catalyst. This result indicates that the catalyst HPW/Nb$_2$O$_5$-300 °C has high stability, good recyclability, and low active phase leaching, thus confirming its effectiveness to be used in HMF production from glucose.

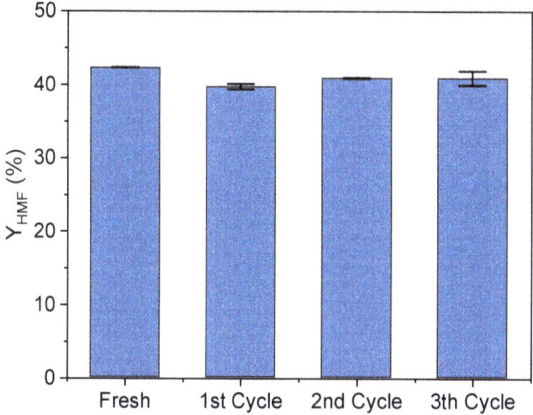

Figure 6. Catalytic performance of HPW/Nb$_2$O$_5$-300 °C during a recycling test. Reaction conditions were 160 °C, 5% *w/v* of catalyst, 50 g/L of glucose, 1:1 *v/v* acetone-to-water, 300 rpm, and reaction time of 30 min.

4. Conclusions

This study revealed that the conditions used for preparing the heterogeneous catalysts based on heteropolyacids, i.e., type of support, active phase, and calcination temperature, strongly affects catalyst performance to convert glucose into HMF. The heterogeneous catalyst prepared by using HPW as active phase, supported in Nb$_2$O$_5$, and calcined at 300 °C presented optimal performance at obtaining HMF yield (9.5%) with 1:1 *w/v* acetone-to-water ratio, 160 °C, 1% *w/v* of catalyst, 100 g/L of glucose and reaction time of 30 min. After optimizing reaction conditions (temperature at 160 °C, 1:1 (*v/v*) acetone-to-water ratio, glucose concentration of 50 g/L, and catalyst concentration of 5% *w/v*), HMF yield increased to 40.8%. The catalyst presented high stability and is capable of maintaining HMF yield at around 40%, even after four runs. This was the first time that a catalyst based on HPW and Nb$_2$O$_5$ was used for HMF production from glucose. The promising results here obtained will open up new opportunities to expand HMF production from lignocellulosic biomass. The possibility of recycling the catalyst is also a point to be highlighted contributing to the sustainability aspect of the proposed process.

Author Contributions: Conceptualization, J.P.A.S., S.I.M. and L.M.C.; Methodology, J.S.M.N., J.P.A.S. and L.M.C.; Software, J.S.M.N., J.P.A.S. and L.M.C.; Investigation, J.S.M.N., J.P.A.S and L.M.C.; Resources, J.P.A.S, S.I.M. and L.M.C.; Writing—Original Draft, J.S.M.N. and L.M.C.; Writing—Review and Editing, J.P.A.S, S.I.M. and L.M.C.; Supervision, J.P.A.S, S.I.M. and L.M.C.; Project Administration, J.P.A.S, S.I.M. and L.M.C.; Funding Acquisition, J.P.A.S, S.I.M. and L.M.C. All authors have read and agreed to the published version of the manuscript.

Funding: This research was funded by Fundação de Amparo à Pesquisa do Estado de São Paulo (FAPESP), project number 2017/24050-8 and 2018/03714-8, Brazil; Conselho Nacional de Desenvolvimento Científico e Tecnológico (CNPq), Brazil; Coordenação de Aperfeiçoamento de Pessoal de Nível Superior (CAPES), Brazil; and Novo Nordisk Foundation, grant number: NNF10CC1016517, Denmark.

Acknowledgments: We thank Companhia Brasileira de Metalurgia e Mineração (CBMM) for donations the materials used for experiments.

Conflicts of Interest: The authors declare no conflict of interest.

References

1. Mussatto, S.I. (Ed.) *Biomass Fractionation Technologies for a Lignocellulosic Feedstock Based Biorefinery*; Elsevier Inc.: Waltham, MA, USA, 2016; p. 674. ISBN 9780128023235.
2. Werpy, T.; Petersen, G.; National Renewable Energy Laboratory (NREL). Top Value Added Chemicals from Biomass. Volume I—Results of Screening for Potential Candidates from Sugars and Synthesis Gas. 2004. Available online: https://www.nrel.gov/docs/fy04osti/35523.pdf (accessed on 3 September 2019).
3. Mukherjee, A.; Dumont, M.J.; Raghavan, V. Review: Sustainable production of hydroxymethylfurfural and levulinic acid: Challenges and opportunities. *Biomass Bioenergy* **2015**, *72*, 143–183. [CrossRef]
4. Jong, E.; Dam, M.A.; Sipos, L.; Gruter, J.M. Furandicarboxylic acid (FDCA), a versatile building block for a very interesting class of polyesters. In *Biobased Monomers, Polymers, and Materials*; Smith, P.B., Gross, R.A., Eds.; American Chemical Society: Midland, MI, USA, 2012; Volume 1105, pp. 1–13. [CrossRef]
5. Sweygers, N.; Alewaters, N.; Dewil, R.; Appels, L. Microwave effects in the dilute acid hydrolysis of cellulose to 5-hydroxymethylfurfural. *Sci. Rep.* **2018**, *8*, 1–11. [CrossRef] [PubMed]
6. Candu, N.; Fergani, M.E.; Verziu, M.; Cojocaru, B.; Jurca, B.; Apostol, N.; Teodorescu, C.; Parvulescu, V.I.; Coman, S.M. Efficient glucose dehydration to HMF onto Nb-BEA catalysts. *Catal. Today* **2019**, *325*, 109–116. [CrossRef]
7. Lopes, M.; Dussan, K.; Leahy, J.J.; Da Silva, V.T. Conversion of d-glucose to 5-hydroxymethylfurfural using Al_2O_3-promoted sulphated tin oxide as catalyst. *Catal. Today* **2017**, *279*, 233–243. [CrossRef]
8. Rusanen, A.; Lahti, R.; Lappalainen, K.; Kärkkäinen, J.; Hu, T.; Romar, H.; Lassi, U. Catalytic conversion of glucose to 5-hydroxymethylfurfural over biomass based activated carbon catalyst. *Catal. Today* **2019**. [CrossRef]
9. Shahangi, F.; Chermahini, A.N.; Saraji, M. Dehydration of fructose and glucose to 5-hydroxymethylfurfural over Al-KCC-1 silica. *J. Energy Chem.* **2018**, *27*, 769–780. [CrossRef]
10. Zhang, T.; Li, W.; Xin, H.; Jin, L.; Liu, Q. Production of HMF from glucose using an Al^{3+}-promoted acidic phenol-formaldehyde resin catalyst. *Catal. Commun.* **2019**, *124*, 56–61. [CrossRef]
11. Hou, Q.; Zhen, M.; Li, W.; Liu, L.; Liu, J.; Zhang, S.; Nie, Y.; Bai, C.; Bai, X.; Ju, M. Efficient catalytic conversion of glucose into 5-hydroxymethylfurfural by aluminum oxide in ionic liquid. *Appl. Catal. B Environ.* **2019**, *253*, 1–10. [CrossRef]
12. Daorattanachai, P.; Khemthong, P.; Viriya-empikul, N.; Laosiripojana, N.; Faungnawakij, K. Effect of calcination temperature on catalytic performance of alkaline earth phosphates in hydrolysis/dehydration of glucose and cellulose. *Chem. Eng.* **2015**, *278*, 92–98. [CrossRef]
13. Gomes, F.N.D.C.; Pereira, L.R.; Ribeiro, N.F.P.; Souza, M.M.V.M. Production of 5-hydroxymethylfurfural (HMF) via fructose dehydration: Effect of solvent and salting-out. *Braz. J. Chem. Eng.* **2015**, *32*, 119–126. [CrossRef]
14. Yuan, B.; Guan, J.; Peng, J.; Guang-Zhou, Z.; Ji-Hong, J. Green hydrolysis of corncob cellulose into 5-hydroxymethylfurfural using hydrophobic imidazole ionic liquids with a recyclable, magnetic metalloporphyrin catalyst. *Chem. Eng. J.* **2017**, *330*, 109–119. [CrossRef]
15. Yan, D.; Xin, J.; Shi, C.; Lu, X.; Ni, L.; Wang, G.; Zhang, S. Base-free conversion of 5-hydroxymethylfurfural to 2,5-furandicarboxylic acid in ionic liquids. *Chem. Eng. J.* **2017**, *323*, 473–482. [CrossRef]
16. Boisen, A.; Christensen, T.B.; Fu, W.; Gorbanev, Y.Y.; Hansen, T.S.; Jensen, J.S.; Klitgaad, S.K.; Pedersen, S.; Riisager, A.; Stahlberg, T.; et al. Process integration for the conversion of glucose to 2,5-furandicarboxylic acid. *Chem. Eng. Res. Des.* **2009**, *87*, 1318–1327. [CrossRef]
17. Kale, S.S.; Armbruster, U.; Eckelt, R.; Bentrup, U.; Umbarkar, S.B.; Dongare, M.K.; Martin, A. Understanding the role of Keggin type heteropolyacid catalysts for glycerol acetylation using toluene as an entrainer. *Appl. Catal. A Gen.* **2016**, *527*, 9–18. [CrossRef]
18. Mansir, N.; Taufiq-Yap, Y.H.; Rashid, U.; Lokman, I.M. Investigation of heterogeneous solid catalyst performance on low grade feedstocks for biodiesel production: A review. *Energy Convers. Manag.* **2017**, *141*, 171–182. [CrossRef]

19. Méndez, F.J.; Llanos, A.; Echeverría, M.; Jáuregui, R.; Villasana, Y.; Díaz, Y.; Liendo-Polanco, G.; Ramos-García, M.A.; Zoltan, T.; Brito, J.L. Mesoporous catalysts based on Keggin-type heteropolyacids supported on MCM-41 and their application in thiphene hydrodesulfurization. *Fuel* **2013**, *110*, 249–258. [CrossRef]
20. Marcí, G.; García-Lopez, E.; Vaiano, V.; Sarno, G.; Sannino, D.; Palmisano, L. Keggin heteropolyacid supported on TiO_2 used in gas-solid (photo)catalytic propene hydration and in liquid-solid photocatalytic glycerol dehydration. *Catal. Today* **2017**, *281*, 60–70. [CrossRef]
21. Han, X.; Yan, W.; Chen, K.; Hung, C.T.; Liu, L.L.; Wu, P.H.; Huang, S.J.; Liu, S.B. Heteropolyacid-based ionic liquids as effective catalysts for the synthesis of benzaldehyde glycol acetal. *Appl. Catal. A Gen.* **2014**, *485*, 149–156. [CrossRef]
22. Pinto, T.; Dufaud, V.; Lefebvre, F. Isomerization of n-hexane on heteropolyacids supported on SBA-15. 1. Monofunctional impregnated catalysts. *Appl. Catal. A Gen.* **2014**, *483*, 103–109. [CrossRef]
23. Hu, L.; Sun, Y.; Lin, L.; Liu, S. 12-Tungstophosphoric acid/boric acid as synergetic catalysts for the conversion of glucose into 5-hydroxymethylfurfural in ionic liquid. *Biomass Bioenergy* **2012**, *47*, 289–294. [CrossRef]
24. Zhao, P.; Zhang, Y.; Wang, Y.; Cui, H.; Song, F.; Sun, X.; Zhang, L. Conversion of glucose into 5-hydroxymethylfurfural catalyzed by acid–base bifunctional heteropolyacid-based ionic hybrids. *Green Chem.* **2018**, *20*, 1551–1559. [CrossRef]
25. Fan, C.; Guan, H.; Zhang, H.; Wang, J.; Wang, S.; Wang, X. Conversion of fructose and glucose into 5-hydroxymethylfurfural catalyzed by a solid heteropolyacid salt. *Biomass Bioenergy* **2011**, *35*, 2659–2665. [CrossRef]
26. Huang, F.; Su, Y.; Tao, Y.; Sun, W.; Wang, W. Preparation of 5-hydroxymethylfurfural from glucose catalyzed by silicasupported phosphotungstic acid heterogeneous catalyst. *Fuel* **2018**, *226*, 417–422. [CrossRef]
27. Martín, C.; Solana, G.; Malet, P.; Rives, V. Nb_2O_5-supported WO_3: A comparative study with WO_3/Al_2O_3. *Catal. Today* **2003**, *78*, 365–376. [CrossRef]
28. Kozhevnikov, I.V. Sustainable heterogeneous acid catalysis by heteropoly acids. *J. Mol. Catal. A Chem.* **2007**, *262*, 86–92. [CrossRef]
29. Alsalme, A.M.; Wiper, P.V.; Khimyak, Y.Z.; Kozhevnikova, E.F.; Kozhevnikov, I.V. Solid acid catalysts based on $H_3PW_{12}O_{40}$ heteropoly acid: Acid and catalytic properties at a gas–solid interface. *J. Catal.* **2010**, *276*, 181–189. [CrossRef]
30. Keggin, J.F. Structure and formula of 12-phosphotungstic acid. *Proc. R. Soc. Lond. Ser. A Math. Phys. Eng. Sci.* **1934**, *144*, 75–100. [CrossRef]
31. Liao, X.; Huang, Y.; Zhou, Y.; Liu, H.; Cai, Y.; Lu, S.; Yao, Y. Homogeneously dispersed HPW/graphene for high efficient catalytic oxidative desulfurization prepared by electrochemical deposition. *Appl. Surf. Sci.* **2019**, *484*, 917–924. [CrossRef]
32. Caliman, E.; Dias, J.A.; Dias, S.C.L.; Garcia, F.A.C.; Macedo, J.L.D.; Almeida, L.S. Preparation and characterization of $H_3PW_{12}O_{40}$ supported on niobia. *Microporous Mesoporous Mater.* **2010**, *132*, 103–111. [CrossRef]
33. Conceição, L.R.V.D.; Carneiro, L.M.; Giordani, D.S.; Castro, H.F.D. Synthesis of biodiesel from macaw palm oil using mesoporous solid catalyst comprising 12-molybdophosphoric acid and niobia. *Renew. Energy* **2017**, *113*, 119–128. [CrossRef]
34. Peela, N.R.; Yedla, S.K.; Velaga, B.; Kumar, A.; Golder, A.K. Choline chloride functionalized zeolites for the conversion of biomass derivatives to 5-hydroxymethykfurfural. *Appl. Catal. A Gen.* **2019**, *580*, 59–70. [CrossRef]
35. Marianou, A.A.; Michailof, C.M.; Pineda, A.; Iliopoulou, E.F.; Triantafyllidis, K.S.; Lappas, A.A. Effect of Lewis and Brønsted acidity on glucose conversion to 5-HMF and lactic acid in aqueous and organic media. *Appl. Catal. A Gen.* **2018**, *555*, 75–87. [CrossRef]
36. García-Sancho, C.; Fúnez-Núñez, I.; Moreno-Tost, R.; Santamaría-González, J.; Pérez-Inestrosa, E.; Fierro, J.L.G.; Maireles-Torres, P. Beneficial effects of calcium chloride on glucose dehydration to5-hydroxymethylfurfural in the presence of alumina as catalyst. *Appl. Catal. B Environ.* **2017**, *206*, 617–625. [CrossRef]
37. Zhang, Y.; Zhang, J.; Su, D. 5-Hydroxymethylfurfural: A key intermediate for efficient biomass conversion. *J. Energy Chem.* **2015**, *24*, 548–551. [CrossRef]

38. Teimouri, A.; Mazaheri, M.; Chermahini, A.N.; Salavati, H.; Momenbeik, F.; Fazel-Najafabadi, M. Catalytic conversion of glucose to 5-hydroxymethylfurfural (HMF) using nano-POM/nano-ZrO₂/nano-γ-Al₂O₃. *J. Taiwan Inst. Chem. Eng.* **2015**, *49*, 40–50. [CrossRef]
39. Shen, Y.; Kang, Y.; Sun, J.; Wang, C.; Wang, B.; Xu, F.; Sun, R. Efficient production of 5-hydroxymethylfurfural from hexoses using solid acid SO₄⁻²/In₂O₃-ATP in a biphasic system. *Chin. J. Catal.* **2016**, *37*, 1362–1368. [CrossRef]
40. Moreno-Recio, M.; Santamaría-González, J.; Maireles-Torres, P. Brönsted and Lewis acid ZSM-5 zeolites for the catalytic dehydration of glucose into 5-hydroxymethylfurfural. *Chem. Eng. J.* **2016**, *303*, 22–30. [CrossRef]

© 2020 by the authors. Licensee MDPI, Basel, Switzerland. This article is an open access article distributed under the terms and conditions of the Creative Commons Attribution (CC BY) license (http://creativecommons.org/licenses/by/4.0/).

Article

Production of Itaconic Acid from Cellulose Pulp: Feedstock Feasibility and Process Strategies for an Efficient Microbial Performance

Abraham A. J. Kerssemakers, Pablo Doménech, Marco Cassano, Celina K. Yamakawa, Giuliano Dragone and Solange I. Mussatto *

Novo Nordisk Foundation Center for Biosustainability, Technical University of Denmark, Kemitorvet, Building 220, 2800 Kongens Lyngby, Denmark; aajker@biosustain.dtu.dk (A.A.J.K.); pablo.domenech@ciemat.es (P.D.); marcocassano08@gmail.com (M.C.); celinayamakawa@biosustain.dtu.dk (C.K.Y.); giudra@biosustain.dtu.dk (G.D.)
* Correspondence: smussatto@biosustain.dtu.dk or solangemussatto@hotmail.com; Tel.: +45-93-511-891

Received: 27 February 2020; Accepted: 31 March 2020; Published: 2 April 2020

Abstract: This study assessed the feasibility of using bleached cellulose pulp from Eucalyptus wood as a feedstock for the production of itaconic acid by fermentation. Additionally, different process strategies were tested with the aim of selecting suitable conditions for an efficient production of itaconic acid by the fungus *Aspergillus terreus*. The feasibility of using cellulose pulp was demonstrated through assays that revealed the preference of the strain in using glucose as carbon source instead of xylose, mannose, sucrose or glycerol. Additionally, the cellulose pulp was easily digested by enzymes without requiring a previous step of pretreatment, producing a glucose-rich hydrolysate with a very low level of inhibitor compounds, suitable for use as a fermentation medium. Fermentation assays revealed that the technique used for sterilization of the hydrolysate (membrane filtration or autoclaving) had an important effect in its composition, especially on the nitrogen content, consequently affecting the fermentation performance. The carbon-to-nitrogen ratio (C:N ratio), initial glucose concentration and oxygen availability, were also important variables affecting the performance of the strain to produce itaconic acid from cellulose pulp hydrolysate. By selecting appropriate process conditions (sterilization by membrane filtration, medium supplementation with 3 g/L $(NH_4)_2SO_4$, 60 g/L of initial glucose concentration, and oxygen availability of 7.33 (volume of air/volume of medium)), the production of itaconic acid was maximized resulting in a yield of 0.62 g/g glucose consumed, and productivity of 0.52 g/L·h.

Keywords: lignocellulosic biomass; cellulose pulp; hydrolysis; oxygen availability; C:N ratio; fermentation; biorefinery; itaconic acid; *Aspergillus terreus*

---

1. Introduction

The development of new process technologies using lignocellulosic feedstock as a carbon source for the production of fuels and chemicals is, currently, one of the main drivers of society to move towards a more sustainable future [1]. Second-generation biofuel plants are already a reality at a commercial scale and, to become truly sustainable and circular, industry is also increasingly viewing the production of chemicals from renewable resources as an attractive area for investment.

Biofuels and biochemicals can be produced in single product processes; however, their production in an integrated biorefinery is seen as a more efficient and interesting approach to solve economic challenges related to biomass conversion processes since, currently, the cost of single biobased production processes, in many cases, still exceeds the cost of petrochemical production [2]. One of the main reasons for these high costs is the recalcitrant nature of biomass that, therefore requires a two-step processing to obtain sugars for fermentation as follows: a pretreatment step to fractionate the material and solubilize especially hemicellulose sugars, and a subsequent hydrolysis step to recover glucose from cellulose. Pretreatment, in particular, is an energy intensive step and significantly contributes to the final cost of the process [3].

Itaconic acid stands out as one of the most relevant among the variety of chemicals that can be produced from lignocellulosic biomass, since it is a platform chemical with extensive applications in different fields. Some of the main interests around itaconic acid arise from its potential to substitute petrochemically produced acrylic acid. However, it can also be used to produce biodegradable polymers, paints, varnishes, and different organic compounds. Moreover, itaconic acid and its derivatives support the synthesis of a wide range of innovative polymers through crosslinking, with applications in special hydrogels for water decontamination, drug delivery, nanohydrogels for food applications, coatings, and elastomers [4].

Currently, itaconic acid is produced industrially from aerobic fungal fermentation using pure glucose as a carbon source, which is not the cheapest or the most sustainable substrate option. Moreover, although the production is done by fermentation, at present, the cost to produce itaconic acid is high and has been a bottleneck preventing its application in different sectors [5]. With a market in expansion due to the increased number of potential applications (its market was worth USD 126.4 million in 2014 and with an expected growth rate of 60% it is predicted to reach around USD 204.6 million by 2023 [6]), the establishment of a more sustainable and cost-competitive process for the production of itaconic acid from renewable feedstock has been strongly encouraged. The present study aims to contribute with new knowledge to advance this area by using industrially produced bleached cellulose pulp as a feedstock for itaconic acid production.

Bleached cellulose pulp, which is the material used for paper manufacturing, is one of the most abundant raw materials worldwide. With a huge volume of production, and a weak demand from the paper industry in the last years, the stocks of bleached cellulose pulp have been extremely high and are posing a major problem for the entire pulp market, according to industry experts [7]. To overcome this problem, different alternative uses for the pulp have been explored with the aim of promoting innovation and new business opportunities [8], including the production of biofuels, nanocellulose, and biocomposites.

Recently, attempts have been done to produce itaconic acid from different types of biomass including beech wood [9], corn stover [10], wheat chaff [11], rice husks [12], and corn cobs [13]. A comprehensive examination of the itaconic acid production from these different feedstocks clearly demonstrates an important impact of biomass pretreatment steps, presence of inhibitor compounds, and fermentation conditions on itaconic acid yield and on the feasibility of the process in general.

This paper is the first study on the use of bleached cellulose pulp for the production of itaconic acid. In this study, the composition of cellulose pulp, as well as its degradation by enzymes and fermentability were some of the points explored to evaluate its feasibility for application in the production of itaconic acid. Then, efforts were done to select process conditions able to result in an improved bioconversion efficiency. Sterilization of the cellulose pulp hydrolysate through different techniques, medium composition in terms of carbon-to-nitrogen ratio (C:N ratio) and initial glucose concentration, and oxygen availability were evaluated and discussed in detail. At the end, the process conditions that maximize the production of itaconic acid were selected and the results were compared with literature data from other feedstocks to conclude on the potential of this new bioprocess.

2. Materials and Methods

2.1. Microorganism and Inoculum Preparation

The filamentous fungus *Aspergillus terreus* NRRL 1960 was used in the experiments. The strain was obtained from the ARS Culture Collection (Peoria, IL, USA) and preserved in the form of spores in 20% (*v/v*) glycerol stock solution at −80 °C.

For inoculum preparation, the stock culture was activated on 2.4% (w/v) potato dextrose broth (PDB) medium at 35 °C for 3 days, and subsequently on 3.9% (w/v) potato dextrose agar (PDA) plates at 35 °C for 7 days. Then, spores were collected from plates by using a sterilized solution of 4% (w/v) Tween 80. The spore suspension was diluted with sterile MilliQ water in order to obtain a concentration of 10^6 spores/mL at the beginning of the fermentation.

2.2. Cellulose Pulp Characterization and Hydrolysis

Bleached cellulose pulp from Suzano S/A (Brazil) was used as raw material for the production of itaconic acid. The cellulosic material, which was produced from Eucalyptus wood and had a moisture content of approximately 5% (w/w), was ground to particle size ≤ 1.0 mm by means of a mill Polymix PX-MFC 90D (Kinematica, Switzerland) and its composition was determined by following standard methods [14–16].

Enzymatic hydrolysis of the cellulose pulp was carried out using the enzyme concentrate Cellic® CTec2, kindly supplied by Novozymes (Bagsværd, Denmark). The cellulase activity of the concentrate, which was measured according to standard protocol [17] and expressed in filter paper units (FPU), was 217.5 FPU/mL. One unit of FPU was defined as the amount of enzyme required to liberate 1 µmol of glucose from Whatman no.1 filter paper per minute at 50 °C.

For the experiments, an enzyme load of 10 FPU/g cellulose was added to 0.1 M sodium citrate buffer (pH 4.7), and then mixed with the cellulose pulp in a concentration of 12% (w/v). The reactions were carried out in 2-L Duran laboratory bottles with vertical baffles containing 0.6 L of working volume. The bottles were accommodated horizontally in a Bottle/Tube Roller system (Thermo Scientific, USA) placed inside an incubator, and kept at 50 °C and 20 rpm for 96 h. Afterwards, the hydrolysate was separated by centrifugation (10,000 rpm, 5 °C, 20 min).

2.3. Hydrolysate Sterilization

Three different methods (membrane filtration, autoclave at 112 °C for 15 min, and autoclave at 121 °C for 20 min) were tested for sterilization of the hydrolysate prior its use as fermentation medium. The autoclave assays were carried out in an autoclave MultiControl 2 (CertoClav, Austria); while for the membrane method, Nalgene RapidFlow™ PES-membrane filters with a pore size of 0.2 µm (Thermo Fisher Scientific, USA) were used.

2.4. Fermentation Media and Conditions

Initially, different synthetic media were tested for the production of itaconic acid by *A. terreus*, which contained only one type of carbon source (glucose, xylose, sucrose, mannose, or glycerol) at a concentration of 50 g/L. Later, the cellulose pulp hydrolysate was used as fermentation medium, which contained around 53 g/L of glucose as carbon source. For all the experiments, the initial pH of the media was adjusted to 3.0.

All the fermentation media, synthetic and hydrolysate, were supplemented with the following nutrients (in g/L): KH_2PO_4 (0.2), $(NH_4)_2SO_4$ (3.0), $MgSO_4 \cdot 7H_2O$ (3.0), $CaCl_2 \cdot 1H_2O$ (0.2), $ZnSO_4 \cdot 7H_2O$ (0.15), $FeSO_4 \cdot 7H_2O$ (0.16), and $CuSO_4 \cdot 5H_2O$ (0.015). To assess the effect of nitrogen concentration on itaconic acid production from cellulosic hydrolysate, the following three different concentrations of $(NH_4)_2SO_4$ were evaluated: 1, 3, and 5 g/L. For comparison, hydrolysate without any nutrient supplementation was also used as fermentation medium.

Fermentation experiments were carried out in 250-mL Erlenmeyer flasks at 35 °C and 200 rpm for 3 to 5 days (72 h to 120 h). A working volume of 50 mL was used in the experiments with pure carbon sources. Experiments performed to assess the effect of aeration on itaconic acid production from cellulosic hydrolysate were carried out with different working volumes varying from 20 to 50 mL in order to result in different air-to-liquid ratios (V_{air}/V_m) as shown in Table 1. All other fermentations from cellulosic hydrolysate medium were performed using a working volume of 30 mL. All experiments were carried out in duplicate.

Table 1. Different air-to-liquid ratios used for the fermentation experiments.

Volume of Medium (mL)	Air Column (mL)	V_f/V_m	V_{air}/V_m
20	230	12.5	11.5
30	220	8.33	7.33
50	200	5.00	4.00

V_f, volume of the flask; V_m, volume of medium; V_{air}, volume of air.

2.5. Analytical Methods and Statistical Analysis

Nitrogen content in the hydrolysates was determined by using an elemental analyzer Vario MACRO cube (Elementar Analysensysteme GmbH, Germany), following the Dumas method.

Cell mass concentration during the fermentations was estimated by dry weight measurement. The fermentation broth was centrifuged at 10,000 rpm for 10 min and the biomass pellet was rinsed two times with deionized water and dried at 60 °C for 48 h. The supernatant of centrifuged samples was used for pH measurement and determination of sugars, itaconic acid, and potential by-products.

The concentrations of glucose, cellobiose, xylose, sucrose, mannose, glycerol, organic acids (itaconic, acetic and formic), 5-hydroxymethylfurfural (5-HMF) and furfural were determined by high-performance liquid chromatography (HPLC) using a Dionex Ultimate 3000 HPLC equipment (Thermo Scientific, USA) coupled with a Biorad Aminex® HPX-87H column (300 × 7.8 mm). For analysis, the column was maintained at 65 °C and a 5 mM H_2SO_4 solution was used as mobile phase at a flow rate of 0.5 mL/min. Sugars, glycerol, and organic acids were detected using a Shodex RI-101 refractive index detector, whereas 5-HMF and furfural were detected using an ultraviolet detector at 254 nm.

Statistical analysis including graphs and quantitative information such as mean and standard deviation was performed using the software OriginPro 9.1.0 (OriginLab Corporation, USA).

2.6. Fermentation Parameters

To evaluate the performance of itaconic acid production by *A. terreus* the following fermentation parameters were considered:

1. Itaconic acid yield per sugar consumed as $Y_{P/S}$ (g/g) = $-\frac{\Delta P}{\Delta S}$;
2. Biomass yield per sugar consumed as $Y_{X/S}$ (g/g) = $-\frac{\Delta X}{\Delta S}$;
3. Itaconic acid productivity as Q_P (g/L·h) = $\frac{\Delta P}{\Delta t}$.

where P is the concentration of itaconic acid (g/L), S is the concentration of sugar (g/L), and t is the fermentation time (h).

3. Results and Discussion

3.1. Carbon Source Assessment

With the aim of identifying potential raw materials for the production of itaconic acid, initial experiments were carried out to evaluate the performance of *A. terreus* when cultivated in different carbon sources. As can be seen in Table 2, hexoses, especially glucose, were the preferred carbon sources used by the strain to produce itaconic acid. Glucose has also been reported as being the preferred carbon source for other filamentous fungi, and this could be explained by the fact that this hexose enables the most direct conversion pathway, eliminating the need for additional biochemical steps [18]. In the present study, the production of itaconic acid from other carbon sources including xylose, sucrose, mannose, and glycerol were clearly lower that that observed from glucose (Table 2). The relevance of glucose as compared with the other carbon sources is also highlighted by the values of itaconic acid yield per substrate consumed, which was of 0.61 g/g of glucose consumed, more than three times higher than that observed for mannose, which was the second best carbon source evaluated in our study.

Table 2. Itaconic acid concentration and yield and biomass yield for the fermentations with *A. terreus* using different carbon sources.

Carbon Source	Itaconic Acid (g/L)	$Y_{P/S}$ (g/g)	$Y_{X/S}$ (g/g)
Glucose	24.85	0.61	0.22
Xylose	1.85	0.09	0.61
Sucrose	2.15	0.05	0.28
Mannose	9.16	0.18	0.40
Glycerol	2.20	0.05	0.43

$Y_{P/S}$, itaconic acid yield per sugar consumed; $Y_{X/S}$, biomass yield per sugar consumed.

It is also interesting to note that the strain presented a very good ability to consume all the carbon sources, including C3 (glycerol), C5 (xylose), and C12 (sucrose), being a potential candidate for use in bioprocesses. However, unlike glucose, the other carbon sources were mainly used for biomass growth instead of itaconic acid formation, which is evidenced by the higher values of biomass yield ($Y_{X/S}$) obtained (Table 2). The biomass formation in glucose was the lowest as compared with the other carbon sources, which was due to the main use of this carbon source for product formation. These results play a crucial role in selecting novel lignocellulosic biomass sources for use on the production of itaconic acid, opening up good possibilities for integrating the production of this acid in a biorefinery.

3.2. Cellulose Pulp as Feedstock for Itaconic Acid Production

Since glucose was the best carbon source for itaconic acid production by *A. terreus*, a cellulose-rich material can be considered to be the ideal candidate for use in this bioprocess as an alternative to replace pure glucose as the carbon source. During recent years, studies have reported the use of cellulose pulp as a substrate for the production of different compounds by fermentation, including ethanol and hydrogen [19,20]. However, there are no studies reporting the use of bleached cellulose pulp for the production of itaconic acid. This study confirms that bleached cellulose pulp can be a feasible feedstock for use in the production of itaconic acid since this material is produced in high amounts in the pulp and paper industries and has attracted great interest for use in the production of valuable compounds (rather than for use in the production of paper only). In addition to its great availability, other important advantages of using bleached cellulose pulp for itaconic acid production include its high content of cellulose (which can be converted into glucose) and the possibility of applying an enzymatic hydrolysis directly, without previous pretreatment, saving time and energy, which can positively impact on the final costs of the itaconic acid production process, making it more economical.

The cellulose pulp used in this study was produced from Eucalyptus wood. The chemical composition of this material was cellulose 89.7% and hemicellulose 10.3% (dry weight). Lignin was present in trace amount. Enzymatic hydrolysis of this material under the conditions used in this study yielded a hydrolysate containing 72.3 g/L of glucose and 14.8 g/L of xylose, representing a rich carbon source for use as fermentation medium by *A. terreus*. Fermentation results from this medium are discussed in the next sections.

3.3. Hydrolysate Sterilization

Sterilization is an important step to be performed when a pure culture has to be used in a bioprocess. However, since sterilization conditions can affect the composition of the hydrolysate, three different methods were evaluated in this study with the aim of selecting the option that most favors the production of itaconic acid by fermentation. Sugar degradation with consequent formation of toxic compounds (5-HMF and furfural) and carbon-to-nitrogen ratio (C:N ratio) were the responses considered to select the best sterilization method.

Although the temperatures used for autoclaving could promote some degradation of glucose and xylose into 5-HMF and furfural, respectively, the formation of these compounds was low for all three sterilization methods evaluated (Figure 1). For the membrane sterilization, the 5-HMF obtained was mostly likely residual and was already present after the enzymatic hydrolysis. It is well known that 5-HMF and furfural are toxic compounds that can affect the microbial performance during fermentation [21]. In the case of itaconic acid production, 5-HMF and furfural concentrations of 0.4 g/L have been reported as being toxic for *A. terreus*, inhibiting the growth, sugar utilization, and product formation [22]. These values are much higher than the concentrations found in the cellulose pulp hydrolysate, independent of the sterilization method applied (Figure 1). Therefore, it can be concluded that none of the sterilization methods was able to promote significant degradation of sugars and formation of toxic compounds at a level unsuitable for fermentation.

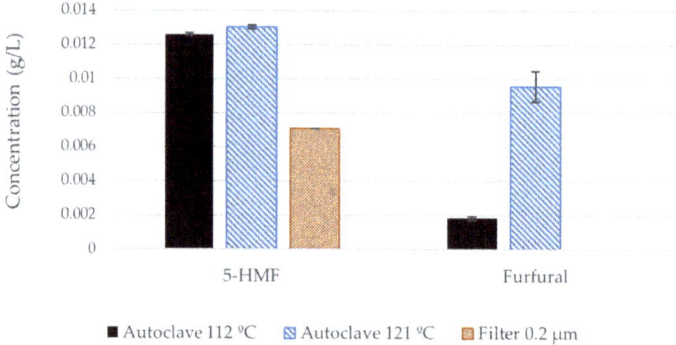

Figure 1. Concentration of 5-HMF and furfural in the cellulose pulp hydrolysate after sterilization by three different methods.

The carbon-to-nitrogen ratio (C:N ratio) is another important characteristic of the hydrolysate that can strongly affect the microbial performance during fermentation, being of high importance to define a suitable C:N ratio to obtain high product yield during fermentation [23]. Analyses of the carbon and nitrogen contents in the cellulose pulp hydrolysate revealed that the carbon composition was not affected by any of the sterilization methods evaluated in this study. However, the nitrogen content was changed, leading to hydrolysates with different C:N ratios (Table 3). Sterilization by filtration clearly resulted in a medium with lower content of nitrogen, which was also visually cleaner and more translucent than the hydrolysates sterilized by autoclaving (figure not shown). According to the literature, nitrogen limitation can be beneficial for the production of organic acid [18]. Since nitrogen is required for biomass production, lack of nitrogen can slow down cell growth, to which some fungi respond by increasing the organic acid production [18]. In addition, high C:N ratios would direct more carbon into the tricarboxylic acid (TCA) cycle, allowing for higher productivities [24]. Ratios that are too high, however, could lead to reduced productivity due to substrate inhibition [18].

Table 3. Carbon and nitrogen contents in the hydrolysate after sterilization by different methods.

Sterilization Method	Carbon (% w/w)	Nitrogen (% w/w)
Membrane filter 0.2 µm	36.16 ± 0.04	0.10 ± 0.01
Autoclave 112 °C	36.35 ± 0.14	0.17 ± 0.00
Autoclave 121 °C	36.30 ± 0.07	0.25 ± 0.06

In order to select the sterilization method more suitable for the production of itaconic acid by *A. terreus*, fermentation assays were performed using the sterilized hydrolysates, with and without nutrient supplementation. No biomass growth or glucose consumption were observed from media without nutrient addition, thus, confirming the necessity of adding nutrients to all the hydrolysates. The results obtained from fermentation of sterilized hydrolysates with nutrient supplementation showed a clear advantage for the method of sterilization by membrane filtration (Figure 2), which provided the highest itaconic acid yield (0.52 g/g) and productivity (0.40 g/L·h) after 72 h (Table 4). It is also interesting to note that glucose consumption and itaconic acid production were maximum at 72 h of fermentation but decreased afterwards, indicating possible consumption of the product when glucose, the main carbon source, was exhausted.

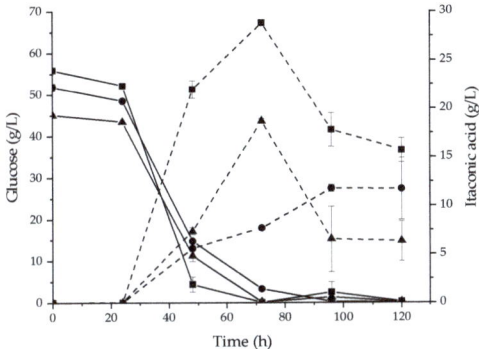

Figure 2. Glucose (solid lines) and itaconic acid (dashed lines) concentrations during the fermentation of the cellulose pulp hydrolysate sterilized by different methods. Membrane filter 0.2 µm (■); autoclave at 112 °C (●); and autoclave at 121 °C (▲). All the hydrolysates were supplemented with nutrients to be used as fermentation medium.

Analysis of the fermentation parameters (Table 4) clearly indicated that the hydrolysate sterilized by the filtration method promoted the best fermentation performance, resulting in the highest values of itaconic acid yield ($Y_{P/S}$) and productivity (Q_P). Such a result could be attributed to the C:N ratio present in the hydrolysate. The filtration method resulted in lower nitrogen content in the hydrolysate, and, as a consequence, in a higher C:N ratio that could have changed the metabolism towards acid production as opposed to fungal growth, thus, explaining the highest production of itaconic acid and the lowest production of biomass obtained from this medium. It is also worth noting that the production of itaconic acid from cellulose pulp hydrolysate sterilized by membrane filtration obtained in the present study compares very well to other studies on the production of this acid from different biomass hydrolysates [25,26]. When corn starch and wheat bran hydrolysates were used as fermentation medium for *A. terreus*, itaconic acid yields of 0.41–0.42 (g/g) were obtained [26].

Table 4. Fermentation parameters obtained for the production of itaconic acid by *A. terreus* from cellulose pulp hydrolysate sterilized by different methods.

Sterilization Method	$Y_{P/S}$ (g/g)	$Y_{X/S}$ (g/g)	Q_P (g/L·h)
Membrane filter 0.2 µm	0.52 ± 0.01	0.13 ± 0.02	0.40 ± 0.01
Autoclave 112 °C	0.23 ± 0.04	0.23 ± 0.17	0.10 ± 0.03
Autoclave 121 °C	0.42 ± 0.02	0.28 ± 0.03	0.26 ± 0.01

$Y_{P/S}$, itaconic acid yield per sugar consumed; $Y_{X/S}$, biomass yield per sugar consumed; and Q_P, itaconic acid productivity.

On the basis of the above, sterilization by membrane filtration was selected as the most suitable sterilization technique as it provided the best results of itaconic acid titer, yield and productivity, and therefore was the sterilization method used in all the subsequent experiments.

3.4. Effect of Aeration on the Production of Itaconic Acid

A sufficient oxygen supply is a fundamental requirement for a successful performance of the microbial strain during fermentation processes. To better understand its effect on the fermentation of cellulose pulp hydrolysate by *A. terreus*, three different aeration conditions were tested in a shake flask setup (Table 1), which were promoted by varying the working volume used in the flasks. A similar setup has been used and discussed in other studies to understand the effect of aeration during fermentation in flasks [27,28]. Results of these experiments are summarized in Table 5.

Table 5. Fermentation parameters obtained during the production of itaconic acid by *A. terreus* from cellulose pulp hydrolysate under different aeration conditions.

V_{air}/V_m	$Y_{P/S}$ (g/g)	$Y_{X/S}$ (g/g)	Q_P (g/L·h)
11.5	0.20 ± 0.02	0.28 ± 0.02	0.17 ± 0.03
7.33	0.52 ± 0.01	0.13 ± 0.02	0.40 ± 0.06
4.00	0.21 ± 0.06	0.29 ± 0.04	0.13 ± 0.06

$Y_{P/S}$, itaconic acid yield per sugar consumed; $Y_{X/S}$, biomass yield per sugar consumed; and Q_P, itaconic acid productivity.

Interestingly, the two boundary conditions, V_{air}/V_m of 11.5 and 4, showed a decreased fermentation performance as compared with that observed for the intermediate condition, V_{air}/V_m of 7.33 (Table 5). In addition, the biomass yield was lower for a V_{air}/V_m of 7.33, revealing that an increased flux of carbon was deviated to the product formation under this oxygen condition. These results indicate that oxygen plays an important role in the production of itaconic acid by *A. terreus* from cellulose pulp hydrolysate. Therefore, selecting the ideal condition is highly important to maximize the product formation since conditions of excess or limitation of oxygen did not provide the best results. According to some authors, interrupting aeration can completely stop the production of itaconic acid by *A. terreus* [29]. Moreover, experiments using different shaking speeds in flasks showed that lowering the RPMs had a negative effect on the production of itaconic acid [22]. On the other hand, research with *Aspergillus niger* revealed that a reduced level of dissolved oxygen has a positive effect on the production of itaconic acid since high levels of dissolved oxygen increase the production of other organic acids such as citric and oxalic acid, which redirects carbon away from itaconic acid production [30]. Therefore, it is important to manage the aeration of the system carefully according to the strain and medium conditions used. Low and high concentrations of dissolved oxygen could both have an adverse effect on the production of itaconic acid. Research is, therefore, required to establish the best oxygen level to be used during fermentation. This is also of great importance for upscaling experiments in bioreactors.

Since a V_{air}/V_m of 7.33 was the oxygen condition that provided the best results of itaconic acid production, this condition was selected and used in the subsequent experiments.

3.5. Effect of C:N Ratio on the Production of Itaconic Acid

Considering that the previous experiments on the sterilization method suggested a significant influence of the C:N ratio on the production of itaconic acid, additional experiments were performed at this step to explore such effect with the aim of selecting conditions able to improve the production of itaconic acid from cellulose pulp hydrolysate. As a first approach, experiments consisted in changing the nitrogen availability in the medium by varying the concentration of $(NH_4)_2SO_4$ added to it. As can be seen in Table 6, the addition of 1 g/L $(NH_4)_2SO_4$ did not provide sufficient nitrogen for the microorganism to properly metabolize the carbon source and convert it into itaconic acid. Better results were obtained for the other two nitrogen concentrations tested, 3 and 5 g/L. From these, supplementation of the medium with 3 g/L $(NH_4)_2SO_4$ gave the best results of itaconic acid production, with yield and productivity of 0.52 g/g and 0.40 g/L·h, respectively. These results confirm that the production of itaconic acid can be improved by using an appropriate C:N ratio. Nitrogen limitation or excess are both non ideal conditions for the metabolism of *A. terreus* go through the itaconic acid formation.

The influence of different nitrogen sources and concentrations on the production of itaconic acid has also been reported in other studies using different microbial strains and fermentation media. For example, the production of itaconic acid by the fungus *Ustilago maydis* in medium containing 200 g/L of glucose was improved when the concentration of NH_4Cl added as nitrogen source was increased from 15 to 75 mM [31]; while the production of itaconic acid by *A. terreus* ATCC 10020 from rice husk hydrolysate containing 15 g/L of glucose was improved when the medium was supplemented with 1.3 g/L NaNO$_3$ and 1.1 g/L (NH$_4$)$_2$SO$_4$ [12]. This makes it possible to conclude that different strains have different nitrogen requirements for their metabolism and, according to the medium used for fermentation, different concentrations of nitrogen should be added to promote the best performance of the strain towards product formation.

Table 6. Effect of the nitrogen concentration on the fermentation parameters obtained during the production of itaconic acid by *A. terreus* from cellulose pulp hydrolysate.

Nitrogen (g/L)	$Y_{P/S}$ (g/g)	$Y_{X/S}$ (g/g)	Q_P (g/L·h)
1	0.01 ± 0.08	0.30 ± 0.25	0.002 ± 0.08
3	0.52 ± 0.01	0.13 ± 0.02	0.40 ± 0.06
5	0.49 ± 0.01	0.21 ± 0.01	0.33 ± 0.00

$Y_{P/S}$, itaconic acid yield per sugar consumed; $Y_{X/S}$, biomass yield per sugar consumed; and Q_P, itaconic acid productivity.

As a second approach to explore the effect of the C:N ratio on the performance of *A. terreus* to produce itaconic acid from cellulose pulp hydrolysate, small changes in the carbon composition were made for a fixed medium supplementation of 3 g/L (NH$_4$)$_2$SO$_4$ (which gave the best results of itaconic acid production in the previous experiments). According to the results, when the initial concentration of glucose in the medium was increased from 45 to 60 g/L, a significant increase in the production of itaconic acid could be observed, which resulted in 2.3 times and 3 times higher values of yield and productivity, respectively (Table 7). These results reinforce that increasing the initial concentration of carbon source is an important strategy to result in a higher production of itaconic acid by *A. terreus*. However, within the scope of this study, higher concentrations of initial glucose were not evaluated since, as the nitrogen supplementation was fixed, increased carbon sources would lead to much higher C:N ratios, which could negatively impact on the production of itaconic acid. For future experiments, higher concentrations of glucose should be tested using an appropriate nitrogen supplementation to offer the ideal C:N balance required by the strain to maximize the formation of itaconic acid from cellulose pulp hydrolysate.

Table 7. Effect of the substrate concentration on the fermentation parameters obtained during the production of itaconic acid by *A. terreus* from cellulose pulp hydrolysate.

Glucose (g/L)	$Y_{P/S}$ (g/g)	$Y_{X/S}$ (g/g)	Q_P (g/L·h)
45	0.27 ± 0.02	0.24 ± 0.02	0.17 ± 0.00
53	0.52 ± 0.01	0.13 ± 0.18	0.40 ± 0.01
60	0.62 ± 0.02	0.18 ± 0.04	0.52 ± 0.02

$Y_{P/S}$, itaconic acid yield per sugar consumed; $Y_{X/S}$, biomass yield per sugar consumed; and Q_P, itaconic acid productivity.

Finally, considering the different strategies evaluated in the present study, sterilization of the cellulose pulp hydrolysate by membrane filtration, medium supplementation with 3 g/L (NH$_4$)$_2$SO$_4$, 60 g/L of initial glucose concentration, and oxygen availability of 7.33 (volume of air/volume of medium) were the most suitable to maximize the production of itaconic acid by *A. terreus*, resulting in a production of 37.5 g/L, corresponding to a yield of 0.62 g/g glucose consumed, and productivity of 0.52 g/L·h. These values compare very well to other recent studies on the production of itaconic acid by *A. terreus* from different lignocellulosic feedstocks (Table 8) and confirm the feasibility of using bleached cellulose pulp for this application. These results can still be improved by optimization of the fermentation conditions using a bioreactor setup, which will be investigated in a next study.

Table 8. Production of itaconic acid by *A. terreus* from different lignocellulosic feedstocks. All the values correspond to experiments in flasks.

Feedstock, Glucose Concentration (g/L)	Biomass Processing	*Aspergillus terreus* Strain	Itaconic Acid (g/L)	Reference
Corn stover, 54	Steam explosion pretreatment + Enzymatic hydrolysis	CICC 2452, mutant AT-90	19.3	[10]
Wheat chaff, ≈ 50	Alkaline pretreatment + Enzymatic hydrolysis	DSM 23081	27.7	[11]
Rice husks, 35	Acid pretreatment + detoxification	ATCC 10020	1.9	[12]
Corn cobs, 7.7	Enzymatic hydrolysis	DSM 826	0.9	[13]
Bleached cellulose pulp, 60	Enzymatic hydrolysis	NRRL 1960	37.5	Present study

4. Conclusions

To accelerate the use of lignocellulosic feedstocks in fermentative processes it is crucial to select the right biomass for the desired process. This study demonstrated that bleached cellulose pulp is a potential candidate for use on the production of itaconic acid by fermentation since it is highly rich in cellulose that can easily be converted into glucose by enzymatic hydrolysis without requiring a previous step of pretreatment. This is in fact an important aspect contributing to the economic feasibility of the fermentation process for itaconic acid production, since pretreatment is usually a very energy-intensive step and impacts significantly on the final costs of the process and the product.

Other important findings of this study were related to the fermentation of the glucose-rich hydrolysate produced from cellulose pulp. Due to the presence of glucose as the main component, no lignin or sugar degradation products in the medium that could negatively affect the strain performance, the only concern is to establish conditions that can direct the metabolism of the strain towards the product formation with minimum use of carbon source for biomass growth. Within this study, it was demonstrated that the C:N ratio and the oxygen availability play important roles in the production of itaconic acid by *A. terreus* from cellulose pulp hydrolysate and should be carefully considered in subsequent studies in a bioreactor setup. Increasing the initial carbon source was also a strategy able to result in better production of itaconic acid and should be further explored taking into account the use of an appropriate C:N ratio during the experiments. Finally, sterilization of the hydrolysate before fermentation is a required step that can also affect the medium composition leading to an unbalance in the C:N ratio, being the sterilization by membrane filtration the most recommended method to result in a better fermentation performance.

Author Contributions: Conceptualization, S.I.M.; Methodology, S.I.M., G.D., and C.K.Y.; Investigation, M.C. and P.D.; Resources: S.I.M.; Data curation, A.A.J.K., M.C., P.D., G.D., C.K.Y., and S.I.M.; Writing—original draft preparation, A.A.J.K., G.D., and S.I.M.; Writing—review and editing, S.I.M.; Supervision, S.I.M., G.D. and C.K.Y.; Project administration: S.I.M.; Funding Acquisition, S.I.M. All authors have read and agreed to the published version of the manuscript.

Funding: This work was supported by the Novo Nordisk Foundation, Denmark (grant number NNF10CC1016517).

Acknowledgments: Special thanks to Suzano S/A (Brazil) for supplying the cellulose pulp used in this research, and to Novozymes (Denmark) for providing the enzyme concentrate Cellic® CTec2.

Conflicts of Interest: The authors declare no conflict of interest.

References

1. Dragone, G.; Kerssemakers, A.A.J.; Driessen, J.L.S.P.; Yamakawa, C.K.; Brumano, L.P.; Mussatto, S.I. Innovation and strategic orientations for the development of advanced biorefineries. *Bioresour. Technol.* **2020**, *302*, 122847. [CrossRef] [PubMed]
2. De Jong, E.; Stichnothe, H.; Bell, G.; Jørgensen, H. *Bio-Based Chemicals. A 2020 Update.* IEA Bioenergy Task 42. Available online: https://www.academia.edu/42073867/Bio-Based_Chemicals_A_2020_Update (accessed on 26 February 2020).

3. Mussatto, S.I.; Dragone, G.M. Biomass pretreatment, biorefineries, and potential products for a bioeconomy development. In *Biomass Fractionation Technologies for a Lignocellulosic Feedstock Based Biorefinery*; Elsevier: Amsterdam, The Netherlands, 2016; pp. 1–22. ISBN 978-0-12-802323-5.
4. Teleky, B.-E.; Vodnar, D.C. Biomass-derived production of itaconic acid as a building block in specialty polymers. *Polymers* **2019**, *11*, 1035. [CrossRef] [PubMed]
5. Analysts Global Industry Inc. Itaconic Acid (IA)—A Global Strategic Business Report. Available online: https://www.strategyr.com/pressMCP-6465.asp (accessed on 26 February 2020).
6. Transparency Market Research. Rising Demand in Manufacturing of Superabsorbent Polymers to Inundate Itaconic Acid Market. Available online: https://www.transparencymarketresearch.com/pressrelease/itaconic-acid-market.htm (accessed on 26 February 2020).
7. Matthis, S. Pulp Price Erosion Persist in October. Available online: https://www.pulpapernews.com/20191028/10851/pulp-price-erosion-persists-october (accessed on 26 February 2020).
8. Suzano. NDR Boston and Itaú BBA 14th Annual LatAm CEO Conference. Available online: http://ri.suzano.com.br/ptb/7597/ApresentaoNDR_Conferncia_USA_EN.pdf (accessed on 26 February 2020).
9. Regestein, L.; Klement, T.; Grande, P.; Kreyenschulte, D.; Heyman, B.; Maßmann, T.; Eggert, A.; Sengpiel, R.; Wang, Y.; Wierckx, N.; et al. From beech wood to itaconic acid: Case study on biorefinery process integration. *Biotechnol. Biofuels* **2018**, *11*, 279. [CrossRef] [PubMed]
10. Li, X.; Zheng, K.; Lai, C.; Ouyang, J.; Yong, Q. Improved itaconic acid production from undetoxified enzymatic hydrolysate of steam-exploded corn stover using an *Aspergillus* terreus mutant generated by atmospheric and room temperature plasma. *BioResources* **2016**, *11*, 9047–9058. [CrossRef]
11. Krull, S.; Eidt, L.; Hevekerl, A.; Kuenz, A.; Prüße, U. Itaconic acid production from wheat chaff by *Aspergillus terreus*. *Process Biochem.* **2017**, *63*, 169–176. [CrossRef]
12. Pedroso, G.B.; Montipó, S.; Mario, D.A.N.; Alves, S.H.; Martins, A.F. Building block itaconic acid from left-over biomass. *Biomass Convers. Biorefinery* **2017**, *7*, 23–35. [CrossRef]
13. Jimenez-Quero, A.; Pollet, E.; Zhao, M.; Marchioni, E.; Averous, L.; Phalip, V. Itaconic and fumaric acid production from biomass hydrolysates by *Aspergillus* strains. *J. Microbiol. Biotechnol.* **2016**, *26*, 1557–1565. [CrossRef]
14. Sluiter, A.; Hames, B.; Ruiz, R.; Scarlata, C.; Sluiter, J.; Templeton, D. *Determination of Ash in Biomass*; Technical Report NREL/TP-510-42622; National Renewable Energy Laboratory: Golden, CO, USA, 2008.
15. Sluiter, A.; Hames, B.; Ruiz, R.; Scarlata, C.; Sluiter, J.; Templeton, D.; Crocker, D. *Determination of Structural Carbohydrates and Lignin in Biomass*; Technical Report NREL/TP-510-42618; National Renewable Energy Laboratory: Golden, CO, USA, 2012.
16. Sluiter, A.; Ruiz, R.; Scarlata, C.; Sluiter, J.; Templeton, D. *Determination of Extractives in Biomass*; Technical Report NREL/TP-510-42619; National Renewable Energy Laboratory: Golden, CO, USA, 2008.
17. Adney, B.; Baker, J. *Measurement of Cellulase Activities*; Technical Report NREL/TP-510-42628; National Renewable Energy Laboratory: Golden, CO, USA, 2008.
18. Mondala, A.H. Direct fungal fermentation of lignocellulosic biomass into itaconic, fumaric, and malic acids: Current and future prospects. *J. Ind. Microbiol. Biotechnol.* **2015**, *42*, 487–506. [CrossRef]
19. Aierkentai, G.; Liang, X.; Uryu, T.; Yoshida, T. Effective saccharification and fermentation of kraft pulp to produce bioethanol. *J. Fiber Sci. Technol.* **2017**, *73*, 261–269. [CrossRef]
20. Moreau, A.; Montplaisir, D.; Sparling, R.; Barnabé, S. Hydrogen, ethanol and cellulase production from pulp and paper primary sludge by fermentation with *Clostridium thermocellum*. *Biomass Bioenergy* **2015**, *72*, 256–262. [CrossRef]
21. Mussatto, S.I.; Roberto, I.C. Alternatives for detoxification of diluted-acid lignocellulosic hydrolyzates for use in fermentative processes: A review. *Bioresour. Technol.* **2004**, *93*, 1–10. [CrossRef] [PubMed]
22. Saha, B.C.; Kennedy, G.J.; Bowman, M.J.; Qureshi, N.; Dunn, R.O. Factors affecting production of itaconic acid from mixed sugars by *Aspergillus terreus*. *Appl. Biochem. Biotechnol.* **2019**, *187*, 449–460. [CrossRef] [PubMed]
23. Liu, Z.; Feist, A.M.; Dragone, G.; Mussatto, S.I. Lipid and carotenoid production from wheat straw hydrolysates by different oleaginous yeasts. *J. Clean. Prod.* **2020**, *249*, 119308. [CrossRef]
24. Casas López, J.L.; Sánchez Pérez, J.A.; Fernández Sevilla, J.M.; Acién Fernández, F.G.; Molina Grima, E.; Chisti, Y. Production of lovastatin by *Aspergillus terreus*: Effects of the C:N ratio and the principal nutrients on growth and metabolite production. *Enzyme Microb. Technol.* **2003**, *33*, 270–277. [CrossRef]

25. Reddy, C.S.K.; Singh, R. Enhanced production of itaconic acid from corn starch and market refuse fruits by genetically manipulated *Aspergillus terreus* SKR10. *Bioresour. Technol.* **2002**, *85*, 69–71. [CrossRef]
26. Kuenz, A.; Krull, S. Biotechnological production of itaconic acid—Things you have to know. *Appl. Microbiol. Biotechnol.* **2018**, *102*, 3901–3914. [CrossRef]
27. Klöckner, W.; Büchs, J. Advances in shaking technologies. *Trends Biotechnol.* **2012**, *30*, 307–314. [CrossRef]
28. Silva, J.P.A.; Mussatto, S.I.; Roberto, I.C. The influence of initial xylose concentration, agitation, and aeration on ethanol production by *Pichia stipitis* from rice straw hemicellulosic hydrolysate. *Appl. Biochem. Biotechnol.* **2010**, *162*, 1306–1315. [CrossRef]
29. Gyamerah, M.H. Oxygen requirement and energy relations of itaconic acid fermentation by *Aspergillus terreus* NRRL 1960. *Appl. Microbiol. Biotechnol.* **1995**, *44*, 20–26. [CrossRef]
30. Li, A.; Pfelzer, N.; Zuijderwijk, R.; Brickwedde, A.; van Zeijl, C.; Punt, P. Reduced by-product formation and modified oxygen availability improve itaconic acid production in *Aspergillus niger*. *Appl. Microbiol. Biotechnol.* **2013**, *97*, 3901–3911. [CrossRef]
31. Maassen, N.; Panakova, M.; Wierckx, N.; Geiser, E.; Zimmermann, M.; Bölker, M.; Klinner, U.; Blank, L.M. Influence of carbon and nitrogen concentration on itaconic acid production by the smut fungus *Ustilago maydis*. *Eng. Life Sci.* **2014**, *14*, 129–134. [CrossRef]

© 2020 by the authors. Licensee MDPI, Basel, Switzerland. This article is an open access article distributed under the terms and conditions of the Creative Commons Attribution (CC BY) license (http://creativecommons.org/licenses/by/4.0/).

Article

Temporal Aspects in Emission Accounting—Case Study of Agriculture Sector

Lelde Timma [1],*, Elina Dace [2,3] and Marie Trydeman Knudsen [1]

[1] Department of Agroecology, Aarhus University, Blichers Allé 20, DK-8830 Tjele, Denmark; mariet.knudsen@agro.au.dk
[2] Institute of Microbiology and Biotechnology, University of Latvia, 1 Jelgavas Street, LV1004 Riga, Latvia; elina.dace@lu.lv
[3] Research Department, Riga Stradins University, 16 Dzirciema Street, LV1007 Riga, Latvia
* Correspondence: lelde.timma@agro.au.dk

Received: 15 January 2020; Accepted: 10 February 2020; Published: 12 February 2020

Abstract: Complex relations link climate change and agriculture. The vast majority of the studies that are looking into the quantification of the climate impacts use the Global Warming Potential (GWP) for a 100-year time horizon (GWP100) as the default metrics. The GWP, including the Bern Carbon Cycle Model (BCCM), was proposed as an alternative method to take into consideration the amount and time of emission, and the fraction of emissions that remained in the atmosphere from previous emission periods. Thus, this study aims to compare two methods for GHG emission accounting from the agriculture sector: the constant GWP100 and the time dynamic GWP100 horizon obtained by using the BCCM to find whether the obtained results will lead to similar or contradicting conclusions. Also, the effect of global temperature potential (GTP) of the studied system is summarized. The results show that the application of the BCCM would facilitate finding more efficient mitigation options for various pollutants and analyze various parts of the climate response system at a specific time in the future (amount of particular pollutants, temperature change potential). Moreover, analyze different solutions for reaching the emission mitigation targets at regional, national, or global levels.

Keywords: climate modelling; climate change; climate policy; emission accounting; global warming potential; global temperature change potential; greenhouse gas emissions; impulse response function; Bern Carbon Cycle model; climate impacts of agriculture system

1. Introduction

The growing global population leads to an increasing need for resources—food, energy, and materials. This expanding demand forces the shift from a fossil-based linear economy to a sustainable biobased economy. Bioeconomy demands biological feedstock that has the potential to generate a spectrum of bio-based products by involving multidisciplinary areas of science, management, and engineering [1]. Agriculture is the primary supply of nutrition and bioenergy and a substantial contributor to the bioeconomy. Yet, agriculture is also linked with environmental, economic, and social aspects of climate change. For example, climate change affects the productivity of the agriculture sector, and thus change in the agricultural practices feedback to the greenhouse gas (GHG) balance. Therefore, climate change and agriculture are linked by complex relations, which can be difficult to define or measure [2].

The vast majority of the studies that are looking into the quantification of the climate impacts use the Global Warming Potential (GWP) for a 100-year time horizon as the default metrics. Since the development of GWP metric in the early nineties, there have been updates only on the numerical value of this metric, rather than the development of the assessment methodology itself [3].

The use of GWP is an accepted measure within the Kyoto Protocol to the United Nations Framework Convention on Climate Change as a measure to weigh the impact of climate due to the emissions of GHGs. Although the use of GWP has received various criticism due to underlying assumptions, it became widely accepted measure because of transparency and ease of use [4]. There were numerous alternative methods developed to substitute use of GWP, such as Global Temperature Change Potential (GTP) by Shine et al. [4], Global Warming Potential using cumulative CO2 forcing-equivalent (GWP*) by Allen et al. [5] and other normalized point and integration metrics (see the review by Levasseur et al. [6]). Current studies on climate science show that processes occurring in the natural environment sometimes cannot be reasonably well quantified using a single value for measuring the impact created in the 100-year perspective. This quantification using a single value cannot be done due to the non-linear nature of the emission dissipation in various environments that leads to spatial and temporal heterogeneities. Misinterpretation of these effects can lead to policy decisions that underestimate the impacts of the emissions with a short lifetime and with a dominating local pollution effect. An example of this phenomenon is given in the thesis work by Shimako [7] and published in the paper by Shimako et al. [8]. This work shows that the same amount of emissions might have different influence if the timing of emissions is considered. Another limitation of GWP is that it estimates the forcing of the climate but does not characterize the impact of climate dynamics. Although climate dynamics are included in the global temperature change potentials (GTPs), they are not intended to illustrate the influence of radiative forcing and enable a qualitative interpretation of causes [9].

The GWP, including the Bern Carbon Cycle Model (BCCM), was proposed as an alternative method to take into consideration both amount and time of emission, as well as the fraction of emissions remaining in the atmosphere from previous emission periods. Furthermore, BCCM considers the effect of GHG emissions estimated as a continuous pattern that handles removals (via sinks) and addition of new emissions to the "stock" of the atmosphere hence also considering the climate system response to emissions.

Thus, this study aims to compare two methods for GHG emission accounting from the agriculture sector. Firstly, the constant GWP values for a 100-year time horizon (GWP100) and, secondly, the time dynamic GWP values for a 100-year time horizon obtained by using the BCCM to find whether the obtained results will lead to similar or contradicting conclusions. Also, the effect of global temperature potential (GTP) of the studied system is summarized.

The agriculture sector is the world's leading source of non-CO_2 GHG emissions and the second-largest GHGs emission source overall. On the global scale, in 2010, the non-CO_2 GHG emissions from agriculture accounted for 10–12% of the total annual anthropogenic emissions or 5.2–5.8 Gt CO_2 eq. [10]. The same share of the GHGs emissions from agriculture is also evident in the European Union (EU), where 0.442 Gt CO_2 eq. originated from agriculture that corresponds to around 10% of the total annual GHG emissions in the EU. Based on the EU strategy for a low-carbon economy by 2050 [11], non-CO_2 GHG emissions or GHG emissions not covered by the EU Emissions Trading Scheme (non-ETS) should be cut down by 30% in the comparison to the emission in 2005 [12]. Thus, these emission reductions should also substantially relay on the emission cutbacks in the agriculture sector. Therefore, a lot of research is put into the evaluation of emission mitigation potential in the EU Member states, including a detailed analysis of the agriculture sector [13,14].

2. Materials and Methods

In this study, we compare the method for "traditionally" accounted for GHG emissions and the GWP method that is used in the BCCM. By "traditional" accounting, we mean the use of GWP and multiplying this value of GWP with the corresponding amount of GHG emission of the specific pollutant, given in IPCC guidelines for national GHG inventories and Section 2.1 of this article. As a method using the BCCM, we refer to the use of impulse response function and decay of pulse emissions, which is also covered in the latest (fifth) IPCC assessment report [15] given in detail in Section 2.2 of this article.

2.1. Method Used in IPCC Guidelines for the National GHG Inventories

The Intergovernmental Panel on Climate Change (IPCC) has created an internationally agreed methodology for the assessment of GHG emissions from numerous sectors, including agriculture [16]. The assessed GHG emissions (inventories) are used for approximate anthropogenic emissions by sources and removals by sinks of GHGs. Each year, countries, including the EU Member States, submit individual reports on the inventory of national emissions to the United Nations Framework Convention on Climate Change (UNFCCC). These reports are used to account for the current state of the GHG emissions, see global trends, and make forecasts. Moreover, IPCC has accomplished that based on the reports, governments take action towards the mitigation of climate change [2].

The GHGs included in the IPCC guidelines are carbon dioxide (CO_2), nitrous oxide (N_2O), methane (CH_4), and fluorinated gases (HFCs, PFCs, and SF6). To quantify, compare, and analyze the emissions, promote mitigation options, and design sustainable policy strategies, a default emission metric, CO_2 equivalent (eq.), has been developed. The CO_2 eq. is obtained by multiplying the estimated amount of non-CO_2 GHG emission (component i) by a coefficient of that specific non-CO_2 emission for a fixed time horizon (usually 20 or 100 years) and summing the obtained individual CO_2 eq. values into an aggregated emission metric. The coefficient used is known as the global warming potential (GWP)—"an index, based on radiative properties of greenhouse gases, measuring the radiative forcing following a pulse emission of a unit mass of a given greenhouse gas in the present-day atmosphere integrated over a chosen time horizon, relative to that of CO_2" [15].

Based on new scientific and technical knowledge, the guidelines have had two major revisions since their 1996-version: "2006 IPCC Guidelines for National Greenhouse Gas Inventories" and the recently adopted "2019 Refinement to the 2006 IPCC Guidelines for National Greenhouse Gas Inventories". As a result of refinement, new sources, and pollutants, and updates to the previously published methods, have been included in the guidelines. Also, the numeric value of the global warming potential has been updated since its first introduction in the early nineties (see Table 1). Changes in the GWP value have been made due to improved scientific knowledge and updated estimates of the energy absorption, lifetime, impulse response functions. Estimates on impulse response functions or radiative efficiencies of GHGs vary because of changing atmospheric concentrations of GHGs that result in a change in the energy absorption of one additional ton of a gas relative to another [15]. Since GWP of CO_2 is used as a reference, GWP of CO_2 equals one and remains constant regardless of the used time frame. Therefore, any parameter adjustments for CO_2 will affect all results of the assessments done for other GHGs [15].

Table 1. Global Warming Potential (GWP) values since the first assessment report, unitless [15,17].

Time Horizon	Gas	IPCC 1990	IPCC 1996	IPCC 2000	IPCC 2006	IPCC 2013	IPCC 2013 [1]
20 years	CH_4	63	56	62	72	84	86
20 years	N_2O	270	280	275	289	264	268
100 years	CH_4	21	21	23	25	28	34
100 years	N2O	290	310	296	298	265	298

[1] With climate-carbon feedbacks [15].

The selection of time horizon has a substantial influence on the GWP values; hence, the estimated contribution to the total emissions by the component i [15]. The majority of studies and agreements use the 100-year time horizon. Also, the Kyoto Protocol and the Paris Agreement are based on GWPs from pulse emissions over a 100-year time horizon. And, indeed, for many objectives, the 100-year time horizon is applicable, especially taking into account the long-term effects of GHG emissions and the need for long-term modelling of future temperatures. Yet, the 20-year time horizon might be a more appropriate choice for regional/national strategies or mid-term modelling applications as their focus typically is on a much shorter time frame [18]. In addition, tropospheric temperatures that are more relevant for regional/national decision makers may show more rapid changes in radiative forcing, which creates a situation when the choice of a short time horizon is more fitting [19].

Although the use of the GWP is considered a relatively simple and easy-to-use method, in recent years, it has received some criticism. For example, Peters et al. [20] and Ledgard and Reisinger [21] point to the contrary outcomes that frequently are the result of a negligent and only implicit value judgment of the time horizon to be selected. Joos et al. [22] and Tanaka et al. [23] emphasize the importance of the proper selection of the time horizon in the determination of metrics variability, especially for the GWP value given to gases with a relatively short lifetime (e.g., CH_4) compared to gases with long lifetimes (e.g., CH_2). Cherubini et al. [3] state that uses of GWP values omits impacts from short-term gases and biophysical factors arising from changes in land cover, as well as overlooks the temporal and spatial heterogeneities of the climate system response to GHG emissions. Also, Skytt et al. [18] state that with GWP, a difficulty exists on how to value CH_4 emissions with CO_2 and that the use of GWP values provides "static" information expressed as the radiative forcing potential at a specific time horizon. Finally, the phenomenon of overall acceptance and use of the GWP as a metric for GHG emission accounting by policy developers and scientific community is surprising, considering that since its first introduction by the IPCC in the first assessment report [17], it has had no direct estimation of any climate system responses or direct link to policy goals (Myhre et al. [15]; Cherubini et al. [3]). Hence, the potential use of alternative methods have been proposed and extensively discussed in the literature (see, e.g., Cherubini et al. [3], Levasseur et al. [6], and Skytt et al. [18]). In Section 2.2., a method based on the BCCM is presented as an alternative approach for estimating the temporal effects of GHGs on the climate system.

2.2. Method Used in the Bern Carbon Cycle Model (BCCM)

The BCCM describes the decay pattern of GHGs in the atmosphere [24], i.e., both the amount and time of emission are considered, as well as the fraction of emissions remaining in the atmosphere from previous emission periods. Moreover, not constant values are taken (as 20 or 100 years), but the effect of emissions is estimated as a continuous pattern that considers removals (via sinks) and addition of new emissions to the "stock" of the atmosphere, hence also considering the climate system response to emissions. The BCCM targets several aspects of the climate impact cause-effect loop (see Figure 1).

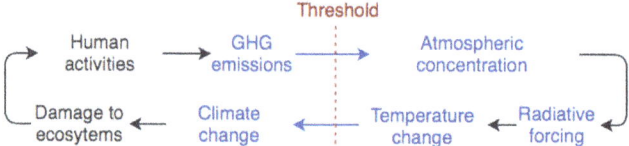

Figure 1. The cause-effect loop of greenhouse gas (GHG) emissions and climate change in blue given segments of the loop that are studied in this article. Figure adapted from Fuglestvedt et al. [25].

The application of climate impulse response models for GHG emissions has been developed by Levasseur et al. [26] and Joos et al. [22]. The impulse response function (IRFs) are usually used in two ways: to describe the decay of atmospheric concentration of pulse emissions or to express global temperature changes due to pulse radiative forcing [27]. The effect of IRF inclusion in the climate response model is graphically represented in Figure 2.

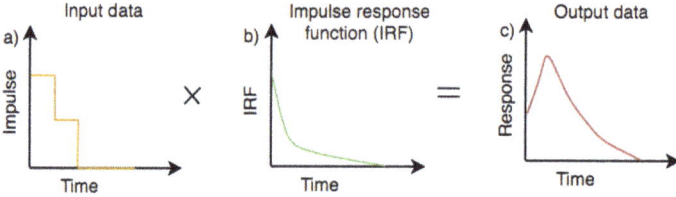

Figure 2. Conversion of input data using impulse response function (*IRF*), adapted from Shimako [7].

Pulse emission is the emission of 1 kg of pollutants at the time $t = 0$. When pulse emission is released to the atmosphere, it serves as an impulse to the complex set of behavioural reactions that occur in climate systems.

These climate responses are condensed into simplified mathematical models that use impulse response function (IRF) [27] given as a response of the temporal temperature to a sudden unit pulse of radiative forcing [28],

$$y_i(t) = \int_0^t x_i(t) IRF_i(t) dt \qquad (1)$$

where $y_i(t)$ is the environmental impact of the pollutant i at the time step t, x is the emitted amount of the pollutant i, and IRF_i is the impulse response function of the pollutant i [29],

$$AGWP_i(H) = \int_0^t RF_i(t) dt = A_i R_i \qquad (2)$$

where $AGWP_i$ is the absolute global warming potential of pollutant i (W·m^{-2}kg^{-1}·year), RF_i is the radiative forcing occurring due to a pulse emission of pollutant i emitted to the atmosphere at time horizon H (W·m^{-2}). RF is the function of specific radiative forcing (A_i, W·m^{-2}kg^{-1})—the ability to increase RF when the unit of the specific pollutant's i mass increases in the atmosphere (see Table 2 for numerical values), and the fraction of pollutant's mass remaining in the atmosphere after the pulse emission of the pollutant i (R_i). The fraction of pollutant's i mass remaining in the atmosphere at the time moment t ($R_i(t)$) is given as a simple exponential decay function:

$$R_i(t) = \exp(-t/\tau_i) \qquad (3)$$

where τ_i is the time needed for the pulse emission of pollutant i to converge to zero concentration, known as perturbation lifetime (years) [22], for CO_2, CH_4, and N_2O emissions, the pattern of R is substantially different over a 1000 years' perspective (see Figure 3).

Table 2. Specific radiative forcing (A_i), perturbation lifetimes (τ_i), and parameter a_i values for the calculation of the pollutant's fraction remaining in the atmosphere (R_i) [3,15,22,29].

Pollutant	Constants		
	A_i, W·m^{-2}kg^{-1}	τ_i, Years	a_i, Unitless
CH_4	1.82·10^{-13}	12.4	-
N_2O	3.88·10^{-13}	121.0	-
CO_2	1.7517·10^{-15}	- 394.40, τ_1 36.54, τ_2 4.304, τ_3	0.2173, a_0 0.2240, a_1 0.2824, a_2 0.2763, a_3

Most of the pollutants follow single exponential decay, while for the CO_2, the behaviour is given with more complex equations [22]. Hence, also the fraction of various GHGs remaining in the air varies by nature. As seen in Figure 3, for CH_4, the decay is much faster, while almost a quarter of the CO_2 emitted in the year 0 is still present in the atmosphere even after 1000 years.

While the perturbation lifetime and specific radiative forcing are known constants for some of the emissions, such as CO_2, CH_4, N_2O, and others (see Table 2 for numerical values). The fraction of CO_2 pulse emission remaining in the atmosphere cannot be represented by a single constant and a simple exponential decay function, as in the case of CH_4 and N_2O. The fraction of CO_2 pulse emission remaining in the atmosphere follows approximation by a sum of exponential functions:

$$R_{CO_2}(t) = a_0 + \sum_{i=1}^{3} a_i \exp(-t/\tau_i) \qquad (4)$$

Global warming potential for pollutant i at time t ($GWP_i(t)$) is calculated by referring absolute global warming potential of the pollutant i ($AGWPi$), to the $AGWP$ of the reference gas, usually CO_2, and integrating it over time period t:

$$GWP_i(t) = AGWP_i(t)/AGWP_{CO_2}(t) = \int_0^t RF_i(t)dt / \int_0^t RF_{CO_2}(t)dt \tag{5}$$

The change of normalized GWP values and absolute GWP values over 100 years in the case of pulse emissions of CO_2, N_2O, and CH_4 are given in Figures 4 and 5, respectively.

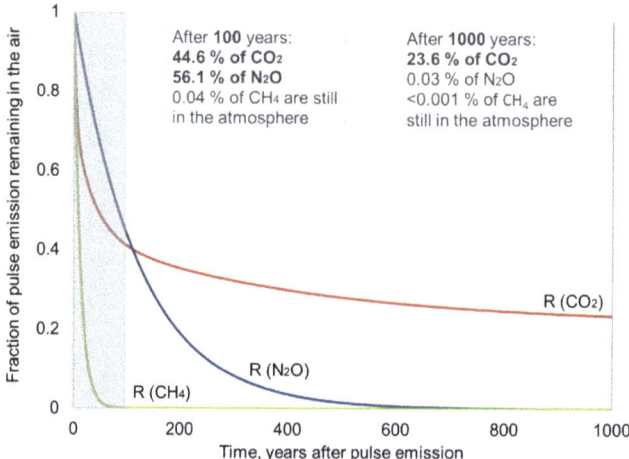

Figure 3. The fraction of pulse emissions at year zero remaining for greenhouse gas emissions of CO_2, N_2O, and CH_4 in 1000 years' time frame.

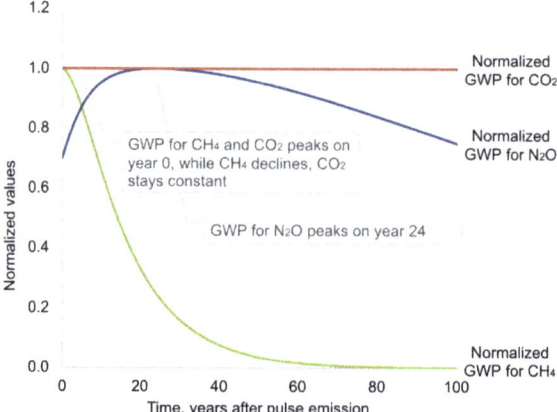

Figure 4. Normalized GWP values as a response to emission of CO_2, N_2O, and CH_4 at year zero. The values are normalized to the maximum value of the corresponding GWP of each gas, unitless.

As can be seen in Figures 4 and 5, the trendlines for the emissions of CH_4 and N_2O are both of different natures and different magnitude, while CO_2 constant values of 1 are assumed.

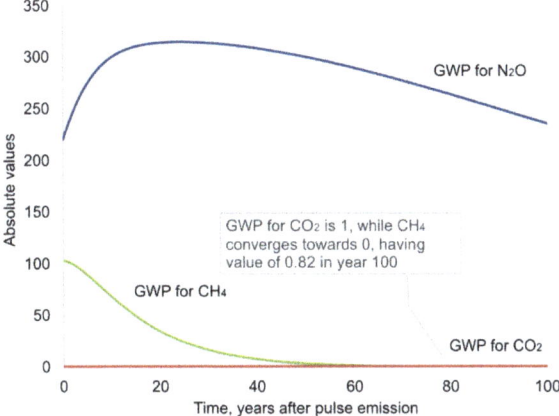

Figure 5. Absolute GWP values as a response to emission of CO_2, N_2O, and CH_4 at year zero, unitless.

Global temperature potential for pollutant i at time t ($GTP_i(t)$) is calculated by referring absolute global temperature potential of the pollutant i ($AGTPi$), to the $AGTP$ of the reference gas, usually CO_2, and integrating it over time period t:

$$GTP_i(t) = AGTP_i(t)/AGTP_{CO_2}(t) \quad (6)$$

where the absolute global temperature change potential of pollutant i in the time horizon H ($AGTP_i(H)$, K·kg^{-1}) [4,30] is calculated as:

$$AGTP_i(H) = \int_0^H RF_i(t) R_T(H-t) dt \quad (7)$$

where R_T is the climate response (K·m^2·W^{-1}·kg^{-1}), H is the time horizon over which the absolute global temperature change potential is calculated (years). R_T is given by the sum of exponentials:

$$R_T(t) = \sum_{j=1}^{M} (c_j/d_j) exp(-t/d_j) \quad (8)$$

where c_j is climate sensitivity (K·(W·m^{-2})$^{-1}$), and d_j is response time (years) (see Table 3 for numerical values). In this equation, the first term is the reaction of the mixed layer in the ocean to a forcing; the second term is the reaction of the deep layer in the ocean. Two exponential terms based on Boucher and Reddy for the non-CO_2 greenhouse gases and CO_2 are given in Equations (9) and (10), respectively.

$$AGTP_i(H) = A_i \sum_{j=1}^{2} \tau c_j/(\tau - d_j) \left(exp(-H/\tau) - exp(-H/d_j) \right) \quad (9)$$

$$AGTP_{CO_2}(H) = A_{CO_2} \sum_{j=1}^{2} \left[a_0 c_j \left(1 - exp\left(-\frac{H}{d_j}\right)\right) \right. \\ \left. + \sum_{i=1}^{3} \frac{a_i \tau_i c_j}{\tau_i - d_j} \left(exp\left(-\frac{H}{\tau_i}\right) - exp\left(-\frac{H}{d_j}\right) \right) \right] \quad (10)$$

Table 3. Values of the climate sensitivity and response time coefficients [15].

Coefficient	1st Term	2nd Term
c_j, K(W·m^{-2})$^{-1}$	0.631	0.429
d_j, years	8.4	409.5

The concept of GTP was first introduced by Shine et al. [4] and further discussed in Shine et al. [31]. The change of normalized GTP values and absolute GTP values over 100 years in the case of pulse emissions of CO_2, N_2O, and CH_4 are given in Figures 6 and 7, respectively.

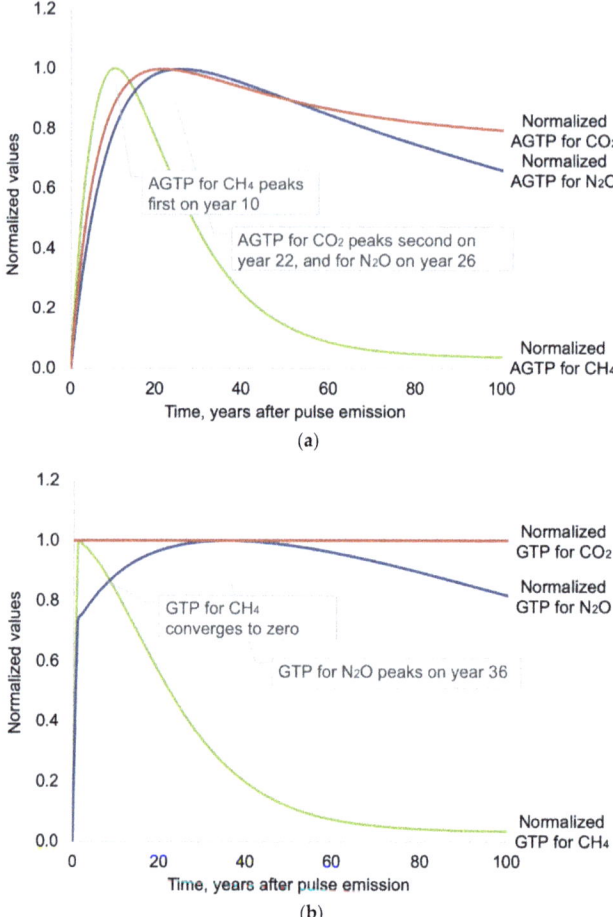

Figure 6. Normalized (**a**) absolute global temperature change potential (AGTP) (**b**) GTP values as a response to emission of CO_2, N_2O, and CH_4 at year zero. The values are normalized to the maximum value of the corresponding AGTP or GTP of each gas, unitless.

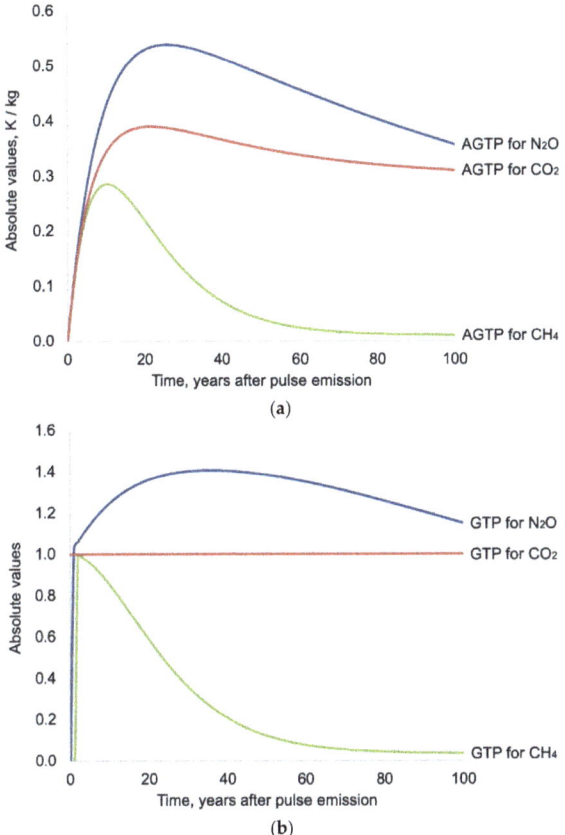

Figure 7. Absolute (**a**) AGTP and (**b**) GTP (unitless) values as a response to emission of CO_2, N_2O, and CH_4 at year zero.

2.3. Case Study—Agriculture Sector in Latvia

In the agriculture sector, aggregated annual GHG emissions is a commonly used measure to characterize pressures and risks that GHGs produced on an ecosystem. The total rate of GHG emissions given as t CO_2 eq. from agriculture per country per year is estimated by following the IPCC guidelines for national GHG inventory [32]. The main contributors to GHG emissions from the agriculture sector are methane (CH_4) and nitrous oxide (N_2O). Livestock enteric fermentation and addition of fertilizers to soils represent the largest emission sources, livestock manure management being a smaller source. In this study, the agricultural GHG emission results obtained and presented by Dace et al. [2] are used. In their study, Dace et al. [2] developed a system dynamics model of the Latvian agriculture sector and followed the IPCC guidelines for national GHG inventories [16] to calculate the sectoral GHG emissions.

The model included the following elements that usually create agricultural systems in the majority of countries: land management, production of livestock and crops, management of manure, fertilization of soil, also various decisions, such as choice of the practices of manure management and the type of crops produced. Thus, the interlinkages and complexity of the sector were simulated. The model was validated against the historic data and used for making GHG emission projections until 2030. In this study, we use the amount of GHG emissions estimated by Dace et al. [2] (see Figure 8) and apply the two methods provided in Section 2 to compare the obtained results expressed as aggregated GHG emissions in CO_2 eq.

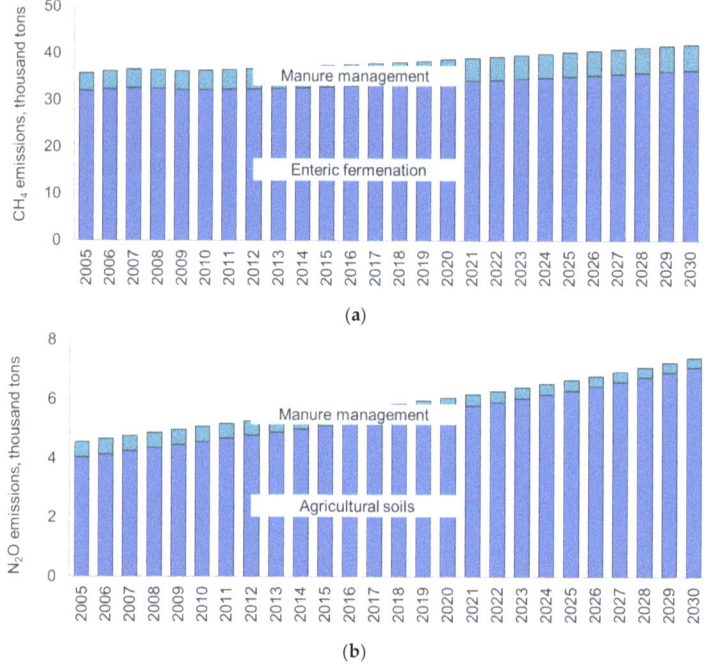

Figure 8. Annual (**a**) methane and (**b**) nitrous oxide emissions from the agriculture sector in Latvia, 2005–2030 (data from Dace et al. [2]).

3. Results and Discussion

The temporal effects of the annual emissions on the current year basis and the cumulative effect of this flow in the atmosphere for both continued emission flow and the eliminated emission flow are given in Figure 9.

In Figure 9, each curve represents the annual pulse emissions. Since the pollutants emitted in previous years are still in the atmosphere and decaying (see Figure 4 for individual emission decay trendlines), these pollutants are added up to emissions of the current year. The accumulation of these pulse emissions is continued until the pollutant has converged to zero concentration.

As can be seen in Figure 9, the flow rate of emission on the current year basis is relatively lower than the amount of the total cumulative emission in the air from the previous years. From this observation, two essential conclusions can be given. First, the importance to account and reduce also "small" amounts of emissions since these "small" amounts add up to a more significant cumulative effect. And second, the interference in the flow of the emissions will have a "visible" affect only a couple of decades later, since it takes time to decay emissions already accumulated in the atmosphere.

The cumulative GHG emissions calculated using the GWP100 values from the IPCC fourth [16] and fifth [15] assessment report and the emissions calculated using the decay function from BCCM show more substantial disparities in a shorter time horizon. At the same time, these disparities reduce in a longer time horizon (see Figure 10).

The use of the latest GWP100 values from IPCC 2013, which include carbon cycle feedback [15], results in emission curves located closer to the curves obtained with BCCM than the use of GWP100 values taken from IPCC 2006 [16]. Especially significant convergence towards numbers obtained by BCCM is evident for CH_4 emissions using IPCC 2013 values instead of IPCC 2006 values. As one of the thought-provoking differences between results in IPCC and BCCM methods, the different nature of the line shapes should be stressed out. Since CH_4 decay has an evident non-linear nature, the most significant difference in the results is created in the near-term estimates of the CH$_4$ impacts. Similar findings are given in the work by Allen et al. [5], where the most significant misrepresentation of the impacts using GWP is evident in the case of methane and aerosol emissions. In Figure 10, this difference is given between the straight lines of IPCC 2006 and IPCC 2018 results and the non-linear line of BCCM results. Where an almost twofold difference is created in the short-term analysis, this difference cannot also be compensated with overestimated N$_2$O impacts, because N$_2$O results by BCCM follow a linear nature until 2030 relatively tightly.

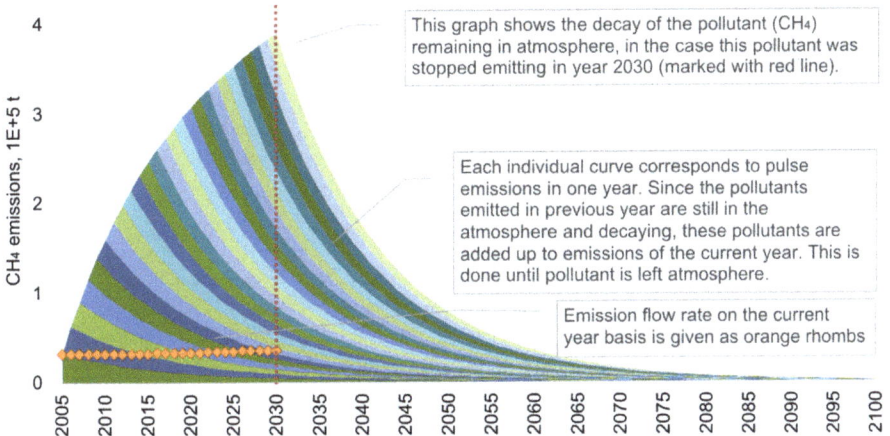

Figure 9. The emission flow rate on the current year basis and the cumulative effect of this flow in the atmosphere for continued emission flow and the elimination of the emission flow: an example of CH$_4$ emissions from enteric fermentation in Latvia, 2005–2100.

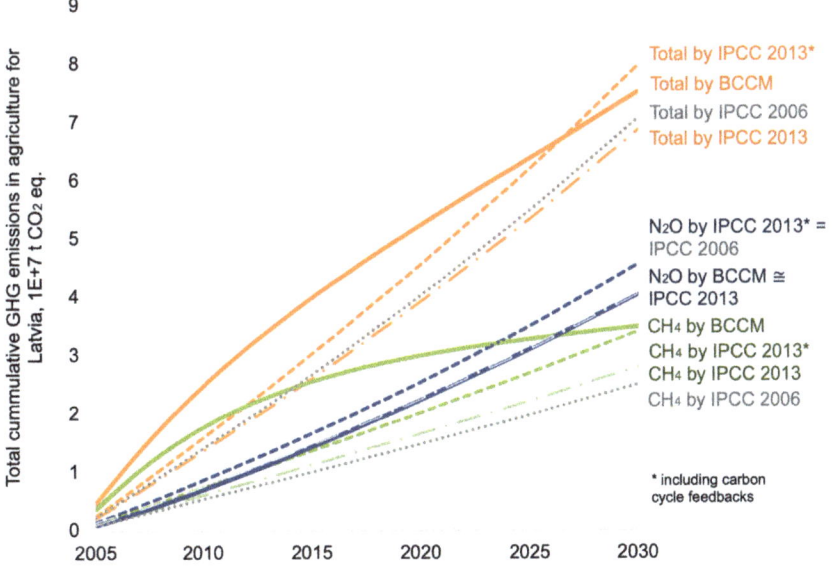

Figure 10. The comparison of the cumulative GHG emissions calculated using the GWP values from IPCC 2006 and IPCC 2013 and the decay function using the Bern Carbon Cycle Model (BCCM).

These differences in the total cumulative amounts of GHG emissions are due to the use of constant GWP values in the IPCC methodology, while in decay function, the GWP value changes with time; see Figure 11 for GWP values for N_2O emissions and Figure 12 for CH_4 emissions.

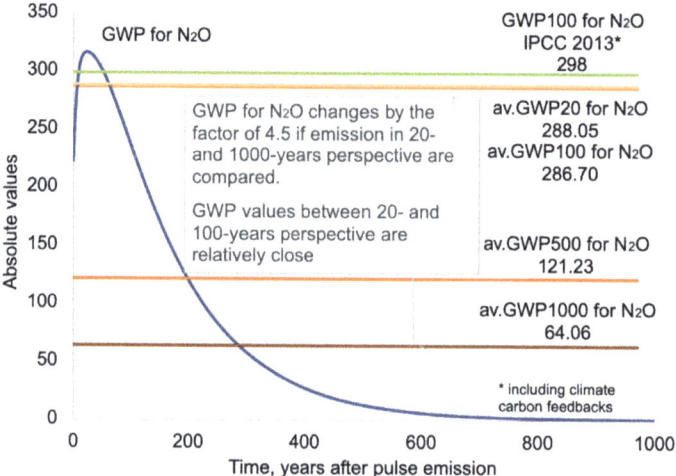

Figure 11. Absolute GWP values for N_2O emissions given using the BCCM "GWP for N_2O" and average GWP values for the first 20 years after pulse emission of N_2O named "av.GWP20 for N_2O" and has the value of 288.05; for the first 100 years—"av.GWP100 for N_2O"; for the first 500 and 1000 years—"av.GWP500 for N_2O" and "av.GWP1000 for N_2O". GWP value for the first 100 years given in the IPCC 2013, including the carbon cycle feedback, has the value of 298 and is given as "GWP100 for N_2O IPCC 2013*", unitless.

The high sensitivity of the GWP values for the chosen time frame for CH_4 and N_2O is due to the non-linear nature and different mathematical functions used for approximation of these decay functions. The difference in the average values of GWP for different time horizons can be substantial, and the obtained conclusions can be misleading. When using these averages, a situation is also possible when a longer time horizon diminishes the importance of local and relatively short lifetime emissions, such as CH_4. For example, the described situation is evident that in the case of N_2O emissions, the time frame of 20 or 100 years does not change the applied GWP values so significantly as they change in the case when CH_4 is assessed either in 20 or 100 years (see Figure 12).

Also, the perturbation time of CH_4 emissions is approximately ten times less than the perturbation time of N_2O emissions. Thus, the use of single metrics for such different behaviour is rather challenging. The difference in the emissions' trendlines, given in Figure 10, between the GWP values from IPCC and the values from BCCM is visually explained in Figures 11 and 12, where GWP100 values used in IPCC2013 for N_2O and CH_4 emissions graphically are located between different segments. In Figure 11, for N_2O emissions, the value of GWP100 from IPCC2013 is 298, including carbon cycle feedbacks. This value of 298 is higher than the mathematical average of GWP20 and GWP100 values for N_2O. While for CH_4 emissions given in Figure 12, the value of GWP100 from IPCC2013 is 34, including carbon cycle feedbacks. This value of 34 is in between the mathematical average of GWP20 and GWP100 values for CH_4. Thus, these selected GWP100 values from IPCC describe different segments of the emission decay period for N_2O and CH_4 and, therefore, create differences in the obtained results.

Also, longer time horizons, in general, give lower GWP values. Thus, it is always better to select a longer time horizon to have a smaller impact, but does it provide a reasonable picture of these impacts? It can be further discussed, how can the assessment of the impact at the government or enterprise-level reasonably give an interpretation for the values of the possible impacts in 100 or 500 years? And how realistic is it that the cost of these created impacts in 100 or even in the next 20 years will be adequately attributed to the producer or consumer of today? In Figure 10, the total cumulative GHG emissions calculated using GWP values suggested by the IPCC and decay functions are compared until 2030. Usually, the comparison is used to show the created impact. In fact, the effect of the emissions occurring in the last year (2030) and a couple of years before 2030 are not included in the calculations when using the decay function. For the more realistic impacts, see Figure 13a.

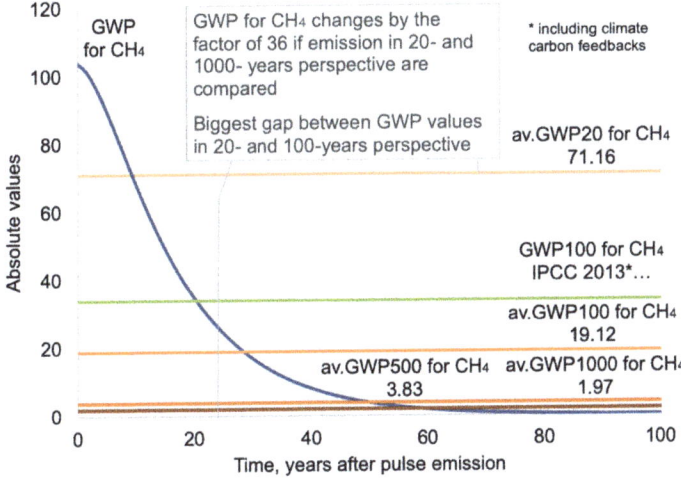

Figure 12. Absolute GWP values for CH_4 emissions given using the BCCM named "GWP for CH_4" and average GWP values for the first 20 years after pulse emission of CH_4 named "av.GWP20 for CH_4" and has the value of 71.16, for the first 100 years—"av.GWP100 for CH_4"; for first 500 and 1000 years—"av.GWP500 for CH_4" and "av.GWP1000 for CH_4". GWP value for the first 100 years given in the IPCC 2013, including carbon cycle feedback, has the value of 34 and is given as "GWP100 for CH_4 IPCC 2013*", unitless.

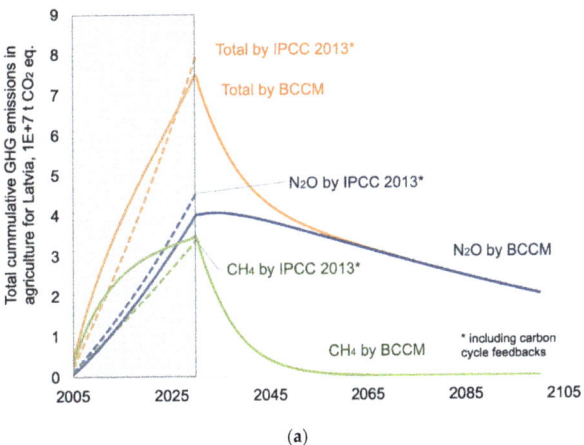

(a)

Figure 13. *Cont.*

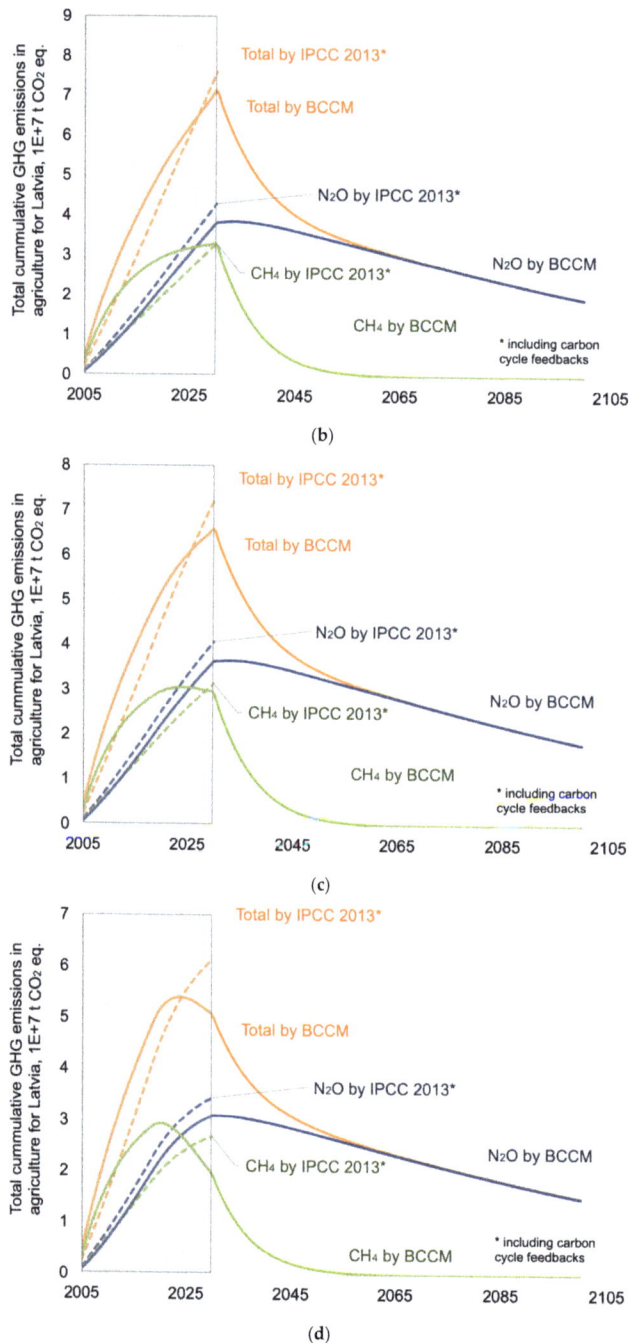

Figure 13. The comparison of the total cumulative GHG emissions calculated using IPCC methodology and decay function before and after year 2030, Latvia (**a**) reference case or business as usual based on Dace et al. [2] (**b**) constant emission after 2020, (**c**) declining emissions by 2% after 2020, (**d**) declining emissions by 10% after 2020.

Assuming that after the year 2030, no more emissions are occurring, Figure 13a shows how the amounts of the emissions released to the atmosphere until 2030 slowly decay. The figure shows that the fraction remaining in the air of various GHG emissions varies by nature. For CH_4, the decay is much faster, while N_2O has not even halved since its release. Similar findings of the lack of appropriate comparison between short-lived GHGs (in this case, CH_4) and long-lived GHGs (N_2O) are discussed in work by Boucher et al. [30]. The authors explain that difficulty in using GWP for short-lived GHGs is because the GWP value does not consider that the radiative forcing of these short-lived GHGs has time to relax and reach equilibrium in the analyzed time horizon. Thus, Boucher et al. [30] have introduced the GTP concept that generalizes climate impacts and considers different climate responses for both short and long-lived GHG emissions.

The faster decay of CH_4 is usually used as the argument to reduce the importance of this emission, especially in the agriculture sector, where CH_4 emissions have a significant share in the total impact categories; see Figure 14a,b.

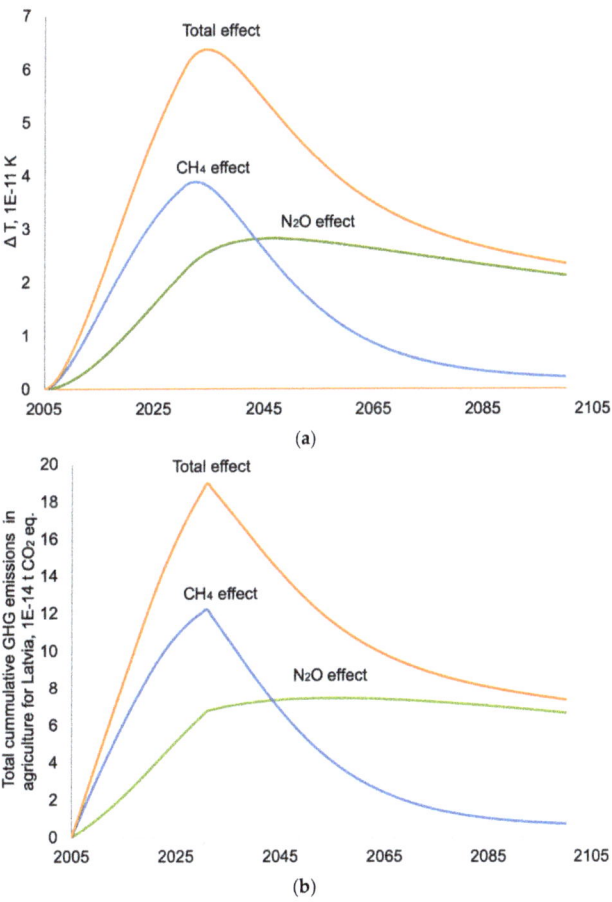

Figure 14. (**a**) Absolute global temperature change potential (AGTP) and (**b**) global temperature change potential (GTP) calculated for the emissions from agriculture in Latvia.

If the impact on the GTP from CH_4 is assessed, it can be seen that CH_4 shows a more obvious temperature change effect in a shorter run and, in total, contributes to more than half of the temperature change effect created by the agriculture in Latvia, as given in Figure 14.

In this case, various different scenarios of the agriculture emission are modelled, such as constant emissions or decreasing emissions; see Figure 15. As can be seen, Figure 13a–c depict related trends very precisely. For example, the total emissions from agriculture in Latvia in the year 2030 between scenarios given in Figure 15a, business as usual or 2% growth of emissions and Figure 15b, constant emissions after the year 2020 calculated based on BCCM will differ by 4% only. This example shows how hard it is to reach the emission reductions due to accumulation and long perturbation time of these emissions in the atmosphere. In work by Olivié and Peters [27], the common characteristics and differences between GWP and GTP are discussed. Both methods are designed to be simple tools for comparing impacts to climate from several types of GHG emissions. Both methods refer to the pulse emission of some specific GHG in comparison to the pulse emission of the same quantity of reference CO_2 emissions, while the difference is in the used mathematical model—where GWP is based on comparing the changes in radiative forcing overtime, and GTP on global mean temperature changes over time [4].

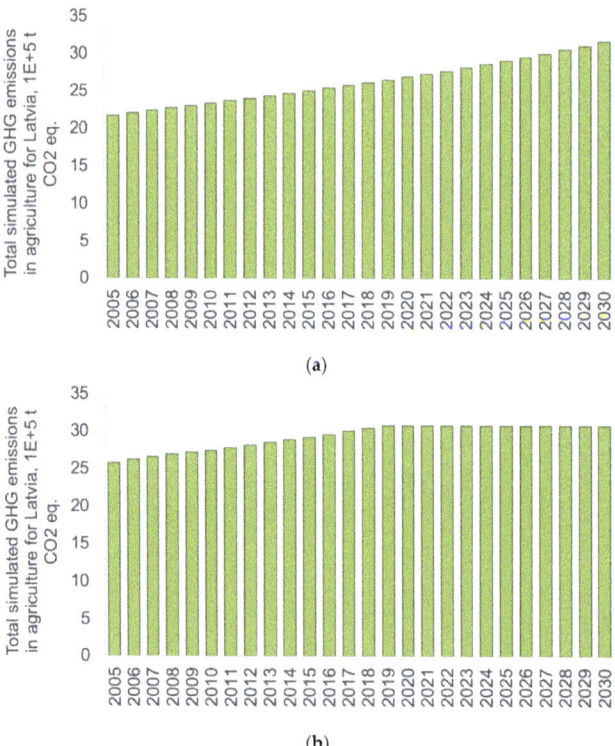

Figure 15. *Cont.*

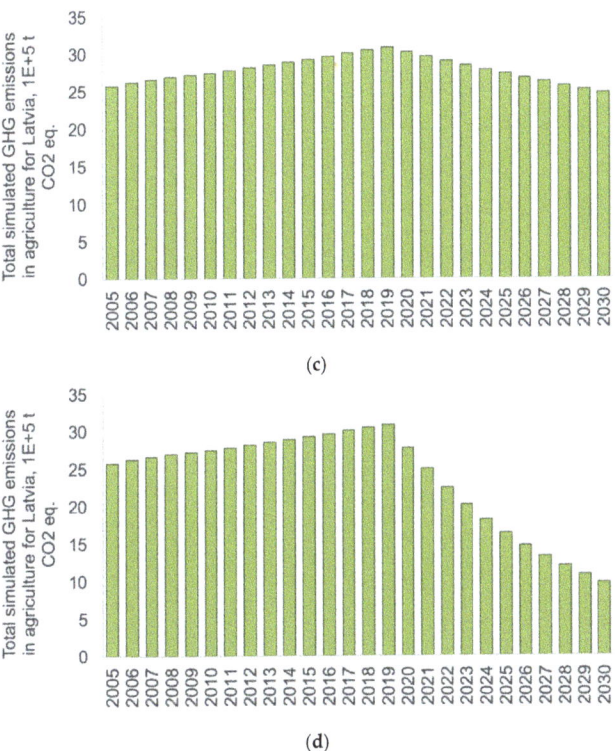

Figure 15. Modelled scenarios for the emissions from agriculture in Latvia, (**a**) reference case or business as usual based on Dace et al. [2], (**b**) constant emission after 2020, (**c**) declining emissions by 2% after 2020, and (**d**) declining emissions by 10% after 2020.

The most significant difference between GWP and GTP is that GTP is an end-point metrics, while GWP is a cumulative measure of climate change. Thus, the value of radiative forcing is of great importance in the analysis of GTP, and more weight is given to climate effects of radiative forcing that come later in the perturbation time of the analyzed pollutants. Thus, for the emissions with a shorter perturbation time, there will be a more significant difference between the results of GWP and GTP results. This theory implies that the GWP assessment gives an overestimation of the short-lived pollutants for the mitigation of climate change. In work by Boucher and Reddy [30], the difference between black carbon emissions for 100 years perspective in the case of using GTP gave seven times smaller impact than the corresponding GWP assessment. These findings are also in agreement with the work of Shine et al. [4]. Also, the choice of the time horizon to evaluate the impacts of the emissions is of significant importance. By far, the most common practice is to use a 100-year time horizon, since it is used in the Kyoto Protocol [27], but there is no scientific justification in using this particular time horizon. The larger the time horizon is chosen, the fewer effects can be attributed to short-lived pollutants [30]. On the other hand, sustainability cannot be achieved if only long-lived pollutants are accounted and restricted, while short-lived continue to degrade local ecosystems. Therefore, here, both short-term and long-term impacts on sustainability should be assessed and balanced.

Moreover, the choice of the time horizon of 20 years or 100 years or some other time horizon is still not scientifically justified by any concrete evidence [30]. Thus, the obtained results are also sensitive to the assumed time horizon and can lead to contradicting conclusions.

Our findings are also in agreement with work published by Shimako [7] and Shimako et al. [8]. The authors of the research studied the same total amount of emissions but taken with two different emission timing profiles. One emission profile was constant through the simulation, another emission profile peaked in the beginning and then was zero for 4/5 of the simulation. Due to this difference in the emission profiles, the temporal effects of these two emission profiles also differ. In contrast, the total amount of emissions at the end of the simulation was the same for both profiles. In case GWP would be multiplied by the total amount of emissions, the obtained results would be the same for both profiles. This phenomenon is also evident in our findings—the cumulative emissions have different impact profiles when the same amount of total emissions is considered using temporal impacts.

Since the short-lived GHG emissions affect local environments in more apparent patterns, such as effects on air quality, human health, and local ecosystems, in work by Rypdal et al. [33], it was proposed to regulate short-lived emission in regional policy contexts. The influence of the emissions on local metrics is reviewed in detail in work by Rypdal et al. [34] and by Levasseur et al. [6].

Work by Olivié and Peters [27] also explains that in coupled systems, the temperature changes will also affect the ocean in two ways. Firstly, the absorption of CO_2 directly by the ocean will increase if the temperature will rise. Secondly, ocean circulation patterns will be changing, due to direct effects from increased respiration and photosynthesis or indirectly by changes in precipitation. Also, the authors discuss how various changes in these coupled systems might influence the numerical values used in IRF. Nevertheless, we would like to stress that, in this work, the precise numerical values were not of such high importance as the depiction of overall dynamics and different results that can be obtained using two different methodological approaches.

Also, work by Jardine et al. [35] shows that various feedbacks exist when CH_4 emissions are analyzed. For example, the global atmospheric lifetime of CH_4 is defined by the amount of atmospheric concentration of CH_4 (CH_4 burden) divided by the amount of annual removal of CH_4 from the atmosphere (CH_4 sink). Thus, increasing the concentration of CH_4 in the atmosphere will lead to longer global atmospheric lifetimes. In work by Holmes [36], the strength of chemical feedback for CH_4 was analyzed using meteorological, chemical, and emissions factors. The research shows that this feedback depends weakly (likely in the 10% range) on temperature, insolation, water vapour, and emissions of NO. While perturbation time of CH_4 might rise as high as 40% and more, this means that close accounting of the balance in CH_4 is needed in order to have valid assumptions about the time when "constant" values cannot be treated as constants anymore.

To sum up, we believe that both types of emission accounting (the constant GWP values for a 100-year time horizon (GWP100) and the time dynamic GWP values for a 100-year time horizon obtained by using the BCCM) are valuable to use; each has its strength and weaknesses. Thus, we propose to also look on the temporal impacts of emissions, since it might help in designing more precisely targeted policy measures appropriate for the chosen mitigation priorities.

As given in the report by Jardine et al. [35], CH_4 emissions decay about ten times faster than N_2O, but from the other point of view, policy measures targeting CH_4 reduction will also show the effect on the reduction of climate change ten times faster than measures targeting N_2O. Also, the report by the United Nations Environment Programme and the World Meteorological Organization [37] shows that the adoption of policies targeting short-lived GHGs would allow reaching climate change mitigation targets with higher confidence. The report also shows that CO_2 reduction measures alone would exceed the set temperature thresholds anyway already in the near term. In contrast, only CH_4 reduction measures would keep the temperature below the threshold in the near term while exceeding the limit later because of the effect of other GHGs. As the opposing argument for stricter control of long-term pollutants is the fact that any technological solution and policy measure will be implemented only for the finite time. Usually, the selected time horizon is much shorter than the consequences of the pollution associated with these technologies.

4. Conclusions

In this paper, two methods for GHG emission accounting were compared. First, using the constant GWP values for a 100-year time horizon (GWP100) and second, the time dynamic GWP values for a 100-year time horizon obtained by using the BCCM that takes into consideration the climate system response to the amount, time and decay rate of the emitted pollutant. The GWP100 values are the default emission metric suggested by the IPCC for the annual emission accounting, and it is considered a relatively simple and easy to use the method. Although no scientific evidence backs the use of GWP100 and, more importantly, GWP100 has "no direct estimation of any climate system responses or direct link to policy goals" (Myhre et al. [15]; Cherubini et al. [38]), policymakers widely use the GWP100 values in designing GHG emission mitigation strategies and international agreements, like the Kyoto protocol and Paris agreement.

The results of our study show that the cumulative emissions have different impact profiles when the same amount of total emissions is considered using temporal impacts. The obtained results are also sensitive to the assumed time horizon and can lead to contradicting conclusions. The high sensitivity of the GWP values for the chosen time frame for CH_4 and N_2O is due to the non-linear nature and different mathematical functions used for approximation of these decay functions. The difference in the average values of GWP for different time horizons can be substantial, and obtained conclusions can be misleading. When using these averages, a situation is also possible when a longer time horizon diminishes the importance of local and relatively short lifetime emissions, such as CH_4. For example, the described situation is evident that in the case of N_2O emissions, the time frame of 20 or 100 years does not change the applied GWP values so significantly as they change in the case when CH_4 is assessed either in 20 or 100 years.

If the impact on the GTP from CH_4 is assessed, it can be seen that CH_4 shows a more obvious temperature change effect in a shorter run and, in total, contributes to more than half of the temperature change effect created by the agriculture in Latvia.

The BCCM facilitates the selection of the time horizon needed for the specific purpose and expresses the results of policy decisions as to the effect of emissions on the global temperature change potential. The use of GWP100 is still useful and needed as (at least) two purposes of the emission accounting should be separated—one is for the emission inventory, the other is for the policy planning. The inventory is needed to keep track of the annual emission rates and assess the trends and success achieved in the emissions mitigation in the past, and GWP100 is useful for the purpose.

Meanwhile, policy strategies and instruments aim to achieve some desirable behaviour that may effectively govern a system in the future. And the GWP values obtained by using the BCCM would be much more useful for the purpose. Although "countries and the international community have made significant investments in inventory systems" [16] reconsideration and use of other methodology by the policymakers might eventually be less costly than dealing with consequences of climate change and wrong decisions (sub-optimal policies). Application of the BCCM would facilitate finding more efficient mitigation options for various pollutants, analyze multiple parts of the climate response system at a specific time in the future (amount of particular pollutants, temperature change potential), or analyze different solutions for reaching the emission mitigation targets at regional, national, or global levels.

Author Contributions: Conceptualization, L.T., E.D., M.T.K.; methodology, L.T.; Software, E.D.; validation, L.T., E.D.; formal analysis, L.T., E.D., M.T.K.; investigation, L.T.; resources, L.T., E.D.; data curation, L.T.; writing—original draft preparation, L.T.; writing—review and editing, L.T., E.D., M.T.K.; visualization, L.T.; supervision, M.T.K.; project administration, L.T., M.T.K.; funding acquisition, L.T. All authors have read and agreed to the published version of the manuscript.

Funding: This research was funded by European Union's Horizon 2020 research and innovation programme under the Marie Sklodowska-Curie grant agreement No 798365.

Acknowledgments: This publication is the selected article from the European Biomass Conference and Exhibition (EUBCE) 2019 conference. And the methodology used in the article was initially presented in the EUBCE 2019 conference for the case study on biorefineries under the title Timma L. and Parajuli R. 2019, Time Dynamics in Life Cycle Assessment—Exemplified by a Case Study on Biorefineries. In European Biomass Conference and Exhibition Proceedings., ETA-Florence Renewable Energies, s. 1599–1603, Lisbon, Portugal, 27/05/2019. https://doi.org/10.5071/27thEUBCE2019-4DO.5.2.

Conflicts of Interest: The authors declare no conflict of interest.

References

1. Mussatto, S.I. Challenges in Building a Sustainable Biobased Economy. *Ind. Crops Prod.* **2017**, *106*, 1–2. [CrossRef]
2. Dace, E.; Muizniece, I.; Blumberga, A.; Kaczala, F. Searching for solutions to mitigate greenhouse gas emissions by agricultural policy decisions—Application of system dynamics modeling for the case of Latvia. *Sci. Total Environ.* **2015**, *527–528*, 80–90. [CrossRef] [PubMed]
3. Cherubini, F.; Fuglestvedt, J.; Gasser, T.; Reisinger, A.; Cavalett, O.; Huijbregts, M.A.J.; Johansson, D.J.A.; Jørgensen, S.V.; Raugei, M.; Schivley, G.; et al. Bridging the gap between impact assessment methods and climate science. *Environ. Sci. Policy* **2016**, *64*, 129–140. [CrossRef]
4. Shine, K.P.; Fuglestvedt, J.S.; Hailemariam, K.; Stuber, N. Alternatives to the Global Warming Potential for comparing climate impacts of emissions of greenhouse gases. *Clim. Chang.* **2005**, *68*, 281–302. [CrossRef]
5. Allen, M.R.; Shine, K.P.; Fuglestvedt, J.S.; Millar, R.J.; Cain, M.; Frame, D.J.; Macey, A.H. A solution to the misrepresentations of CO2-equivalent emissions of short-lived climate pollutants under ambitious mitigation. *NPJ Clim. Atmos. Sci.* **2018**, *1*, 1–8. [CrossRef]
6. Levasseur, A.; Cavalett, O.; Fuglestvedt, J.S.; Gasser, T.; Johansson, D.J.A.; Jørgensen, S.V.; Raugei, M.; Reisinger, A.; Schivley, G.; Strømman, A.; et al. Enhancing life cycle impact assessment from climate science: Review of recent findings and recommendations for application to LCA. *Ecol. Indic.* **2016**, *71*, 163–174. [CrossRef]
7. Shimako, A. *Contribution to the Development of a Dynamic Life Cycle Assessment Method*; INSA de Toulouse: Toulouse, France, 2017.
8. Shimako, A.H.; Tiruta-Barna, L.; Bisinella de Faria, A.B.; Ahmadi, A.; Spérandio, M. Sensitivity analysis of temporal parameters in a dynamic LCA framework. *Sci. Total Environ.* **2018**, *624*, 1250–1262. [CrossRef]
9. Seshadri, A.K. Fast–slow climate dynamics and peak global warming. *Clim. Dyn.* **2017**, *48*, 2235–2253. [CrossRef]
10. Intergovernmental Panel on Climate Change. *Drivers, Trends and Mitigation*; IPCC: Cambridge, UK; New York, NY, USA, 2014.
11. EC (European Commission). *Green Paper: A 2030 Framework for Climate and Energy Policies COM (2013) 169*; EC: Brussels, Belgium, 2013.
12. EC (European Commission). *23/24 October 2014—Conclusions, EUCO 169/14*; EC: Brussels, Belgium, 2014.
13. Leip, A.; Weiss, F.; Wassenaar, T.; Perez, I.; Fellmann, T.; Loudjani, P.; Tubiello, F.; Grandgirard, D.; Monni, S.; Biala, K. *Evaluation of the Livestock Sector's Contribution to the EU Greenhouse Gas Emissions—Final Report*; EC: Brussels, Belgium, 2010.
14. Domínguez, I.P.; Fellmann, T.; Witzke, H.-P.; Jansson, T.; Oudendag, D.; Gocht, A.; Verhoog, D. *Agricultural GHG Emissions in the EU: An Exploratory Economic Assessment of Mitigation Policy Options*; EC: Brussels, Belgium, 2012.
15. Myhre, G.; Shindell, D.; Breéon, F.-M.; Collins, W.; Fuglestvedt, J.; Huang, J.; Koch, D.; Lamarque, J.-F.; Lee, D.; Mendoza, B.; et al. Anthropogenic and Natural Radiative Forcing, Supplementary Material. In *Climate Change 2013: The Physical Science Basis. Contribution of Working Group I to the Fifth Assessment Report of the Intergovernmental Panel on Climate, Change*; Stocker, T.F., Qin, D., Plattner, G.-K., Tignor, M., Allen, S.K., Boschung, J., Nauels, A., Xia, Y., Bex, V., Midgley, P.M., Eds.; Cambridge University Press: Cambridge, UK; New York, NY, USA, 2013.
16. IPCC. *Guidelines for National Greenhouse Gas Inventories*; Eggleston, H.S., Buendia, L., Miwa, K., Ngara, T., Tanabe, K., Eds.; IGES: Kanagawa, Japan, 2006.
17. Houghton, J.T.; Jenkins, G.J.; Ephraums, J.; IPCC. *Climate Change: The IPCC Scientific Assessment*; Cambridge University Press: Cambridge, UK, 1990.
18. Skytt, T.; Nielsen, S.N.; Jonsson, B.G. Global warming potential and absolute global temperature change potential from carbon dioxide and methane fluxes as indicators of regional sustainability—A case study of Jämtland, Sweden. *Ecol. Indic.* **2020**, *110*, 105831. [CrossRef]

19. Shine, K.P. The global warming potential-the need for an interdisciplinary retrial. *Clim. Chang.* **2009**, *96*, 467–472. [CrossRef]
20. Peters, G.P.; Aamaas, B.; T. Lund, M.; Solli, C.; Fuglestvedt, J.S. Alternative "global warming" metrics in life cycle assessment: A case study with existing transportation data. *Environ. Sci. Technol.* **2011**, *45*, 8633–8641. [CrossRef] [PubMed]
21. Ledgard, S.; Reisinger, A. Implications of alternative greenhouse gas metrics for life cycle assessments of livestock food products. In Proceedings of the 9th International Conference on Life Cycle Assessment in the Agri-Food Sector (LCA Food 2014), San Francisco, CA, USA, 8–10 October 2014.
22. Joos, F.; Roth, R.; Fuglestvedt, J.S.; Peters, G.P.; Enting, I.G.; Von Bloh, W.; Brovkin, V.; Burke, E.J.; Eby, M.; Edwards, N.R.; et al. Carbon dioxide and climate impulse response functions for the computation of greenhouse gas metrics: A multi-model analysis. *Atmos. Chem. Phys.* **2013**, *13*, 2793–2825. [CrossRef]
23. Tanaka, K.; Johansson, D.J.A.; O'Neill, B.C.; Fuglestvedt, J.S. Emission metrics under the 2 °C climate stabilization target. *Clim. Chang.* **2013**, *117*, 933–941. [CrossRef]
24. Petersen, B.M.; Knudsen, M.T.; Hermansen, J.E.; Halberg, N. An approach to include soil carbon changes in life cycle assessments. *J. Clean. Prod.* **2013**, *52*, 217–224. [CrossRef]
25. Fuglestvedt, J.S.; Berntsen, T.K.; Godal, O.; Sausen, R.; Shine, K.P.; Skodvin, T. Metrics of climate change: Assessing radiative forcing and emission indices. *Clim. Chang.* **2003**, *58*, 267–331. [CrossRef]
26. Levasseur, A.; Lesage, P.; Margni, M.; Deschênes, L.; Samson, R. Considering time in LCA: Dynamic LCA and its application to global warming impact assessments. *Environ. Sci. Technol.* **2010**, *44*, 3169–3174. [CrossRef]
27. Olivié, D.J.L.; Peters, G.P. Variation in emission metrics due to variation in CO2 and temperature impulse response functions. *Earth Syst. Dyn.* **2013**, *4*, 267–286. [CrossRef]
28. Aamaas, B.; Berntsen, T.K.; Fuglestvedt, J.S.; Shine, K.P.; Collins, W.J. Regional temperature change potentials for short-lived climate forcers based on radiative forcing from multiple models. *Atmos. Chem. Phys.* **2017**, *17*, 10795–10809. [CrossRef]
29. Olivié, D.J.L.; Peters, G.P. The impact of model variation in CO2 and temperature impulse response functions on emission metrics. *Earth Syst. Dyn. Discuss.* **2012**, *3*, 935–977. [CrossRef]
30. Boucher, O.; Reddy, M.S. Climate trade-off between black carbon and carbon dioxide emissions. *Energy Policy* **2008**, *36*, 193–200. [CrossRef]
31. Shine, K.P.; Berntsen, T.K.; Fuglestvedt, J.S.; Skeie, R.B.; Stuber, N. Comparing the climate effect of emissions of short- and long-lived climate agents. *Philos. Trans. R. Soc. A Math. Phys. Eng. Sci.* **2007**, *365*, 1903–1914. [CrossRef] [PubMed]
32. Dace, E.; Blumberga, D. How do 28 European Union Member States perform in agricultural greenhouse gas emissions? It depends on what we look at: Application of the multi-criteria analysis. *Ecol. Indic.* **2016**, *71*, 352–358. [CrossRef]
33. Rypdal, K.; Berntsen, T.; Fuglestvedt, J.S.; Aunan, K.; Torvanger, A.; Stordal, F.; Pacyna, J.M.; Nygaard, L.P. Tropospheric ozone and aerosols in climate agreements: Scientific and political challenges. *Environ. Sci. Policy* **2005**, *8*, 29–43. [CrossRef]
34. Aamaas, B.; Peters, G.P.; Fuglestvedt, J.S. A synthesis of climate-based emission metrics with applications. *Earth Syst. Dyn. Discuss.* **2012**, *3*, 871–934. [CrossRef]
35. Jardine, C.N.; Boardman, B.; Osman, A.; Vowles, J.; Palmer, J. *Methane UK*; The Environmental Change Institute, University of Oxford: Oxford, UK, 2003.
36. Holmes, C.D. Methane Feedback on Atmospheric Chemistry: Methods, Models, and Mechanisms. *J. Adv. Model. Earth Syst.* **2018**, *10*, 1087–1099. [CrossRef]
37. United Nations Environment Programme (UNEP); World Meteorological Organization (WMO). *Integrated Assessment of Black Carbon and Tropospheric Ozone. Summary for Decision Makers*; UNEP: Nairobi, Kenya; WMO: Geneva, Switzerland, 2011.
38. Cherubini, F.; Huijbregts, M.; Kindermann, G.; Van Zelm, R.; Van Der Velde, M.; Stadler, K.; Strømman, A.H. Global spatially explicit CO2 emission metrics for forest bioenergy. *Sci. Rep.* **2016**, *9*, 20186. [CrossRef] [PubMed]

© 2020 by the authors. Licensee MDPI, Basel, Switzerland. This article is an open access article distributed under the terms and conditions of the Creative Commons Attribution (CC BY) license (http://creativecommons.org/licenses/by/4.0/).

Article

Economic and Environmental Analysis of Small-Scale Anaerobic Digestion Plants on Irish Dairy Farms

Sean O'Connor [1,*], Ehiaze Ehimen [1], Suresh C. Pillai [1], Gary Lyons [2] and John Bartlett [1]

1. Department of Environmental Science, Institute of Technology Sligo, F91 YW50 Sligo, Ireland; ehimen.ehiaze@itsligo.ie (E.E.); pillai.suresh@itsligo.ie (S.C.P.); bartlett.john@itsligo.ie (J.B.)
2. Agri-Environment Branch, Agri-Food and Biosciences Institute, Large Park, Hillsborough BT26 6DR, UK; Gary.Lyons@afbini.gov.uk
* Correspondence: sean.oconnor2@mail.itsligo.ie; Tel.: +353-(0)71-91-55222

Received: 29 December 2019; Accepted: 26 January 2020; Published: 3 February 2020

Abstract: The European Union's (EU) climate and energy package requires all EU countries to reduce their greenhouse gas (GHG) emissions by 20% by 2020. Based on current trends, Ireland is on track to miss this target with a projected reduction of only 5% to 6%. The agriculture sector has consistently been the single largest contributor to Irish GHG emissions, representing 33% of all emissions in 2017. Small-scale anaerobic digestion (SSAD) holds promise as an attractive technology for the treatment of livestock manure and the organic fraction of municipal wastes, especially in low population communities or standalone waste treatment facilities. This study assesses the viability of SSAD in Ireland, by modelling the technical, economic, and environmental considerations of operating such plants on commercial Irish dairy farms. The study examines the integration of SSAD on dairy farms with various herd sizes ranging from 50 to 250 dairy cows, with co-digestion afforded by grass grown on available land. Results demonstrate feedstock quantities available on-farm to be sufficient to meet the farm's energy needs with surplus energy exported, representing between 73% and 79% of the total energy generated. All scenarios investigated demonstrate a net CO_2 reduction ranging between 2059–173,237 kg CO_2-eq. yr^{-1}. The study found SSAD systems to be profitable within the plant's lifespan on farms with dairy herds sizes of >100 cows (with payback periods of 8–13 years). The simulated introduction of capital subvention grants similar to other EU countries was seen to significantly lower the plant payback periods. The insights generated from this study show SSAD to be an economically sustainable method for the mitigation of GHG emissions in the Irish agriculture sector.

Keywords: anaerobic digestion; methane production; co-digestion; combined heat and power; farm-scale; technical-economic analysis; life cycle assessment; greenhouse gas emission; Ireland

1. Introduction

The European Union's (EU) climate and energy package sets binding greenhouse gas (GHG) emission reduction targets for all EU states by 2020; these include a 20% cut in GHG emissions, to produce 20% of energy consumed from renewable sources, and a 20% improvement in energy efficiency [1]. The Republic of Ireland, in particular, has struggled to meet its emission targets, with most recent estimates projecting a 14–15% shortfall, resulting in the country projected to pay up to €103 million in carbon credits to compensate for its lack of climate action [2,3]. Ireland's agriculture sector has consistently remained the single largest contributor, accounting for 33% of all GHG emissions in 2017, and 46% of all non-emission trading system (ETS) GHG emissions [4]. The country now faces a dilemma, to either limit or reduce the growth of its agriculture sector (which is vital to Ireland's economy) or to disregard its environmental obligations.

A promising technology with the capacity to provide both renewable energy and GHG reduction, particularly in the agriculture sector, is anaerobic digestion (AD). AD is a natural process in which microorganisms (hydrolytic, fermentative, acetogenic and methanogenic bacteria) break down biodegradable material in the absence of oxygen, producing biogas (a mixture mainly composed of methane and carbon dioxide). These systems are beneficial for improving on-site energy generation, upgrading wastes, and producing a nutrient-rich fertiliser from the digester effluents. They can also reduce pathogenic loads, odours and greenhouse gas emissions emanating from the agricultural processes [5–8]. Furthermore, the technology has received considerable research attention, advancing its potential capability through optimisation strategies [9–14]. Despite the apparent benefits, Ireland has been slow to adopt the technology, ranking 20th in AD penetration among the EU-28 countries [15–17]. A contributing factor to the low deployment is the concentration of "large scale plants", particularly in Europe, where the siting of such centralised facilities has been based on the availability of significant quantities of biomass feedstock [18]. However, the biomass quantities in many Irish farms are currently insufficient to meet the feedstock requirements of medium-and large-scale AD plants. The situation is worsened when considering that the average dairy herd in Ireland only consists of approximately 90 cows in 2018 [19].

The application of small-scale anaerobic digestion (SSAD) plants with an electrical output of 15–100 kW_e, holds promise in overcoming the technical and economic barriers associated with treating lesser biomass quantities [18]. SSAD may be particularly useful for the Irish dairy industry, where there is a large livestock population (1.4 million dairy cows) [20], there are predictable process energy demands and reliable feedstock collection potential, its deployment is promising. Despite the potential of this technology, previous studies have largely focused on the implications of deploying medium to large scale AD plants (>100 kW_e) with relatively little focus on the Irish context [21,22]. Therefore, a lack of understanding is apparent in the applicability of SSAD plants in stand-alone agricultural environments within Ireland [23].

The goal of this study was to provide an initial assessment of the viability of SSAD on commercial Irish dairy farms. Thus, not only benefiting the reported case study but also other countries and regions, especially those with significant agricultural and livestock productivities. To achieve this goal, the following objectives were put forward:

- Examine the technical parameters associated with the operation of an SSAD plant at various capacities.
- Conduct a CO_2 balance to assess the various scenarios investigated.
- Conduct an economic analysis investigating total revenues, expenditures and financial indicators, such as net present value (NPV) and internal rate of return (IRR).

2. Materials and Methods

2.1. System Boundary

This study considered a "cradle-to-grave" system boundary, encompassing both the technical and environmental impacts in the construction and operation of SSAD plants at various scales. The system boundary, as described in Figure 1, was divided into four main parts:

1. Associated agricultural processes: (i) crop production; (ii) crop harvest and transport; (iii) manure collection and transport; (iv) storage; (v) transport to digester;
2. Biogas production: (i) digester feeding (ii) the AD process;
3. Energy conversion: (i) energy generation (production of electricity and heat); (ii) final use of energy produced;
4. End of life of digestate: (i) storage; (ii) transport and digestate spreading.

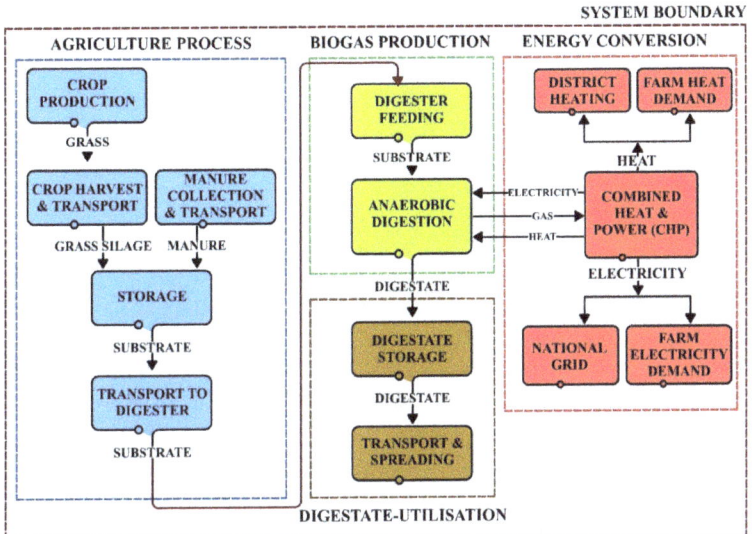

Figure 1. System boundary.

This study did not examine the processes related to the SSAD plant construction, as the material use and key manufacturing processes are unclear. Similarly, the inputs and processes related to the disposal of the plant were also not considered and are outside the study scope. Additionally, the inputs related to the production of farm equipment (e.g., tractors, machines) were not included in the system boundaries, due to the uncertainty regarding their energy input. All simulations were created and run using the software package Microsoft Office Excel (Microsoft Office 2016, Microsoft Corporation, Redmond, WA, USA).

2.2. Feedstock Yield

The farms simulated in this study were selected based on their ability to collectively provide a full representation of the Irish dairy industry, which consists of mainly small to medium-sized farms, as illustrated in Figure 2. The study used a co-digestion feedstock of both dairy cow manure and grass silage. Grass silage was selected because of its popularity in Ireland, where 80% of agriculture land is devoted to pasture, hay and grass silage [24]. Furthermore, Ireland has ideal climate conditions for grass production, experiencing mild and moist conditions, an abundance of rainfall and a lack of extreme temperatures [25,26].

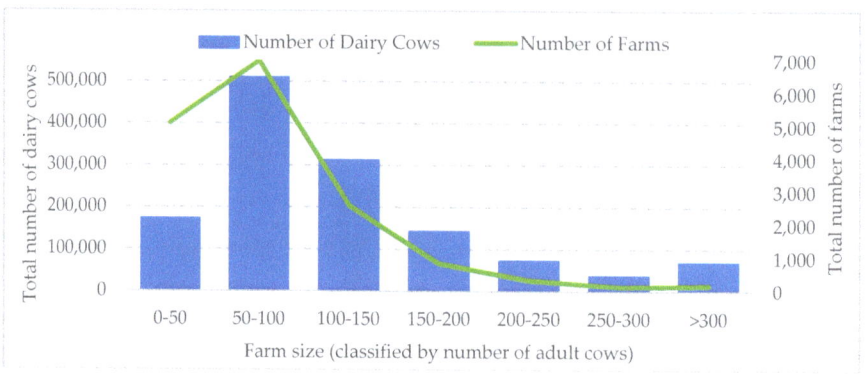

Figure 2. Distribution of total farms and dairy cows by farm size in Ireland. Estimated, based on data received from [27].

Five dairy farm sizes were selected, these relate to the assumed herd sizes of 50 dairy cows (Scenario 1); 100 dairy cows (Scenario 2); 150 dairy cows (Scenario 3); 200 dairy cows (Scenario 4); and 250 dairy cows (Scenario 5). The number of dairy cows refers to the number of female bovine dairy cows, which had already calved and were kept exclusively to produce milk.

The dairy enterprise is based on a self-contained Holstein–Friesian herd [28], retaining pure-bred replacements and selling beef crosses at three weeks. Dairy cows are culled, on average, after five lactations (i.e., annual replacement rate of 18%), which is common in Ireland [29,30]. Manure is predominantly collected from the milking parlour and the cattle housing units (mainly slatted sheds) [31]. The quantity of manure produced per adult cow, heifer, and calve is presented in Table 1. Over the 16 week winter period, it was assumed all manure produced was collected for digestion as the cows are housed [32]. It was more difficult to estimate manure collection over the grazing period (remainder of the year) as collection mainly occurs when the cows are being milked. Based on a milking rate of two times per day and the increased metabolic rate during this period, a 20% manure collection rate was assumed in comparison to Table 1 figures, i.e., 10.4 kg fresh weight (FW) day^{-1} for adult cows, 7.44 kg FW day^{-1} for heifers, and 3.72 kg FW day^{-1} for calves.

Table 1. Characteristics of dairy livestock.

Livestock	Livestock Weight Target	Total Manure Production (FW day^{-1})
Adult cows (<24 months)	550 kg [a]	52.2 kg [b]
Heifers (12 to 24 months)	406 kg [a]	37.2 kg [b]
Calves (>12 months)	175 kg [a]	18.6 kg [b]

[a] Assumed, based on reports for livestock weight by [32]; [b] Ultimate analysis presented in [33] for dairy cow manure production.

In the model, it was assumed that the dairy enterprise was the primary source of income, with revenue from biogas production being a supplementary income stream. Consequently, the needs of the dairy herd were prioritised, with only surplus crops used for biogas production. The area of farmland available to grow feedstock was estimated by subtracting Ireland's mean farm size (based on herd size) from the area of land required to sustain the dairy herd. The mean farm sizes for the scenarios considered corresponded to 43.51 (Scenario 1), 68.74 (Scenario 2), 93.96 (Scenario 3), 119.19 (Scenario 4), and 144.41 hectares (Scenario 5) [27]. The area of farmland required to sustain the dairy herd was based on a recommended ratio of 2.8 cows per hectare with an additional 20% margin of safety added, to account for seasonal variations and unusable land [32]. Silage yields in Ireland are typically between 11 and 15 t dry solids (DS) ha^{-1}; yields are generally higher in the southwest and decrease towards the northeast [25,34]. The model assumed an average yield of 13 t DS ha^{-1} to enable it to represent the majority of Irish dairy farms.

2.3. Pre-Digestion Farm Activities

This study considered the direct and indirect energy inputs for the co-digestion feedstock prior to digestion. For the grass silage feedstock, energy inputs in cultivation, harvesting, recovery and digester feeding were accounted for and are described in Tables 2 and 3. The calculations used in Table 2 were based on the land being ploughed every seven years to maintain grass productivity. For the dairy cow manure feedstock, the energy inputs related to its collection, loading and transportation from the farm's cattle housing and milking parlour to the digester were also accounted. According to Berglund [35], the energy input in loading and transporting liquid manure is 2.5 MJ t^{-1} km^{-1}. The model used this figure and an estimated distance of 500 m between the manure storage and digester to calculate energy consumption. The system boundary assumed that the digestate produced from the AD process was spread as fertiliser on the farms' own land.

Table 2. Fuel consumption by machinery in grass cultivation. Reproduced from [36], Elsevier: 2008.

Operation	Average Diesel Fuel Consumption (l ha^{-1} y^{-1})
Crop production	
Soil ploughing and crumbling	4.67
Sowing and maintenance	6.9
Weed control	0.24
Transport and spreading of fertiliser	18
Crop collection and transport	
Harvest	47.20
Harvest transport	25.49
Silo compaction	8.80
Digester feeding (grass)	23.57

Table 3. Energy consumed and CO$_2$ emitted from raw materials. Reproduced from [36], Elsevier: 2008.

	Application Rate (kg ha^{-1} yr^{-1})	Energy Consumed (MJ kg^{-1})	CO$_2$ Emitted (kg CO$_2$ kg^{-1})
Mineral fertiliser			
Nitrogen	82 [a]	70 ± 34	2.5 ± 0.1
Phosphorus pentoxide	11 [a]	12 ± 4	1.1 ± 0.4
Potassium oxide	29 [a]	7.5 ± 2.5	0.67 ± 0.19
Other raw materials			
Diesel	N/A	56.3 ± 5.6	3.64 ± 3.6
Weed control	0.11 [b]	200 ± 20	15.45 ± 1.5

[a] Assumed, application rate of mineral fertilizer according to [37]; [b] Average pesticide applied to grass reported by [38].

2.4. Operation of the Biogas Plant

The biogas available for potential recovery in an AD plant is largely dependent on the fraction of volatile solids (VS) in the feedstock, high fractions of VS correlate to higher biogas production [39]. The VS content represents the portion of organic solids that can be digested in the feedstock, while the remainder of the solids is fixed [40]. Using the feedstock physical and chemical properties described in Table 4, the biogas flowrates per kg of VS were quantified using the Boyle–Buswell stoichiometric relationship described in Equation (1) [41]. This methodology assesses the biogas potential of organic solids through the AD process. As this methodology considers the total content of VS to be biologically degraded, it can lead to an overestimation of the biogas produced from the feedstock in comparison to real-world case studies [42]. Nevertheless, Boyle–Buswell has been commonly applied in literature as an effective indicator to gauge biogas potential [21,43,44]. The subsequent methane yield was 0.6376 m^3 CH$_4$ kg^{-1} VS from dairy cow manure and 0.822 m^3 CH$_4$ kg^{-1} VS from grass silage.

$$C_aH_bO_cN_dS_e + \left(a - \frac{b}{4} - \frac{c}{2} + \frac{3d}{4} + \frac{e}{2}\right)H_2O \rightarrow \left(\frac{a}{2} + \frac{b}{8} - \frac{c}{4} - \frac{3d}{8} - \frac{e}{4}\right)CH_4 + \left(\frac{a}{2} - \frac{b}{8} + \frac{c}{4} + \frac{3d}{8} + \frac{e}{4}\right)CO_2 + dNH_3 + eH_2S \quad (1)$$

Table 4. Physical and chemical properties for dairy cow slurry and grass silage.

Physical Properties	Dairy Cow Manure	Grass Silage
DS (g kg^{-1}) [a]	87.5 ± 2.1 [c]	292.7 ± 3.4 [c]
VS (g kg^{-1}) [b]	66.9 ± 1.8 [c]	87.5 ± 2.1 [c]
VS DS^{-1} (%) [a,b]	76.5 [c]	91.7 [c]
Carbon (%)	58.62 [d]	43.3 [e]
Hydrogen (%)	7.69 [d]	6.43 [e]
Oxygen (%)	30.50 [d]	44.72 [e]
Nitrogen (%)	2.92 [d]	2.36 [e]
Sulphur (%)	0.27 [d]	0.06 [e]

[a] DS is dry solids; [b] VS is volatile solids; [c] Characteristics of grass and manure are based on [21]; [d] Ultimate analysis of dry and ash free cow manure reported by [41,42]; [e] Ultimate analysis of grass silage as presented in [45].

The plant simulated consisted of a mesophilic continuously stirred tank reactor (CSTR) with all biogas produced used in a combined heat and power (CHP) unit. The annual operating time of the plant was assumed to be 8000 h (91% of the year), allowing for routine maintenance and repair, as reported in the literature [46–48]. The hydraulic retention time of the plant was 25 days [49]. Based on the rate of biogas flow, it was possible to size the required CHP unit using Equation (2) [50]. The CHP unit was assumed to have an electrical efficiency of 30% and a thermal efficiency of 55%, which is typical for similar sized systems [35,48,51,52].

Berglund and Börjesson [35] reported that the primary power consumption in the operation of an AD plant is the pumping and stirring of feedstock (7.2 kWh t^{-1}). The net electricity produced via the CHP unit was first used to meet the electrical demand of the farm, with surplus electricity exported to the national grid. The energy required to heat and maintain the digester's temperature was calculated using Equation (3). The plant's heat losses (hl) were estimated using Equation (4). The heat transfer coefficients of the plant's construction materials correspond to the following: floating cover (1.0 W m^2.°C); 6 mm steel plate "sandwich" with 100 mm insulation (0.35 W m^2.°C); 300 mm concrete floor in contact with earth (1.7 W m^2.°C) (Zhang, 2013). Equation (5) describes the energy required to heat the digester feedstock (q). The operating temperature of the digester was assumed to be constant at 40 °C, with the temperature of the incoming feedstock at 10 °C [53].

$$\text{CHP capacity }(kW_e) = \frac{\text{Biogas production }(m^3) \times [\text{Calorific value of biogas}\left(\frac{MJ}{Nm^3}\right)/3.6]}{\text{Operational full load}\left(\frac{h}{yr}\right)} \times \text{Electrical efficiency (\%)}, \quad (2)$$

$$\text{Total heat requirement for the process} = hl + q, \quad (3)$$

$$hl = U\,A\,\Delta T, \quad (4)$$

where hl is heat loss (kJ s^{-1}); U is the overall coefficient of heat transfer (W m^{-2} K); A is the cross-sectional area through which heat loss occurs (m^2); ΔT is temperature drop across the surface area (°C).

$$q = C\,Q\,\Delta T, \quad (5)$$

where q is the energy required for heating feedstock (kJ s^{-1}); C is the specific heat of the feedstock (kJ kg^{-1} °C^{-1}); Q is the volume to be added (m^3); ΔT is the outside and inside temperature difference (°C).

2.5. Final use of Energy Produced

The energy produced in the form of electricity and heat via the CHP unit was used in four main areas. These include: (i) the operation of AD plant; (ii) satisfying the dairy enterprises energy demand; (iii) exported to the national grid (electricity); (iv) exported to district heating system (thermal energy). The energy demand of the farm was calculated by using the energy requirements per litre of milk, as reported in the literature [54]. The average yield of an Irish dairy cow was assumed to be 5000 litres [55]. The thermal energy generated by the CHP unit was understood to displace kerosene, which is the primary heating fuel on Irish farms [54].

The heat produced that exceeds the needs of the plant and the farm has a number of potential local applications, such as drying woodchips, use in the horticulture sector, or in local industry. Another promising option is its use in a district-heating scheme, where heat generated is distributed from a central location through a network of insulated pipes to nearby residential and commercial energy users. Although these systems are not common in Ireland, this study has selected this technology to demonstrate its potential applicability. The study assumed that the thermal energy supplied to the scheme displaces kerosene, which is commonly used to heat residential homes in Ireland [54]. Equation (6) was used to describe the heat transfer capacity of the pipework utilised, with the subsequent heat losses calculated using Crane's methodology [56]. An average distance of 300 m was assumed between the CHP unit and the residential housing for this study.

$$Q = \pi\, r^2\, v\, \Delta T\, C \tag{6}$$

where Q is heat transfer capacity of pipe (kW); r is internal pipe radius (mm); v is the fluid velocity (m^3 s^{-1}); ΔT is temperature difference between the flow and return (°C); C is the specific heat of fluid (kJ kg^{-1} °C^{-1}).

2.6. Environmental Considerations

As depicted in the system boundary (Figure 1), all energy requirements for the operation of the AD plant were met internally via the CHP engine, where no CO_2 emissions were assumed. Surplus heat was fully used on-site with the understanding that it displaces kerosene, which is a conventional heating fuel on farms in Ireland [54]. According to Upton [57], the energy output from kerosene is 36.4 MJ l^{-1}, with CO_2 related emissions at 0.25 tCO_2 MWh^{-1}. All electricity generated that exceeds the energy demand of the AD plant and farm was exported to the national grid. The subsequent CO_2 savings were calculated based on the average emissions produced by the current energy mix of 0.367 t CO_2 MWh^{-1} [58].

The study accounted for the release of CO_2 in the combustion of biogas, at a rate of 83.6 kg GJ^{-1} [59]. Furthermore, the study included a "do nothing scenario", which incorporated the GHG emission savings in comparison to a no AD plant scenario. This included the emissions released from manure storage and application to land. Calculations follow guidelines from an OECD report, where emissions during storage are based on 20% potential biogas production over a 2-month period. Emissions from land application were calculated based on 10% remaining biogas potential [60]. The emission factor of biogas was calculated to be equivalent to be 11.9 kg CO_2 based on global warming potential (GWP) of 28 for methane [61].

2.7. Establishment and Operating Costs

As a new enterprise, establishment costs have to be accounted for within the model. The capital cost for the AD plant was quantified by compiling the capital costs and associated CHP electrical capacity of several SSAD plants (Figure 3). The data gathered gave an estimation of the average establishment costs for the model. Figure 3 correlates with similar studies [48], seeing a reduction in capital costs as the capacity of the plant increased.

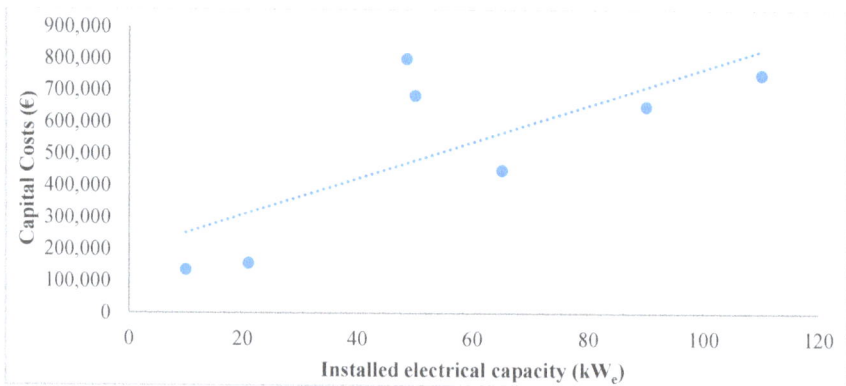

Figure 3. Establishment cost for farm-scale anaerobic digestion plants. Estimated, based on reports from [62–69].

The published data available on the running of Irish farm-scale AD plants are quite limited, mainly due to the relatively low number of plants in operation [16]. Considering these limitations, this study puts forward a list of annual expenditures to provide an appropriate representation of the Irish context.

- The plants incur an annual maintenance cost of 2.5% of the total capital cost, as reported in the literature [70].
- Insurance costs are typically 1% of total capital costs, which was observed in the model [71].
- The time required to operate the AD plant is a minimum of 8.5 working hours (net) per kW$_e$ capacity installed [67]. The cost of labour for a staff member in this position is estimated to be €15 hr^{-1}, which is considered standard in Ireland for this position [67].

Taxes and interest were not considered in the financial assessment of the plants. Taxes are calculated based upon the company's total profits or loss; therefore, including taxes would not reflect the actual revenue generated by the project. Interest was also not considered, as it would give a distorted representation of the cost of financing, because of its reliance on fluctuations in the financial market.

2.8. Revenue Streams and Financial Indicators

Electricity exported to the national grid is sold according to the Renewable Energy Feed-in Tariff (REFIT), introduced by the Irish Government in May 2010 [72]. These tariffs were offered for a period of 15 years with indexation, including a rate of 15.8 c€ kWh^{-1} for electricity exported from an AD plant with a CHP capacity of less than or equal to 500 kW. The current Irish REFIT schemes have since closed as of December 2015. It is presumed that this support will reopen in the coming years with a new funding round at the same rates for a period of 20 years. Revenue is calculated at the point that exported electricity enters the national grid, with subsequent transmission and distribution losses not considered.

Energy used to satisfy the farm's on-site power demand was based upon Ireland's business electricity rates from July to December 2017 [73]. The farm scenarios considered under this study were compatible with two rates: energy users consuming less than 0.02 GWh yr^{-1}, a purchase rate of 19.9 c€ kWh^{-1} applies; for energy users consuming between 0.02 to 0.5 GWh yr^{-1} a rate of 15.1 c€ kWh^{-1}.

The thermal energy produced via the CHP engine was understood to displace kerosene heating oil as a fuel at a cost of 8 c€ l^{-1} [74]. In addition, the simulated plants take advantage of the "*Support Scheme for Renewable Heat*" launched in mid-2019 [75]. The scheme provides a tariff of 2.95 c€ kWh^{-1} for a period of 15 years for AD plants producing less than 300 MWh yr^{-1} [75]. Accounting for the cost of infrastructure, the revenue generated from the sale of thermal energy via the district heating system was estimated to be €0.03 kWh^{-1}.

The financial indicators used to assess and compare the economic performance of the different plant scenarios included the net present value (NPV), internal rate of return (IRR), simple payback period, and discounted payback period. The NPV gives an indication of whether the project is profitable, taking into account the value of cash flows at different times, as shown in Equation (7). The IRR is a discount rate that makes the NPV of all cash flows equal to zero. The discount rate indicates the risk an investor takes in investing in a project. The higher the risk, the larger the discounted rate expected in compensation. This study used a discount factor of 5% and a project lifespan of 20 years, which is deemed appropriate for an AD project of this scale as reported in the literature [76–78]. The payback period refers to the number of years it takes to generate enough revenues to pay the investment back. The discounted payback period makes the same calculation but includes the time value of money.

$$NPV = \sum_{t=0}^{n} \frac{NCF_t}{(1+r)^t}, \qquad (7)$$

where NCF_t is the expected net cash flow, t is time and r is the discount rate.

Government supports through capital subvention grants have proven effective in increasing the deployment of AD plants by significantly lowering establishment costs. Grants of up to 50% have been adopted in countries such as Sweden, France, Wales and England [68]. This study incorporated a government subvention grant of 50% to provide an understanding of its implications.

3. Results

3.1. Technical Results

The technical parameters of the SSAD plants under study are presented in Table 5. These parameters provide an overview of the plant's operation in terms of feedstock used, plant specifications, resulting methane yield and application of energy. The cow manure available increased linearly, as it was directly proportional to the number of livestock on the farm. Interestingly, the farmland available for biogas production increased by just 35.4% between the smallest and largest farm sizes, showing that a larger proportion of farmland is potentially available for biogas production in farms with smaller herd sizes. Consequently, the grass feedstock represented a much larger percentage of total methane production in Scenario 1 (51%) in comparison to Scenario 5 (23%).

Table 5. Technical characteristics of scenarios under study.

	Scenario 1	Scenario 2	Scenario 3	Scenario 4	Scenario 5
Herd Characteristics					
Herd size (adult cows)	50	100	150	200	250
Cow manure yield (t FW yr^{-1})	505	1010	1515	2020	2525
Crop Characteristics					
Land available for energy crops (ha)	21.19	24.10	27.00	29.90	32.81
Grass silage yield (t FW yr^{-1})	941	1070	1199	1328	1457
CHP Specifications					
CHP engine power (kW_e)	17	26	39	46	55
Methane Yield					
Methane yield [a] ($m^3\ yr^{-1}$)	42,316	66,718	91,120	115,521	139,923
Energy Consumption of AD Plant					
Electricity consumption (kWh yr^{-1})	10,414	14,979	19,544	24,109	28,674
Heat consumption (kWh yr^{-1})	48,225	69,212	90,173	111,117	132,048
Farm Energy Demand					
Electricity demand (kWh yr^{-1})	8125	16,250	24,375	32,500	40,625
Heat demand (kWh yr^{-1})	2458	4915	7373	9830	12,288
Final Use of Excess Energy					
Exported electricity to grid (kWh yr^{-1})	102,697	159,917	217,137	274,357	331,577
Equivalent electricity consumption in residential homes (Irish homes $year^{-1}$) [b]	24.5	38.1	51.7	65.3	78.9
Exported heat to district heating system (kWh yr^{-1})	148,193	252,918	357,667	462,434	567,215
Equivalent heat consumption in residential homes (homes $year^{-1}$) [c]	13.5	23.0	32.5	42.0	51.6

[a] Methane yield utilised by the CHP unit annually; [b] Electricity consumption of an average residential house was assumed to be 4200 kWh yr^{-1} [79]; [c] Heat consumption of an average residential house was assumed to be 11,000 kWh yr^{-1} [79].

All scenarios examined exhibited a net energy generation, which was used to supply external applications, as shown in Figure 4. The farm's energy demand represented a relatively small portion of the total energy generated, ranging from 3.08% to 4.66%. The majority of the energy generated was exported off-site, representing between 73.04% and 79.13% of the total energy generated, demonstrating the need for external applications at the plants' planning stage.

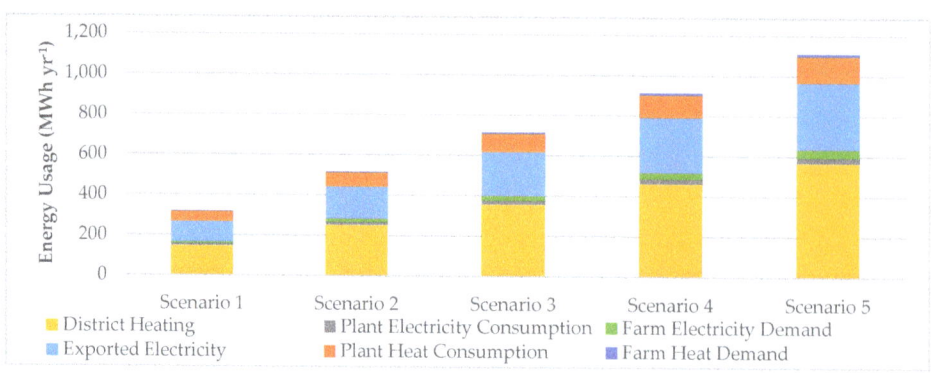

Figure 4. Final electrical and thermal energy usage via the combined heat and power (CHP) unit.

3.2. Environmental Results

A CO_2 balance that fully assesses the CO_2 inputs and outputs of the scenarios under investigation is presented in Table 6. The methodology undertaken was a "cradle-to-grave" approach to provide an accurate representation of the net CO_2- savings for each of the SSAD scenarios per year.

Table 6. Annual CO_2 balance for scenarios under study.

	Scenario 1	Scenario 2	Scenario 3	Scenario 4	Scenario 5
Herd size (adult cows)	50	100	150	200	250
CO_2 Produced (kg CO_2-eq. yr^{-1})					
Crop Production					
Soil ploughing and crumbling	264	300	336	372	408
Sowing and maintenance	300	341	382	423	464
Sowing	90	102	114	126	139
Weed control (fuel)	13	15	17	19	21
Weed control (mineral production)	36	41	46	51	56
Fertiliser spreading (fuel)	381	434	486	538	591
Fertiliser (mineral production)	5013	5699	6386	7073	7760
Feedstock Collection and Transport					
Harvest	2665	3030	3395	3760	4125
Harvest transport	1439	1636	1833	2030	2227
Silo compaction	497	565	633	701	769
Digester feeding (Crops)	1331	1513	1695	1878	2060
Collection and digester feeding (Manure)	92	185	277	370	462
Biogas Production Process					
CO_2 Content	133,652	210,722	287,722	364,863	441,933
Digestate Disposal					
Transport and spreading of digestate	2355	3387	4419	5451	6484
Total CO_2 produced	148,127	227,970	307,813	387,656	467,499
CO_2 reduction (kg CO_2-eq. yr^{-1})					
Do nothing scenario					
Manure storage	51,323	102,646	153,970	205,293	256,616
Manure land application	20,529	41,059	61,588	82,117	102,646
Farm Energy Demand					
On-farm electricity	2982	5964	8946	11,928	14,909
On-farm heating	614	1229	1843	2458	3072
Final Use of Excess Energy					
Electricity exported	37,690	58,689	79,689	100,689	121,689
Heat exported to district heating	37,048	63,229	89,417	115,609	141,804
Total CO_2 reduction	150,187	272,816	395,452	518,093	640,736
Net CO_2 savings (kg CO_2-eq. yr^{-1})	2059	44,846	87,639	130,437	173,237
Equivalent savings in cars displaced (cars per year) [a]	4.36	94.90	185.45	276.01	366.57

[a] Diesel consumption per car is reported to be 1259 litres yr^{-1}, as reported in the literature [80].

All scenarios investigated exhibited a net CO_2 reduction, ranging between 2059–173,237 kg CO_2-eq yr^{-1}. Significant net CO_2 savings were shown for each of the scenarios under investigation, even in the smallest farm size investigated (Scenario 1), with savings of 41,180 kg CO_2-eq. over the lifespan of the plant (equivalent to taking 87 cars off the road). This shows that SSAD can have a meaningful contribution, even at relatively small sizes. The activity which resulted in the largest production of CO_2 emissions was the "Biogas Production Process", where the release of CO_2 in the combustion of biogas contributed approximately 90% to 95% of the total CO_2 emissions released per annum.

3.3. Economic Results

A comprehensive economic analysis was carried out to investigate the revenues, expenditures, and financial indicators of each of the scenarios under investigation over a 20-year life span, as illustrated in Table 7. The results of this analysis showed SSAD plants to be economically feasible and profitable for commercial dairy farms with >100 dairy cows. However, the payback periods of farm sizes between 100 and 200 dairy cows were relatively long, which may dissuade potential investors.

Table 7. Economic results of small-scale anaerobic digestion plants over a 20-year lifespan.

	Scenario 1	Scenario 2	Scenario 3	Scenario 4	Scenario 5
Herd size (adult cows)	50	100	150	200	250
Project Revenues (€)					
On-site electricity savings	€32,338	€64,675	€73,613	€98,150	€122,688
On-site heating savings	€3932	€7864	€11,796	€15,728	€19,660
Sale of exported electricity	€323,727	€504,099	€684,472	€864,844	€1,045,216
Sale of exported heat to district heating	€88,916	€151,751	€214,600	€277,461	€340,329
Support Scheme for Renewable Heat	€66,663	€114,091	€161,530	€208,977	€256,430
Total Revenues	€515,576	€842,480	€1,146,000	€1,465,159	€1,784,322
Project Expenditures (€)					
Investment Costs					
Capital Costs Inc. CHP	€290,099	€345,479	€400,860	€456,241	€511,622
Operating Costs					
Maintenance and Repair Costs incl. CHP	€145,049	€172,740	€200,430	€228,121	€255,811
Insurance	€87,030	€103,644	€120,258	€136,872	€153,487
Labour	€42,625	€67,204	€91,784	€116,363	€140,943
Total Operating Costs	€274,704	€343,588	€412,472	€481,356	€550,241
Financial Indicators					
Profit before tax (€)	€240,872	€498,892	€733,538	€983,803	€1,234,082
NPV at 5% (€)	-€135,418	-€26,758	€67,339	€171,168	€275,006
IRR (%)	-2%	4%	7%	9%	11%
Payback period (Years)	25.65	12.87	10.18	8.66	7.75
Discounted payback period (Years)	N/A	24.02	14.56	11.64	10.05
Payback period Incl. capital grant (Years)	11.03	6.43	5.09	4.33	3.88
Discounted payback period Incl. capital grant (Years)	16.34	7.96	6.02	5.00	4.42

The largest revenue generators were electricity sold to the national grid and thermal energy sold to a nearby district heating system (where available). These two applications should be key considerations in the planning process for any such development considered.

The capital expenditure required decreased significantly as the capacity of the plant increases, primarily due to the economies of scale that occur. In addition to the economic analysis of the scenarios under study, this work also explored the adoption of a capital grant subvention in an attempt to provide a possible political pathway to increase the adoption of SSAD in Ireland. Such subvention has proven successful in countries such as Sweden, France, Wales and England, where capital grants of up to 50% have been applied [5]. As shown in Figures 5 and 6, the addition of a 50% capital subvention grant had a significant impact on the scenarios payback periods, resulting in all scenarios having a discounted payback period of under 17 years, with herd sizes above 100 cows particularly attractive with a payback period of under eight years.

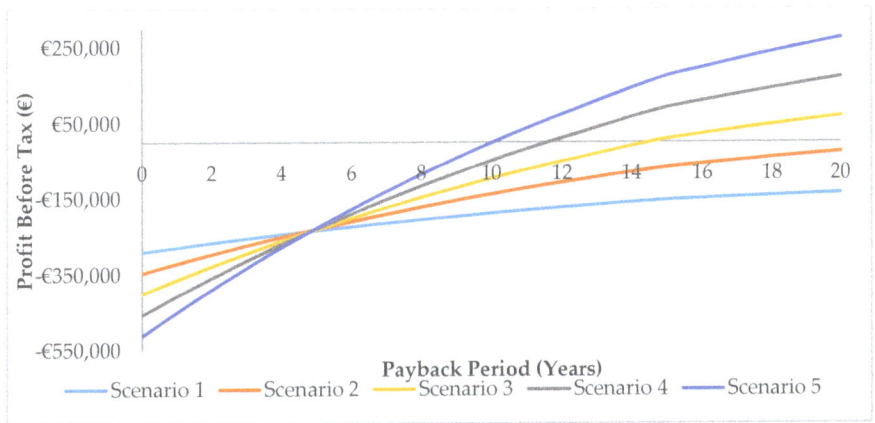

Figure 5. Comparison of discounted payback periods of scenarios.

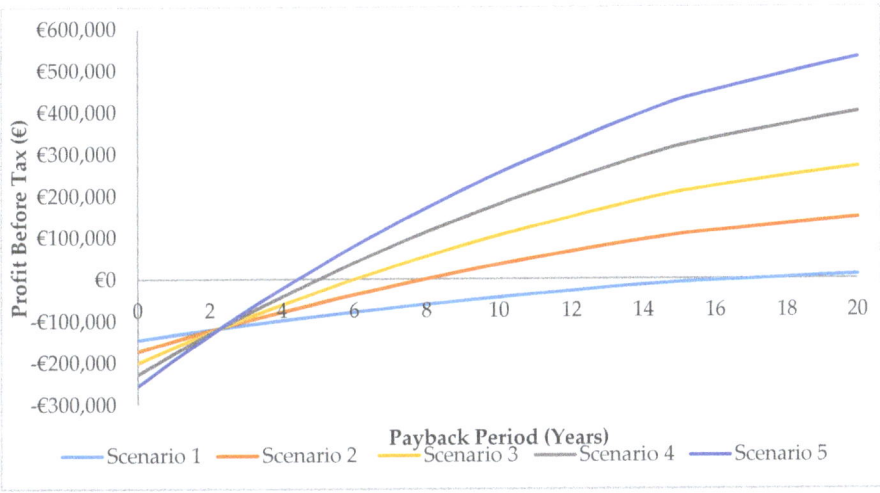

Figure 6. Comparison of discounted payback periods of scenarios (including 50% capital subvention grant).

4. Discussion

4.1. Financial Significance

The scenarios explored in this work showed SSAD plants to be economically feasible and profitable for dairy farms with >100 dairy cows. However, due to the study's boundaries, some costs were not considered, such as grid connection, civil works, etc. Such considerations are deemed important for the overall viability and future implementation of SSADs in practice and should be further investigated.

The need for further government supports and financial incentives is still apparent, where the relatively long payback periods projected may dissuade investors. Incentives available in Ireland, such as the REFIT scheme, have had a significant economic influence in reducing payback periods. Although the scheme provided only two tariffs, at a rate of 15.8 c€ kWh^{-1} for plants with a CHP capacity up to 500 kW and 13.7 c€ kWh^{-1} for plants exceeding this capacity [72]. Consequently, this puts smaller capacity plants at a disadvantage, as they have higher costs due to economies of scale.

Comparing Irish rates to other EU countries, Germany provides a rate of 23.73 c€ kWh^{-1} to plants with a total installed capacity of less than 75 kW$_e$. Likewise, the United Kingdom provides a tariff of 4.50 p£ kWh^{-1} to plants with a capacity of less than 250 kW. The issue reappears with the recently introduced *Support Scheme for Renewable Heat*, which provides a single tariff of 2.95 c€ kWh^{-1} to all plants generating less than 1000 MWh yr^{-1} [75]. To maximise the potential deployment of SSAD plants, government support schemes need to recognise the additional costs associated with smaller capacity plants and, therefore, implement policy that counteracts such expenditures.

Based on the literature and the findings of this study, the cost of finance has been the overriding barrier in the deployment of SSAD plants across Europe [11,63,81]. Issues cited include investors being uneasy with the technology due to limited case studies, the relative newness of the technology, and a lack of expertise within financial institutions to assess such plants. A potential government support explored in this study was the adoption of a capital grant subvention. Such legislation has proven successful in countries such as Sweden, France, Wales and England, where capital grants of up to 50% have been applied [68]. As shown in Table 7, the addition of a 50% capital grant subvention reduced the payback period by 3.88 years to 14.62 years, providing a possible pathway for the Irish government to support the deployment of SSAD.

Over the next few years, it is anticipated that the capital and operational costs of such plants will reduce dramatically. This is based on the most recent technological advancements, where a growing emphasis on smaller capacity plants has led to cost reductions, primarily through the development of modular systems and plug and play design. Several companies are in the testing phase or have fully commercialised such systems in the European market place, with a wide variety of technologies at various sizes now in development [82–87].

4.2. Environmental Outlook

From an environmental perspective, all scenarios examined exhibited a net CO_2 reduction ranging between 2 and 173 tonnes CO_2-eq. per year (Table 6). If the widespread deployment of SSAD were to occur in Ireland, a CO_2 reduction of at least 211,349 tonnes could be achieved if 20% of all farms with >250 dairy cows (61 farm holdings) were to implement the technology [27]. Ireland's expected failure to meet its EU 2020 commitments will put further pressure on the state to undertake a climate action policy, due to the compensation (in the form of carbon credits) it will be forced to pay [88]. In addition, the state has also committed itself to at least a 40% reduction in GHG emissions by 2030, resulting in a need for long-term climate action policy [89].

4.3. Comparison to Other Studies

Although this study reports on a specific case study, the results are relevant worldwide, especially those with significant livestock and agricultural productivities. In the literature, some studies have investigated the economics of implementing AD plants on small- to medium-sized farms in various countries and regions [68,90–92]. The overriding theme has been that the financial viability of a plant often needs to be assessed on a case-by-case bases as it is often highly dependent on local conditions, such as cost of energy, feedstock type and availability, and government incentives. Therefore, careful consideration of such variables needs to be taken at the project planning stages.

SSAD has increasingly become a topic of interest for researchers, mainly driven by the growing emphasis to reduce the negative environmental impact of agriculture waste streams and increasing investment in renewable energy production. Research trends in the topic have included the optimisation of plant design and operations [93–95], feedstock pre-treatments [95–97], the impact of trace compounds [98–100], and biogas cleaning technologies, and its integration to afford further energy generation [101–103]. Further research could expand the potential integration of these technologies with small-scale AD systems, and make its implementation more likely.

4.4. Irelands Future Outlook

Irelands national herd size has grown significantly in the past five years from 1,082,500 dairy cows in 2013 to 1,369,100 in 2018 (+21%) [20,104]. Much of the recent growth has stemmed from the removal of the European-wide milk production quotas in 2015, which saw milk output increase by 8% and 9% in 2016 and 2017, respectively [105]. Growing herd sizes allows SSAD plants to become more feasible because of economies of scale, as shown in Table 7.

The sector is projected to grow further due to Teagasc (the government's semi-state advisory authority) targeting a national herd increase of 19% by 2025 in comparison to 2018 [20,106]. When considering these targets, it is anticipated that the average national farm herd will exceed 100 dairy cows by 2025. The argument for the applicability of SSAD continues to deepen not only for the potential economic benefits but also for its capacity to mitigate GHG emissions.

5. Conclusions

Over the coming years, it is anticipated that the Irish government will come under increased pressure to enact measures to mitigate the negative environmental impact of the agricultural sector. This will be further heightened by the targeted growth of the dairy sector, increasing to 1.7 million dairy cows by 2025, with the average herd size growing to over 100 cows [106]. Of the renewable energy technologies available, SSAD is particularly promising for both the reduction of GHG emissions and the economic value in the form of on-site energy generation. This study uses a non-linear model to determine the technical, environmental, and economic viability of SSAD on Irish dairy farms ranging from 50 to 250 dairy cows. The study found the technology to be profitable within the lifespan of the plant on farms with dairy herds exceeding 100 cows (payback periods of 12.87 to 7.75 years). In addition, all scenarios with dairy herds sizes >100 cows showed a net CO_2 reduction ranging between 2059 and 173,237 kg CO_2-eq. yr^{-1}.

Although SSAD plants were shown to be viable, significant government supports are still needed to achieve financial returns that are attractive to investors. One support explored in this study was the inclusion of a capital subvention grant at rates similar to schemes in other EU countries. Incorporating the result had a significant economic impact, reducing payback periods by 3.88 years to 14.62 years. Furthermore, there is a need for the reintroduction of an electricity feed-in tariff applicable to SSAD plants. Without such a mechanism, the size of plants is limited to the electrical demand of local applications, significantly limiting expansion and financial returns. Both measures provide potential pathways for the government to support and accelerate a domestic biogas industry.

For future research, we suggest the analysis of the seasonal feedstock supply, parasitic energy consumption and net energy production variabilities experienced by farm-scale AD plants. Such seasonal variabilities can negatively affect the sustained operability and economic viability of plants as they often have contractual obligations to provide a consistent energy output year-round with minimum variations in the quantities and quality of energy produced. In addition, a greater understanding of Irish farmer's perception of AD is needed. Key information essential to the long-term success of AD in Ireland is still lacking in the literature, such as characteristics of potential adaptors, uptake rates, and perceived barriers.

Author Contributions: Conceptualization, S.O., E.E., and J.B.; validation, S.O., E.E., and J.B.; writing—original draft preparation, S.O.; writing—review and editing, S.O., E.E., S.C.P., G.L., and J.B.; supervision, E.E., S.C.P., and J.B. All authors have read and agreed to the published version of the manuscript.

Funding: This research was funded by the European Union's INTERREG VA Programme, managed by the Special EU Programmes Body (SEUPB), with match funding provided by the Department for the Economy, and Department of Jobs, Enterprise and Innovation in Ireland, grant number IVA5033.

Acknowledgments: In this section you can acknowledge any support given which is not covered by the author contribution or funding sections. This may include administrative and technical support, or donations in kind (e.g., materials used for experiments).

Conflicts of Interest: The authors declare no conflict of interest. The funders had no role in the design of the study; in the collection, analyses, or interpretation of data; in the writing of the manuscript, or in the decision to publish the results.

References

1. *Decision No. 406/2009/EC of the European Parliament and of the Council of 23 April 2009 on the Effort of Member States to Reduce their Greenhouse Gas Emissions to Meet the Community's Greenhouse Gas Emission Reduction Commitments up to 2020*; European Commission: Strasbourg, France, 2009.
2. Enviromental Protection Agency. *Ireland's Greenhouse Gas Emissions Projections 2018–2040*; Enviromental Protection Agency, Johnstown Castle Estate: Wexford, Ireland, 2020.
3. Parliamentary Budget Office. *An Overview of Carbon Pricing: PBO Publication 35 of 2019*; Tithe an Oireachtais Houses of the Oireachtas: Dublin, Ireland, 2019.
4. Howley, M.; Holland, M. *Energy-Related CO_2 Emissions in Ireland 2005–2016*; Sustainable Energy Authority of Ireland Energy: Cork, Ireland, 2018.
5. Hung, Y.T.; Kajitvichyanukul, P.; Wang, L.K. Advances in anaerobic systems for organic pollution removal from food processing wastewater. In *Handbook of Water and Energy Management in Food Processing*; Klemes, J., Smith, R., Kim, J.K., Eds.; Woodhead Publishing Ltd.: Cambridge, UK, 2008; pp. 755–775.
6. Chadwick, D.; Sommer, S.; Thorman, R.; Fangueiro, D.; Cardenas, L.; Amon, B.; Misselbrook, T. Manure management: Implications for greenhouse gas emissions. *Anim. Feed Sci. Technol.* **2011**, *166–167*, 514–531. [CrossRef]
7. Moral, R.; Bustamante, M.A.; Chadwick, D.R.; Camp, V.; Misselbrook, T.H. N and C transformations in stored cattle farmyard manure, including direct estimates of N2 emission. *Resour. Conserv. Recycl.* **2012**, *63*, 35–42. [CrossRef]
8. Gerber, P.J.; Steinfeld, H.; Henderson, B.; Mottet, A.; Opio, C.; Dijkman, J.; Falcucci, A.; Tempio, G. *Tackling Climate Change Through Livestock—A Global Assessment of Emissions and Mitigation Opportunities*; Food and Agriculture Organization of the United Nations (FAO): Rome, Italy, 2013.
9. Choong, Y.Y.; Norli, I.; Abdullah, A.Z.; Yhaya, M.F. Impacts of trace element supplementation on the performance of anaerobic digestion process: A critical review. *Bioresour. Technol.* **2016**, *209*, 369–379. [CrossRef] [PubMed]
10. Romero-Güiza, M.S.; Vila, J.; Mata-Alvarez, J.; Chimenos, J.M.; Astals, S. The role of additives on anaerobic digestion: A review. *Renew. Sustain. Energy Rev.* **2016**, *58*, 1486–1499. [CrossRef]
11. Kampman, B.; Leguijt, C.; Scholten, T.; Tallat-Kelpsaite, J.; Brückmann, R.; Maroulis, G.; Lesschen, J.P.; Meesters, K.; Sikirica, N.; Elbersen, B. *Optimal Use of Biogas From Waste Streams—An Assessment of the Potential of Biogas from Digestion in the EU Beyond 2020*; European Commission: Strasbourg, France, 2017.
12. Jeguirim, M.; Limousy, L. Strategies for bioenergy production from agriculture and agrifood processing residues. *Biofuels* **2018**, *9*, 541–543. [CrossRef]
13. Muradin, M.; Joachimiak-Lechman, K.; Foltynowicz, Z. Evaluation of Eco-Efficiency of Two Alternative Agricultural Biogas Plants. *Appl. Sci.* **2018**, *8*, 2083. [CrossRef]
14. Chiumenti, A.; Pezzuolo, A.; Boscaro, D.; Borso, F.D. Exploitation of Mowed Grass from Green Areas by Means of Anaerobic Digestion: Effects of Grass Conservation Methods (Drying and Ensiling) on Biogas and Biomethane Yield. *Energies* **2019**, *12*, 3244. [CrossRef]
15. Stambasky, J. *The Potential Size of the Anaerobic Digestion Industry in Ireland by the Year 2030*; Composting & Anaerobic Digestion Association of Ireland and The Irish Bioenergy Association: Meath, Ireland, 2016.
16. Auer, A.; Vande Burgt, N.H.; Abram, F.; Barry, G.; Fenton, O.; Markey, B.K.; Nolan, S.; Richards, K.; Bolton, D.; De Waal, T.; et al. Agricultural anaerobic digestion power plants in Ireland and Germany: Policy and practice. *J. Sci. Food Agric.* **2017**, *97*, 719–723. [CrossRef]
17. Tabassum, M.R.; Xia, A.; Murphy, J.D. Potential of seaweed as a feedstock for renewable gaseous fuel production in Ireland. *Renew. Sustain. Energy Rev.* **2017**, *68*, 136–146. [CrossRef]
18. O'Connor, S.; Ehimen, E.; Black, A.; Pillai, S.C.; Bartlett, J. An overview of biogas production from small-scale anaerobic digestion plants on European farms. In Proceedings of the Energy Technology Partnership (ETP) Annual Conference, University of Strathclyde, Glasgow, UK, 29 October 2018.

19. De Paor Consultancy. *Review of the Irish Agri-food Industry 2017–2018*; Irish Farmers Monthly: Dublin, Ireland, 2018.
20. Central Statistics Office. Livestock Survey: December 2018. Available online: https://www.cso.ie/en/releasesandpublications/er/lsd/livestocksurveydecember2018/ (accessed on 10 December 2019).
21. Wall, D.M.; O'Kiely, P.; Murphy, J.D. The potential for biomethane from grass and slurry to satisfy renewable energy targets. *Bioresour. Technol.* **2013**, *149*, 425–431. [CrossRef]
22. Hijazi, O.; Munro, S.; Zerhusen, B.; Effenberger, M. Review of life cycle assessment for biogas production in Europe. *Renew. Sustain. Energy Rev.* **2016**, *54*, 1291–1300. [CrossRef]
23. IrBEA and Cre. *Biogas Support Scheme-Mobilising an Irish biogas Industry with Policy and Action*; Irish Bioenergy Association (IrBEA), Composting and Anaerobic Digestion Association of Ireland (Cre): Meath, Ireland, 2019.
24. Government of Ireland, Fact Sheet on Irish Agriculture–January 2018. Available online: https://www.agriculture.gov.ie/media/migration/publications/2018/January2018Factsheet120118.pdf (accessed on 10 December 2019).
25. Holdent, N.M.; Brereton, A.J. An Assessment of the Potential Impact of Climate Change on Grass Yield in Ireland over the next 100 years. *Irish J. Agric. Food Res.* **2002**, *41*, 213–226.
26. Smyth, B.M.; Murphy, J.D.; O 'Brien, C.M. What is the energy balance of grass biomethane in Ireland and other temperate northern European climates? *Renew. Sustain. Energy Rev.* **2009**, *13*, 2349–2360. [CrossRef]
27. Central Statistics Office. Farm Structure Survey 2016. Available online: https://www.cso.ie/en/releasesandpublications/ep/p-fss/farmstructuresurvey2016/ (accessed on 10 December 2019).
28. Wickham, B. *Cattle Breeding in Ireland*; Irish Farmers Journal: Dublin, Ireland, 2007.
29. Berry, D.; Shallooa, L.; Cromieb, A.; Olorib, V.; Veerkampc, R.; Dillon, P.; Amer, P.; Evans, R.; Kearney, F.; Wickham, B. *The Economic Breeding Index: A Generation on*; Technical Report to the Irish Cattle Breeding Federation; The Irish Cattle Breeding Federation: Cork, Ireland, 2007.
30. O'Brien, D.; Capper, J.L.; Garnsworthy, P.C.; Grainger, C.; Shalloo, L. A case study of the carbon footprint of milk from high-performing confinement and grass-based dairy farms. *J. Dairy Sci.* **2014**, *97*, 1835–1851. [CrossRef] [PubMed]
31. Ryan, T.; Lenehan, J.J. Chapter 48-Winter accommodation for beef animals. In *Teagasc Beef Manual*; Teagasc: Carlow, Ireland, 2016; pp. 271–284.
32. *Teagasc Dairy Manual—A Best Practice Manual for Ireland's Dairy Farms*; Teagasc: Carlow, Ireland, 2016.
33. Midwest Plan Service. *Livestock Waste Facilities Handbook*; Iowa State University: Iowa, IA, USA, 1985.
34. Ryan, M. Grassland Productivity 1. Nitrogen and soil effects on yield of herbage. *Irish J. Agric. Res.* **1974**, *13*, 275–291.
35. Berglund, M.; Börjesson, P. Assessment of energy performance in the life-cycle of biogas production. *Biomass Bioenergy* **2006**, *30*, 254–266. [CrossRef]
36. Gerin, P.A.; Vliegen, F.; Jossart, J.M. Energy and CO_2 balance of maize and grass as energy crops for anaerobic digestion. *Bioresour. Technol.* **2008**, *99*, 2620–2627. [CrossRef]
37. Dillon, E.; Buckley, C.; Moran, B.; Lennon, J.; Wall, D. *Teagasc National Farm Survey-Fertiliser Use Survey 2005–2015*; Teagasc: Carlow, Ireland, 2018.
38. Pesticide Control Division. *Pesticide Usage in Ireland-Grassland & Fodder Crops Survey Report 2013*; Department of Agriculture, Food and the Marine: Kildare, Ireland, 2014.
39. Manchala, K.R.; Sun, Y.; Zhang, D.; Wang, Z.W. Anaerobic digestion modelling. *Adv. Bioenergy* **2017**, *2*, 69–141.
40. Nijaguna, B.T. *Biogas Technology*; New Age International: Delhi, India, 2002.
41. Boyle, W.C. Energy recovery from sanitary landfills. In *Microbial Energy Conversion*; Schlegel, H.G., Barnea, J., Eds.; Pergamon Press: Oxford, UK, 1977; pp. 119–138.
42. Oreggioni, G.D.; Gowreesunker, B.L.; Tassou, S.A.; Bianchi, G.; Reilly, M.; Kirby, M.E.; Toop, T.A.; Theodorou, M.K. Potential for energy production from farm wastes using anaerobic digestion in the UK: An economic comparison of different size plants. *Energies* **2017**, *10*, 1396. [CrossRef]
43. Jain, S. Cost of Abating Greenhouse Gas Emissions from UK Dairy Farms by Anaerobic Digestion of Slurry. Ph.D. Thesis, University of Southampton, Southampton, UK, 2013.
44. Theofanous, E.; Kythreotou, N.; Panayiotou, G.; Florides, G.; Vyrides, I. Energy production from piggery waste using anaerobic digestion: Current status and potential in Cyprus. *Renew. Energy* **2014**, *71*, 263–270. [CrossRef]

45. Miller, S.F.; Miller, B.G. The Occurrence of Inorganic Elements in Various Biofuels and Its Effect on the Formation of Melt Phases During Combustion. In Proceedings of the International Joint Power Generation Conference, Scottsdale, AZ, USA, 24–26 June 2002; pp. 873–880.
46. *Handreichung-Biogasgewinnung Und-Nutzung*; Fachagentur Nachwachsende Rohstoffe e.V.: Gulzow, Germany, 2006.
47. Akbulut, A. Techno-economic analysis of electricity and heat generation from farm-scale biogas plant: Cicekdagi case study. *Energy* **2012**, *44*, 381–390. [CrossRef]
48. Lantz, M. The economic performance of combined heat and power from biogas produced from manure in Sweden—A comparison of different CHP technologies. *Appl. Energy* **2012**, *98*, 502–511. [CrossRef]
49. Bioenergy Training Center. Introduction to Anaerobic Digestion Course-Types of Anaerobic Digesters. Available online: https://farm-energy.extension.org/types-of-anaerobic-digesters/ (accessed on 28 January 2020).
50. The German Solar Energy Society; Ecofys. *Planning and Installing Bioenergy Systems: A Guide for Installers, Architects and Engineers*; Taylor & Francis: Lindon, UK, 2005.
51. Murphy, J.D.; McKeogh, E.; Kiely, G. Technical/economic/environmental analysis of biogas utilisation. *Appl. Energy* **2004**, *77*, 407–427. [CrossRef]
52. Enerblu Cogeneration. Combined Heat and Power (CHP) Specifications. Available online: http://www.enerblu-cogeneration.com/products/80-impianti-biogas.html (accessed on 12 May 2019).
53. Walsh, S. *A Summary of Climate Averages for Ireland 1981–2010*; Met Éireann: Dublin, Ireland, 2012.
54. Upton, J.; Humphreys, J.; Groot Koerkamp, P.W.G.; French, P.; Dillon, P.; De Boer, I.J.M. Energy demand on dairy farms in Ireland. *J. Dairy Sci.* **2013**, *96*, 6489–6498. [CrossRef]
55. Irish Farming Association. Factsheet on Irish Dairying 2017. Available online: https://www.ifa.ie/sectors/dairy/dairy-fact-sheet/ (accessed on 19 April 2019).
56. Crane, M. Energy efficient district heating in practice-the importance of achieving low return temperatures. In Proceedings of the CIBSE Technical Symposium Edinburgh, Heriot Watt University, Edinburgh, UK, 14–15 April 2016.
57. The Engineering Toolbox. Combustion from Fuels-Carbon Dioxide Emission. Available online: https://www.engineeringtoolbox.com/co2-emission-fuels-d_1085.html (accessed on 26 June 2019).
58. Commission for Regulation of Utilities. *Fuel Mix Disclosure 2016*; Commission for Regulation of Utilities: Dublin, Ireland, 2017.
59. Nielsen, M.; Nielsen, O.K.; Plejdrup, M. *Danish Emission Inventories for Stationary Combustion Plants. Inventories until 2011*; Scientific Report from DCE–Danish Centre for Environment and Energy; Aarhus University, Danish Centre for Environment and Energy: Aarhus, Denmark, 2014.
60. Organisation of Economic Community and Development (OECD). *Estimation of Greenhouse-Gas Emissions and Sinks Final Report from ODED Expererts Meeting, 18–21, February 1991*; OECD: Paris, France, 1991.
61. Myhre, G.; Shindell, D.; Bréon, F.M.; Collins, W.; Fuglestvedt, J.; Huang, J.; Koch, D.; Lamarque, J.F.; Lee, D.; Mendoza, B.; et al. 2013: Anthropogenic and Natural Radiative Forcing. In *Climate Change 2013: The Physical Science Basis. Contribution of Working Group I to the Fifth Assessment Report of the Intergovernmental Panel on Climate Change*; Stocker, T.F., Qin, D., Plattner, G.K., Tignor, M., Allen, S.K., Boschung, J., Nauels, A., Xia, Y., Bex, V., Midgley, P.M., Eds.; Cambridge University Press: Cambridge, UK, 2013.
62. Redman, G. *A Detailed Economic Assessment of Anaerobic Digestion Technology and its Suitability to UK Farming and Waste Systems*; The Andersons Centre: Leicestershire, UK, 2010.
63. Bywater, A. *A Review of Anaerobic Digestion Plants on UK Farms—Barriers, Benefits and Case Studies*; Royal Agricultural Society of England: Stoneleigh Park, Warwickshire, UK, 2011.
64. Heinsoo, K. *Implementation Plan for BioEnergy Farm*; BioEnergy Farm Publication: Tartu, Estonia, 2011; Available online: https://ec.europa.eu/energy/intelligent/projects/sites/iee-projects/files/projects/documents/bioenergy_farm_description_of_best_case_examples_en.pdf (accessed on 28 January 2019).
65. The Wales Centre Of Excellence For Anaerobic Digestion; Landes Energie Verein Steiermark; Vienna University of Technology. *European Case Studies of Anaerobic Digestion Plants Showcasing their Monitoring Practices*; Bio-methane Regions: Brussels, Belgium, 2012; Available online: https://www.severnwye.org.uk/fileadmin/Resources/SevernWye/Projects/Biomethane_Regions/Downloads/BMR_D_5_1_Best_Practice_Monitoring_FINAL_a_Resubmission_Final7.pdf (accessed on 28 January 2020).

66. De Dobbelaere, A.; De Keulenaere, B.; De Mey, J.; Lebuf, V.; Meers, E.; Ryckaert, B.; Schollier, C.; Van Driessche, J. *Small-scale Anaerobic Digestion: Case Studies in Western Europe*; Mia Demeulmeester: Rumbeke, Belgium, 2015.
67. Hjort-Gregersen, K. *Market Overview Micro Scale Digesters*; AgroTEch A/S: Aarhus, Denmark, 2015; Available online: http://www.bioenergyfarm.eu/wp-content/uploads/2015/05/WP2_report_revised_version-FINAL-ENGLISH.pdf (accessed on 19 April 2019).
68. Lukehurst, C.; Bywater, A. *Exploring the Viability of Small Scale Anaerobic Digesters in Livestock Farming*; IEA Bioenergy: Paris, France, 2015; Available online: https://www.iea-biogas.net/files/daten-redaktion/download/Technical%20Brochures/Small_Scale_RZ_web2.pdf (accessed on 28 January 2020).
69. Li, Y.; Samir, K.K. *Bioenergy: Principles and Applications*; John Wiley & Sons: Hoboken, NZ, USA, 2016.
70. Jones, P. *Missing Integrated Systems for Farm Diversification into Energy Production by Anaerobic Digestion: Implications for Rural Development, Land Use and the Environment-Modelling the Commercial Profitability of AD Energy Production at the Farm Level within Arable and Dairy Systems*; University of Reading: Reading, UK, 2010.
71. Jones, P.; Salter, A. Modelling the economics of farm-based anaerobic digestion in a UK whole-farm context. *Energy Policy* **2013**, *62*, 215–225. [CrossRef]
72. Department of Communications Energy and Natural and Resources. *Renewable Energy Feed in Tariff: A Competition For Electricity Generation—From Biomass Technologies 2010–2015*; Department of Communications Energy and Natural and Resources: Dublin, Ireland, 2013.
73. Howley, M.; Barriscale, A. *Electricity & Gas Prices in Ireland—2nd Semester (July–December) 2016*; Sustainable Energy Authority of Ireland: Dublin, Ireland, 2017.
74. SEAI. *Domestic Fuels Comparison of Energy Costs*; Sustainable Energy Authority of Ireland: Dublin, Ireland, 2018. Available online: https://www.seai.ie/resources/publications/Domestic-Fuel-Cost-Comparison.pdf (accessed on 17 November 2019).
75. Department of Communications Climate Action & Environment. *Support Scheme for Renewable Heat Scheme Overview*; Department of Communications Climate Action & Environment: Dublin, Ireland, 2018.
76. Bishop, C.P.; Shumway, C.R. The Economics of Dairy Anaerobic Digestion with Coproduct Marketing. *Rev. Agric. Econ.* **2009**, *31*, 394–41016. [CrossRef]
77. Abu-Orf, M.; Bowden, G.; Pfrang, W. *Wastewater Engineering: Treatment and Resource Recovery*; Tchobanoglous, G., Stensel, H.D., Tsuchihashi, R., Burton, F., Eds.; McGraw Hill Higher Education: New York, NY, USA, 2014.
78. Redican, J.H. *Federal Discount Rate for Fiscal Year 2019: Economic Guidance Memorandum 19-01*; Army Crops of Engineers: Washington, DC, USA, 2018.
79. Commission for Energy Regulation. *Review of Typical Domestic Consumption Values for Electricity and Gas Customers*; Commission for Energy Regulation: Dublin, Ireland, 2017.
80. Central Statistics Office. Fuel Consumption by Sector, Fuel Type and Year. Available online: https://statbank.cso.ie/px/pxeirestat/Statire/SelectVarVal/Define.asp?maintable=SEI06&PLanguage=0 (accessed on 13 December 2019).
81. Ricardo Energy & Environment. *Assessment of Cost and Benefits of Biogas and Biomethane in Ireland*; Sustainable Energy Authority of Ireland: Dublin, Ireland, 2017.
82. Earthlee, Onsite Organic Waste Management & Energy Solution. Available online: https://www.earthlee.com/ (accessed on 29 December 2018).
83. Alchemy Utilities. Creating a Circular Economy. Available online: https://alchemyutilities.ie/ (accessed on 29 December 2019).
84. Demetra. AD Bag-biogas Made Easy. Available online: https://www.demetra.ie/wp-content/uploads/2016/12/ADbag.pdf (accessed on 29 December 2019).
85. Bio Ferm Energy Systems. Range of Anaerobic Digestion Systems. Available online: https://www.biofermenergy.com/ (accessed on 29 December 2019).
86. SEaB Energy. Products. Available online: https://seabenergy.com/ (accessed on 29 December 2019).
87. QUBE Renewables. Innovative Small Scale Anaerobic Digestion. Available online: https://www.quberenewables.co.uk/ (accessed on 29 December 2019).
88. Environmental Protection Agency. *Ireland's Greenhouse Gas Emission Projections 2016–2035*; Environmental Protection Agency: Dublin, Ireland, 2017.
89. Scheer, J.; Clancy, M.; Gaffney, F. *Ireland's Energy Targets—Progress, Ambition & Impacts*; Sustainable Energy Authority of Ireland: Dublin, Ireland, 2016.

90. Wilkinson, K.G. Development of on-farm anaerobic digestion. In *Integrated Waste Management*; Kumar, S., Ed.; InTech.: Rijeka, Croatia, 2011; Volume 1, pp. 179–194.
91. Walker, M.; Theaker, H.; Yaman, R.; Poggio, D.; Nimmo, W.; Bywater, A.; Blanch, G.; Pourkashanian, M. Assessment of micro-scale anaerobic digestion for management of urban organic waste: A case study in London, UK. *Waste Manag.* **2017**, *122*, 221–236. [CrossRef]
92. Department for Business Energy & Industrial Strategy. *Review of Support for Anaerobic Digestion and Micro-combined Heat and Power Under the Feed-in Tariff Scheme*; Department for Business Energy & Industrial Strategy: Dublin, Ireland, 2017.
93. Nguyen, D.; Gadhamshetty, V.; Nitayavardhana, S.; Khanal, S.K. Automatic process control in anaerobic digestion technology: A critical review. *Bioresour. Technol.* **2015**, *193*, 513–522. [CrossRef]
94. Kougias, P.G.; Angelidaki, I. Biogas and its opportunities–A review. *Front. Environ. Sci. Eng.* **2018**, *12*, 14. [CrossRef]
95. Wiese, J.; Haeck, M. Instrumentation, control and automation for full-scale manure-based biogas systems. *Water Sci. Technol.* **2006**, *54*, 1–8. [CrossRef]
96. Carlsson, M.; Lagerkvist, A.; Morgan-Sagastume, F. The effects of substrate pre-treatment on anaerobic digestion systems: A review. *Waste Manag.* **2012**, *32*, 1634–1650. [CrossRef] [PubMed]
97. Ehimen, E.A.; Connaughton, S.; Sun, Z.; Carrington, G.C. Energy recovery from lipid extracted, transesterified and glycerol codigested microalgae biomass. *GCB Bioenergy* **2009**, *1*, 371–381. [CrossRef]
98. Papurello, D.; Tomasi, L.; Silvestri, S.; Santarelli, M. Evaluation of the Wheeler-Jonas parameters for biogas trace compounds removal with activated carbons. *Fuel Process. Technol.* **2016**, *152*, 93–101. [CrossRef]
99. Rasi, S.; Läntelä, J.; Rintala, J. Trace compounds affecting biogas energy utilisation-A review. *Energy Convers. Manag.* **2011**, *52*, 3369–3375. [CrossRef]
100. Papurello, D.; Boschetti, A.; Silvestri, S.; Khomenko, I.; Biasioli, F. Real-time monitoring of removal of trace compounds with PTR-MS: Biochar experimental investigation. *Renew. Energy* **2018**, *125*, 344–355. [CrossRef]
101. Kupeckia, J.; Papurelloc, D.; Lanzinic, A.; Naumovicha, Y.; Motylinskia, K.; Blesznowskia, M.; Santarelli, M. Numerical model of planar anode supported solid oxide fuel cell fed with fuel containing H2S operated in direct internal reforming mode (DIR-SOFC). *Appl. Energy* **2018**, *230*, 1573–1584. [CrossRef]
102. Wasajja, H.; Lindeboom, R.E.F.; Van Lier, J.B.; Aravind, P.V. Techno-economic review of biogas cleaning technologies for small scale off-grid solid oxide fuel cell applications. *Fuel Process. Technol.* **2020**, *197*, 106215. [CrossRef]
103. Papurello, D.; Lanzini, A. SOFC single cells fed by biogas: Experimental tests with trace contaminants. *Waste Manag.* **2018**, *72*, 306–312. [CrossRef]
104. Central Statistics Office. Livestock Survey: December 2013. Available online: https://www.cso.ie/en/releasesandpublications/er/lsd/livestocksurveydecember2013/ (accessed on 13 December 2019).
105. *Food Harvest 2020: A Vision for Irish Agri-food and Fisheries*; Department of Agriculture Fisheries and Food: Dublin, Ireland, 2010.
106. Teagasc. *Sectoral Road Map: Dairying*; Teagasc: Carlow, Ireland, 2016.

© 2020 by the authors. Licensee MDPI, Basel, Switzerland. This article is an open access article distributed under the terms and conditions of the Creative Commons Attribution (CC BY) license (http://creativecommons.org/licenses/by/4.0/).

The objective of the study was to carry out comparative LCA of four agroforestry systems with olive production in Italy and two production systems, specifically one agroforestry and one conventional wheat production system, in Denmark. Comparisons of environmental footprints were carried out among the production systems within the country and between Italy and Denmark.

2. Method

2.1. Study Site

The case studies are four agroforestry systems in Italy and two systems, specifically one agroforestry and one conventional wheat production system, in Denmark. The four agroforestry systems are olive trees under (i) silvopastoral agroforestry, (ii) organic agroforestry, (iii) traditional agroforestry, and (iv) conventional olive system, located in Orvieto in the Umbria region in Italy. The two production systems in Denmark are (v) a combined food and energy agroforestry system and (vi) a conventional wheat production system in Denmark.

The four production systems in Italy have olive trees as one of the components, while management and inputs differ among the systems depending on the production system as described below:

(i) Silvopastoral agroforestry system consists of 135 trees in 1.0 ha. Olive production is 3.6 t ha^{-1}, and natural grass pasture is present between the olive trees. A total of 177 sheep graze the pasture for 150 days a year, producing 0.33 kg dung in dry matter and 2.9 kg urine per day per sheep, with fertilization effects on the grass pasture. The trees were planted in 1956. While no synthetic fertilizer was used, biological copper was applied at 1.7 kg ha^{-1} [16].

(ii) Organic agroforestry system covered an area of 4.5 ha. The system has 200 trees ha^{-1} and olive yield is 2.2 t ha^{-1}. Naturally growing grass are present in between the trees, fertilizer application is 4.0 t ha^{-1} cow manure in dry matter, and no pesticides were used [16].

(iii) In traditional agroforestry, olive trees were planted in 1982. The yield is 7.05 t ha^{-1} olives with tree density of 529 trees ha^{-1}. The cultivation area was drip-irrigated and fertilized with olive prunings and olive pomace. Glyphosate was applied to control weeds [16].

(iv) Conventional olive system has planting density of 250–400 olive trees ha^{-1} with irrigation facility and practice mechanized harvesting of olives. The olive trees are fertilized with nitrogen (90–150 kg ha^{-1}), phosphorus (20–30 kg ha^{-1}), and potassium (70–120 kg ha^{-1}). Olive yield is estimated to be 4.5 t ha^{-1} with 300 trees ha^{-1} [17,18].

In Denmark, two production systems, namely a combined food and energy production agroforestry system and a conventional wheat production system, were investigated and the descriptions of the system are provided below:

(v) The combined food and energy system covers an area of 11.1 ha, of which 10.1 ha is cropped with barley, wheat, and clover in a four-year crop rotation and 1.0 ha of biofuel crops (mix of willow, hazelnut, and alder). The biofuel crops consist of four shelterbelts of short rotation woody crops, spatially placed at 50, 100, 150, and 200 m, forming alleys for the food and fodder crop production. The production yields were 5430 kg wheat ha^{-1}, 3750 barley kg ha^{-1}, 6700 kg clover ha^{-1}, and 4078 kg woodchips ha^{-1} annually [19,20].

(vi) Conventional wheat production systems relate to winter wheat production, which is sown in September or October and harvested the following year in August. A total of 50% of nitrogen (95 kg ha^{-1}), potassium (20 kg ha^{-1}), and phosphorus (60 kg ha^{-1}) is applied at the time of sowing and the remaining 50% of the nitrogen (95 kg ha^{-1}) is applied in the spring. Fungicides and herbicides are applied as per the standard practice at the experimental farm in Taastrup in Denmark. The details of the management and crop production are available from another study at the same experimental farm [19].

2.2. Data Collection

In Italy, the collection of data on each production system was obtained through interviews and surveys. A detailed interview was conducted with the respective farm mangers to obtain initial information on cultivation and management practices at each farm (Figure 1). Information on the description of the production systems, inputs used, and the management practices were gathered from the managers of the respective farms. Data was collected on animal and plant production, seed use, use of fertilizer or plant protection products, field size, irrigation, machinery use, and if any recent changes have been made to the farm practice within the past five years.

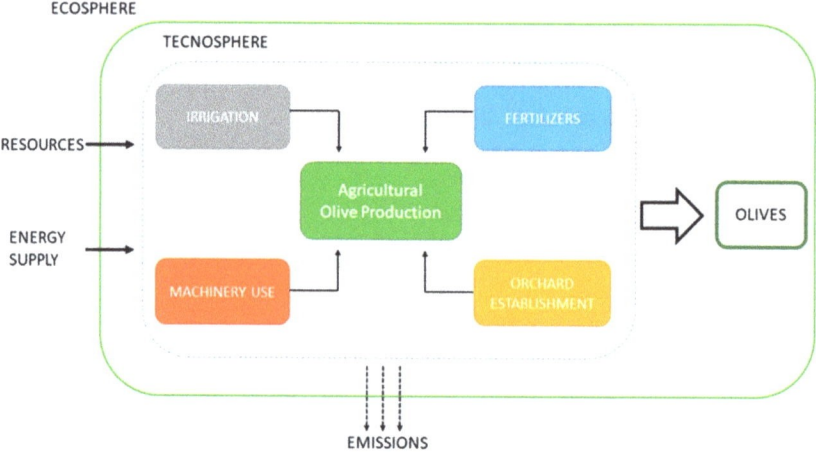

Figure 1. General flow diagram of olive production systems in Italy.

In Italy, three olive-based agroforestry systems were compared with a conventional olive production system in Italy, with a yield of 4300 kg ha^{-1} described in detail in the Ecoinvent database 3.3 [21]. The Ecoinvent database is the world's leading life cycle inventory (LCI) database, which provides well-documented data for thousands of products including agricultural processes.

In Denmark, input and production data were taken for the conventional wheat production system and combined food and energy system in Denmark from a previous study on the system [19]. Data collected included yields, field size, seed use, use of fertilizer or plant protection products, irrigation, machinery use, field preparation and, if any, recent changes made to the farm practice within the past five years (Figure 2). The combined food and energy system was compared with conventional wheat production system with a yield of 7341 kg wheat ha^{-1} and application of fertilizer, pesticides, and herbicides according to Danish standard practice as per the description provided under the study sites.

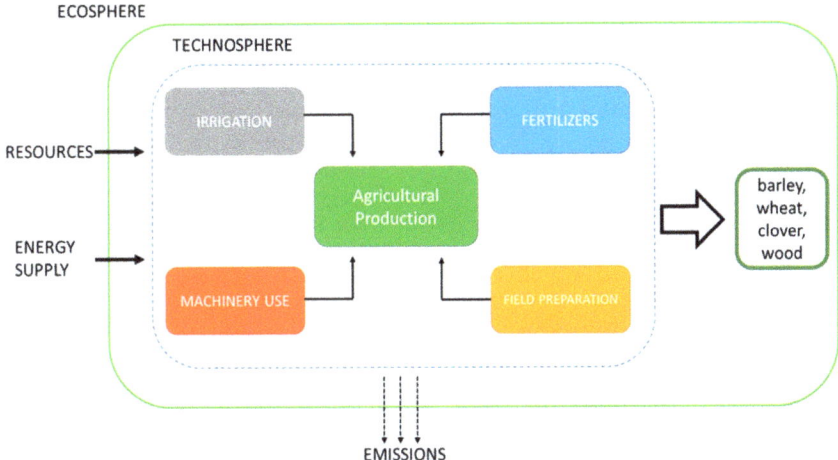

Figure 2. General flow diagram for the combined food and energy production system in Denmark.

2.3. Life Cycle Analysis

There is not one strictly defined methodology for conducting LCA analyses for agricultural production. However, according to the principles set out in ISO PN-EN 14040 [22], the full analysis should include the four phases of PN-EN ISO 14040 (goal and scope definition), PN-EN ISO 14041 (LCI-life cycle inventory), PN-EN ISO 14042 (LCIA-life cycle impact assessment), and PN-EN ISO 14043 (life cycle interpretation). The LCIA is the phase in which the environmental impact assessment (e.g., land use) of the products applies. Data obtained in the previous LCI phase were transformed into impact category indicators. This was done by selecting the impact category and impact indicators, assigning LCI results, and calculating the category indicator values. According to the methodology developed by the Society of Environmental and Chemical Sciences (SETAC), 14 environmental impact categories were taken into account in LCA, of which three were considered in this study, namely, global warming potential (GWP), acidification, and eutrophication.

LCA was done by applying SimaPro 8.4 software [23]. To calculate the emissions of inputs production for the Italian agroforestry systems, the Ecoinvent database 3.3 was used. Dinitrogen monoxide (N2O), ammonia (NH3), and nitric oxide (NO) were modelled based on methodology described in [24]. N_2O follows IPPC guidelines [25] Tier 1 for animal production and Tier 1 for crop production. NH_3 calculations were based on emission factors for NH_3, based on application of mineral N fertilizer and as a function of soil pH. NO is relatively of low importance compared to other sources, for that reason, simple emission factors were used [26]. Carbon dioxide (CO_2) emissions after urea or lime application were calculated based on the factor (1.57 kg CO_2/kg Urea-N[10] for urea and 44 kg CO_2/kg limestone or 48 kg CO_2/kg dolomite [26]. Irrigation quantity was calculated based on Methodological Guidelines for the Life Cycle Inventory of Agricultural Product [26] as consumed water for yield production (m^3 t^{-1}). The emissions related to pesticide use were not included due to low influence on calculated environmental impacts. The impacts of GWP, acidification, and eutrophication were calculated using the CML method [27]. The system boundary was cradle to olive farm gate, i.e., from the extraction of raw materials to the farm gate until the olives were harvested. For the Danish farm, the direct NO2-N and indirect NO2 emissions were calculated based on IPCC 2006 methodology [25], while the ReCiPe method [28] was applied to calculate the potential of GWP, acidification, and eutrophication. Due to the agroforestry systems producing different crops, the yields are not directly comparable. Hence, the yields were converted to monetary values based on the prices indicated: 0.49 \$ kg^{-1} wheat, 0.45 \$ kg^{-1} barley, 0.16 \$ kg^{-1} clover, and 0.14 \$ kg^{-1} woodchips [19].

3. Results

The estimated greenhouse gas (GHG) emissions from fertilizer and irrigation inputs are displayed in Table 1. The results showed that the main GHG emissions from the silvopastoral system was nitric oxide and ammonia, and from the organic systems, it was mainly nitric oxide. The main contribution of the traditional agroforestry system was CO_2 and the largest emission was from irrigation, which was only applied in the traditional agroforestry production system.

Table 1. Estimations of estimated greenhouse gas (GHG) emissions from four agroforestry systems in Italy and two production systems in Denmark. CFE: combined food and energy system.

Agricultural Practice	On Field Emissions	Methodology	Unit	Italy				Denmark	
				Silvo-Pastoral	Organic	Traditional	Conventional	CFE	Conventional
Fertilization	N_2O	EEA/EMEP (2013)	g kg^{-1}	0.00	0.50	0.31	0.4		
	CO_2	Nemecek (2014)	g kg^{-1}	0.00	0.00	31.18	5		
	NH_3	EEA/EMEP (2013)	g kg^{-1}	8.92	0.00	1.03	3		
	NO	EEA/EMEP (2013)	g kg^{-1}	12.35	34.20	0.24	0.7		
	Chemical N		kg ha^{-1}					0.00	190.00
	Manure N		kg ha^{-1}					16.00	35.20
	Crop residues N		kg ha^{-1}					0.00	0.00
	Direct NO_2-N N_2O	IPCC 2006	kg ha^{-1}					0.25	3.54
	Indirect (VOL.) NO_2-N N_2O	IPCC 2006	kg ha^{-1}					0.05	0.41
	Indirect (leaching) NO_2-N N_2O	IPCC 2006	kg ha^{-1}					0.06	0.80
Irrigation	H_2O	Nemecek (2014)	m^3	0.00	0.00	0.14			

In the combined food and energy production system in Denmark, GHG emissions emanated mainly from nitrogen (N) from the applied manure, with a small contribution from direct NO_2-N dinitrogen monoxide (N_2O). Much larger GHG contributions were recorded for conventional wheat production systems due to chemical fertilizer application.

The production system impacts on the environment are displayed in Table 2. In Italy, the highest global warming potential (GWP) was contributed by the traditional production systems, while silvopastoral systems exhibited the lowest GWP. The highest acidification and eutrophication effects originated in the silvopastoral system, whereas the lowest effects were recorded in the traditional system. In Denmark, GWP, acidification, and eutrophication were higher in the conventional wheat system than the combined food and energy system.

Table 2. Environmental impacts of the production systems. Greenhouse gas emissions from the Italian production systems are provided per weight (kg) of olive yield, while the unit of income ($) is used for the Danish production systems. GWP: global warming potential (GWP100a).

Impact Category	Unit	Italy				Denmark	
		Silvo-Pastoral	Organic	Traditional	Conventional Orchard	Combined Food and Energy	Conventional Wheat System
GWP	kg CO_2-eq. kg^{-1} yr^{-1}	0.166	0.266	0.655	0.388	0.615	4.922
Acidification	kg SO_2-eq. kg^{-1} yr^{-1}	0.022	0.018	0.007	0.008	1.290	6.844
Eutrophication	kg PO_4-eq. kg^{-1} yr^{-1}	0.005	0.005	0.002	0.004	2.957	20.446

Evident from Figure 3, the conventional and tradition production systems in Italy exhibited the highest adverse impacts on the environment due to fertilization. Silvopastoral and organic systems emissions mainly emanated from machinery use. GHG emissions due to orchard establishment were high on acidification and eutrophication impact categories in the silvopastoral system.

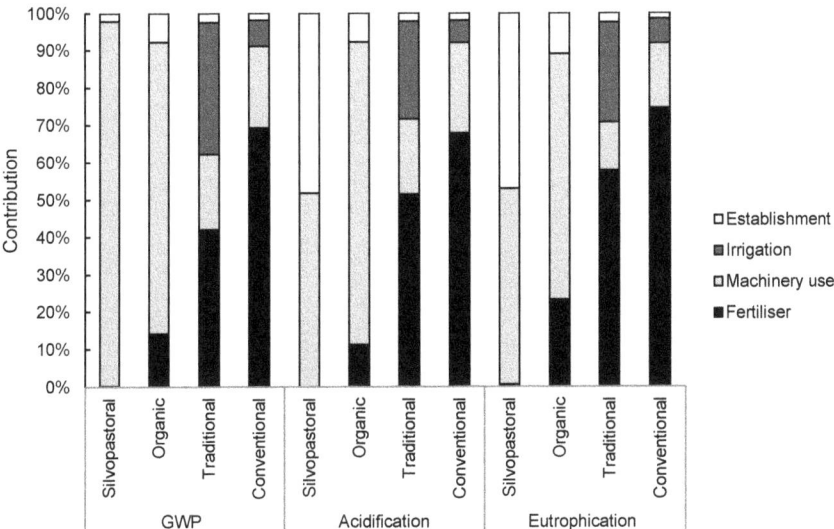

Figure 3. Percentage contribution from fertilizer, machinery use, irrigation, and orchard establishment on environmental impacts categories of global warming potential (GWP), acidification, and eutrophication for four Italian production systems.

Figure 4 shows that the application of fertilizer and manure had the greatest GHG emission contribution on GWP, acidification, and eutrophication for the conventional wheat and the combined food and energy system in Denmark. The second largest contributor to the GWP comes from machinery use, while seeding operation resulted in the second highest contribution to acidification and eutrophication in both production systems.

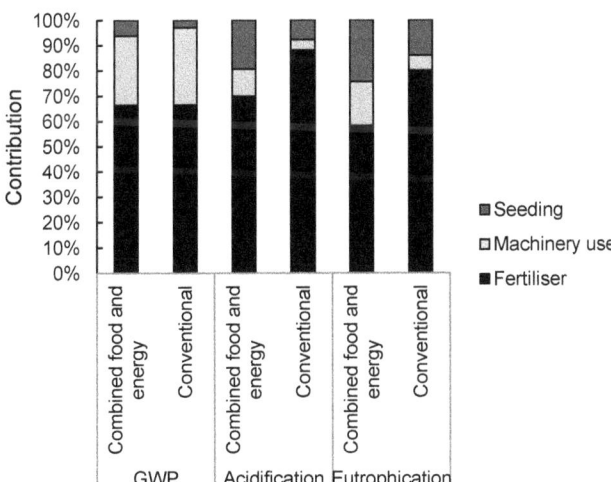

Figure 4. Percentage contribution from fertilizer, machinery use, and seeding on the environmental impact categories of global warming potential (GWP), acidification, and eutrophication for agroforestry and conventional production systems in Denmark.

The income from the diversity of produce in the combined food and energy systems in Denmark is presented in Table 3. Clover contributed to 49.1% of the total income, followed by wheat (30.5%) and barley (19.3%). The sale of willow woodchips amounted to only 1% of the total income.

Table 3. Income from the produce in the conventional wheat and combined food and energy production systems in Denmark.

Crop	Unit	Combined Food and Energy	Conventional	Share of Crop at Farm
Wheat	$ farm^{-1}	16,975.07		30.5
Barley	$ farm^{-1}	10,760.29		19.3
clover	$ farm^{-1}	27,342.72		49.1
Willow	$ farm^{-1}	570.89		1.0
Wheat	$ farm^{-1}		39,028.43	100.0
Total	$ farm^{-1}	55,648.96	39,028.43	

The environmental impacts of the two production systems in Denmark were estimated based on the income from each production system (Table 4). The combined food and energy system had a lower GWP, acidification, and eutrophication impact compared to the conventional wheat production system.

Table 4. Environmental impacts of the conventional wheat and combined food and energy production systems in Denmark. GWP: global warming potential (GWP100a).

Impact Category	Unit	Combined Food and Energy	Conventional
GWP	g CO$_2$-eq. \$$^{-1}$ yr^{-1}	123.89	1413.80
Acidification	g SO$_2$-eq. \$$^{-1}$ yr^{-1}	0.80	6.03
Eutrophication	g PO$_4$-eq. \$$^{-1}$ yr^{-1}	0.02	0.22

From the yield ha^{-1} (kg ha^{-1}) and area (ha) for each of the Italian production systems, and from the income per farm ($ farm^{-1}), yield per area and price of wheat (0.49 $ kg^{-1}) for the Danish production systems, the GHG emission ha^{-1} (Table 5) was calculated based on the results presented in Table 2.

Table 5. GHG emissions ha^{-1} in production systems in Italy and Denmark. CFE: combined food and energy system.

Impact Category	Unit	Italy				Denmark	
		Silvo-Pastoral	Organic	Traditional	Conventional	CFE	Conventional
GWP	kg CO$_2$-eq. ha^{-1} yr^{-1}	606	585	4615	1669	3083	17,705
Acidification	kg SO$_2$-eq. ha^{-1} yr^{-1}	78	39	49	33	6467	24,618
Eutrophication	kg PO$_4$-eq. ha^{-1} yr^{-1}	18	11	16	18	14,825	73,618
Yield ha^{-1}	kg ha^{-1}	3640	2200	7050	4300		7341
Income ha^{-1}	$ ha^{-1}					5013	3597
Area	ha	1.0	4.5	8.5		11.1	

4. Discussion

The comparison between specific types of production systems in two diverse environmental zones is challenging and cumbersome due to the differences in climate, soil, and management. Hence, a simple way to gain an initial overview of the environmental impacts of the studied production systems are presented in Table 5, with GHG emissions calculated per hectare. The data showed that the conventional wheat production system had the largest environmental impact ha^{-1} compared to other studied production systems in Denmark and Italy.

The environmental impacts of the six productions systems were calculated, based on three indicators of GWP, acidification, and eutrophication. Machinery use contributed the highest GHG contribution in the silvopastoral and organic production systems whereas fertilizer contributed the largest GHG in the traditional and conventional production systems in Italy.

The comparison of results with other studies was not straightforward as the farming systems and system boundaries varied between studies [10,11,14]. For Italy, the highest GWP calculated for the traditional farming system (Table 2) was mainly attributed to fertilizer (0.15 kg CO_2-eq.) and irrigation (0.13 kg CO_2-eq.). Romero-Gamez et al. [14] related this with CO2 and NO2 emissions to air caused by the manufacture and application of fertilizers to the cropping systems. In the present study, CO2 and N2O from fertilizer and machinery use were significant contributors to GHG emissions in the production systems. Romero-Gamez et al. [14] found that acidification was dominated by NH3 emissions to the air and those emissions were allocated to fertilizer production, in similarity to the present study, finding that fertilization and machinery use related to NH3 and NO had the highest impact on acidification.

Due to the diversity of products from the combined food and energy system, the environmental impacts were calculated based on the income from the two types of production systems in Denmark. Thus, acidification was found to be more than seven times higher for the conventional wheat production system in comparison to the combined food and energy system (Table 2) with the main impact from fertilizer related to NH3 and NO (Table 1). Likewise, the eutrophication and GWP in the conventional wheat system was 11.0 and 11.4 times higher, respectively, compared to the combined food and energy system, mainly caused by fertilizer use (Table 4). The study by Nemecek et al. [15] found increased environmental impacts by conventional production practice compared to organic agricultural practices by 1.5 times for GWP (4474 vs. 2920 kg CO_2-eq. ha^{-1} yr^{-1}), 1.4 times for acidification (88 vs. 61 kg SO_2-eq. ha^{-1} yr^{-1}), and 1.4 times for eutrophication (123 vs. 88 kg N-eq. ha^{-1} yr^{-1}). This supports the present study's findings of less environmental impacts from practices with reduced application of fertilizers and pesticides. While the GWP for the combined food and energy system is of similar magnitude to the organic system in Nemecek et al. [15], the GWP of the Swiss conventional wheat production system was much higher than the conventional wheat production system in the present study. Likewise, Knudsen et al. [29] found low GWP values of 2032-2599 kg CO_2-eq. ha^{-1} yr^{-1} for conventional wheat production system in Denmark in a four-year barley, potatoes, and winter wheat crop rotation including one year of either faba beans or grass-clover.

5. Conclusions

Among the six production systems, the conventional wheat production system in Denmark accounted for highest global warming potential, acidification, and eutrophication. In Italy, global warming potential was highest in traditional agroforestry and lowest in the silvopastoral system whereas acidification and eutrophication was lowest in the traditional production system with high acidification effects from the silvopastoral system. In Italy, machinery use contributed the highest greenhouse gas emissions in silvopastoral and organic production systems, while the large contribution to greenhouse gas emissions from fertilizer was recorded in the traditional and conventional production systems. In Denmark, the combined food and energy system was found to have lower environmental impacts compared to the conventional wheat production system according to the three indicators. For both systems in Denmark, the main contribution to greenhouse gas emission was due to fertilizer and manure application. Thus, the study demonstrated that the environmental footprint is dependent on the management intensity of the production system. The field-based evidence from the study can contribute to informed decision making by the land managers and policy makers for promotion of environmentally friendly food and non-food production practices to meet the European Union targets of providing biomass-based materials and energy for bio-based economy in Europe and beyond.

Author Contributions: A.P. and G.R. provided data on the Italian production systems, B.B.G. provided data on the Danish production systems, M.B. and K.Ż. conducted the data analysis. L.M.L. drafted the manuscript and B.B.G. and L.M.L revised and improved the scientific content of the manuscript. All authors have read and agreed to the published version of the manuscript.

Funding: Financial support from SustainFARM (grant agreement no. 652615) and WaterFARMING (grant agreement no: 689271) and BioEcon (grant agreement No 669062) projects for supporting the collation of experimental data and manuscript preparation is acknowledged.

Conflicts of Interest: The authors declare no conflicts of interests.

References

1. FAO. Agroforestry. 2017. Available online: http://www.fao.org/forestry/agroforestry/en/ (accessed on 10 May 2019).
2. Rois-Díaz, M.; Lovric, N.; Lovric, M.; Ferreiro-Donmínquez, N.; Mosquera-Losada, M.R.; Den Herder, M.; Graves, A.; Palma, J.H.N.; Paulo, J.A.; Pisanelli, A.; et al. Farmers' reasoning behind the uptake of agroforestry practices: Evidence from multiple case-studies across Europe. *Agrofor. Syst.* **2018**, *92*, 811–828. [CrossRef]
3. Den Herder, M.; Moreno, G.; Mosquera-Losada, M.R.; Palma, J.H.N.; Sidiropoulou, A.; Santiago-Freijanes, J.J.; Crous-Duran, J.; Paulo, J.A.; Tomé, M.; Pantera, A.; et al. Current extent and stratification of agroforestry in the European Union. *Agric. Ecosyst. Environ.* **2017**, *241*, 121–132. [CrossRef]
4. Tsiafouli, M.A.; Thébault, E.; Sgardelis, S.P.; de Ruiter, P.C.; van der Putten, W.H.; Birkhofer, K.; Hemerik, L.; de Vries, F.T.; Bardgett, R.D.; Brady, M.V.; et al. Intensive agriculture reduces soil biodiversity across Europe. *Glob. Chang. Biol.* **2015**, *21*, 973–985. [CrossRef] [PubMed]
5. Danmarks Statistik. Statistikbanken. 2019. Available online: www.statistikbanken.dk (accessed on 8 May 2019).
6. Foereid, B.; Bro, R.; Mogensen, V.O.; Porter, J.R. Effects of windbreak strips of willow coppice-modelling and field experiment on barley in Denmark. *Agric. Ecosyst. Environ.* **2002**, *93*, 25–32. [CrossRef]
7. Jørgensen, U.; Dalgaard, T.; Kristensen, E.S. Biomass energy in organic farming-the potential role of short rotation coppice. *Biomass Bioenergy* **2005**, *28*, 237–248. [CrossRef]
8. FAOSTAT. 2016. Available online: http://www.fao.org/faostat/en/#home (accessed on 26 April 2019).
9. Ramachandran Nair, P.K. (Ed.) *An Introduction to Agroforestry*; Kluwer Academic Publishers: Dordrecht, The Netherlands, 1993; p. 499.
10. Avraamides, M.; Fatta, D. Resource consumption and emissions from olive oil production: A life cycle inventory case study in Cyprus. *J. Clean. Prod.* **2008**, *16*, 809–821. [CrossRef]
11. Hanandeh, A.E.; Gharaibeh, M.A. Environmental efficiency of olive oil production by small and micro-scale farmers in northern Jordan: Life cycle assessment. *Agric. Syst.* **2016**, *148*, 169–177. [CrossRef]
12. Notarnicola, B.; Salomone, R.; Petti, L.; Renzulli, P.A.; Roma, R.; Cerutti, A.K. (Eds.) Life Cycle Assessment in the Agri-Food Sector Case Studies, Methodological Issues and Best Practices; Hardcover; 2015; Volume XXI, 390 p. Available online: www.springer.com/978-3-319--11939-7 (accessed on 26 April 2019).
13. Mohamed, R.S.; Verrastro, V.; Cardone, G.; Bteich, M.R.; Favia, M.; Moretti, M.; Roma, R. Optimization of organic and conventional olive agricultural practices from a Life Cycle Assessment and Life Cycle Costing perspectives. *J. Clean. Prod.* **2014**, *70*, 78–89. [CrossRef]
14. Romero-Gamez, M.; Castro-Rodriguez, J.; Suarez-Rey, E.M. Optimization of olive growing practices in Spain from a life cycle assessment perspective. *J. Clean. Prod.* **2017**, *149*, 25–37. [CrossRef]
15. Nemecek, T.; Dubois, D.; Huguenin-Elie, O.; Gaillard, G. Life cycle assessment of Swiss farming systems: I. integrated and organic farming. *Agric. Syst.* **2011**, *104*, 217–232. [CrossRef]
16. Borzęcka, M.; Żyłowska, K.; Russo, G.; Pisanelli, A.; Freire, F. Life Cycle Assessment of olive cultivation in Italy: Comparison of three management systems. In Proceedings of the 167th EAAE Seminar: European Agriculture and the Transition to Bioeconomy, 24–25 September 2018; Institute of Soil Science and Plant Cultivation-State Research Institute: Pulawy, Poland. Available online: https://ageconsearch.umn.edu/record/281566?ln=en (accessed on 26 April 2019).
17. Olive-4-Climate 2019. Manuale per la Gestione Sostenibile Degli Uliveti. LIFE15 CCM/IT/000141. Available online: www.OLIVE4CLIMATE.eu (accessed on 17 January 2020).
18. Teatro Naturale. Gli Olivi Devono Produrre a Costi Accettabili: Analisi di Quelli Fissi e Variabili. 2018. Available online: https://www.teatronaturale.it/strettamente-tecnico/l-arca-olearia/26289-gli-olivi-devono-produrre-a-costi-accettabili-analisi-dei-costi-fissi-e-variabili.htm (accessed on 17 January 2020).
19. Ghaley, B.B.; Porter, J.R. Emergy synthesis of a combined food and energy production system compared to a conventional wheat (*Triticum aestivum*) production system. *Ecol. Indic.* **2013**, *24*, 534–542. [CrossRef]

20. Ghaley, B.B.; Vesterdal, L.; Porter, J.R. Quantification and valuation of ecosystem services in diverse production systems for informed decision-making. *Environ. Sci. Policy* **2014**, *39*, 139–149. [CrossRef]
21. Wernet, G.; Bauer, C.; Steubing, B.; Reinhard, J.; Moreno-Ruiz, E.; Weidema, B. The ecoinvent database version 3 (part I): Overview and methodology. *Int. J. Life Cycle Assess.* **2016**, *21*, 1218–1230. [CrossRef]
22. ISO. ISO. ISO 14040 International Standard. In *Environmental Management-Life Cycle Assessment-Principles and Framework. International Organisation for Standardization*; ISO: Geneva, Switzerland, 2006; Available online: http://www.iso.org/iso/catalogue_detail?csnumber=37456 (accessed on 26 April 2019).
23. Pré Consultants. *SimaPro Version 7*; Pré Consultants: Amersfoort, The Netherlands, 2006.
24. EEA/EMEP. *EMEP/EEA Air Pollutant Emission Inventory Guidebook 2013*; ISBN 978-92-9213-403-7. Available online: https://www.eea.europa.eu/publications/emep-eea-guidebook-2013 (accessed on 26 April 2019).
25. Intergovernment Panel on Climate Change (IPCC). N2O emissions from managed soils, and CO_2 emissions from lime and urea applications. In *2006 IPCC Guidelines for National Greenhouse Gas Inventories, Volume 4, Agriculture Forestry and Other Land Use'*; Eggleston, H.S., Buendia, L., Miwa, K., Ngara, T., Tanabe, K., Eds.; Institute for Global Environmental Strategies: Hayama, Japan, 2006; pp. 11.1–11.54.
26. Nemecek, T.; Bengoa, X.; Lansche, J.; Mouron, P.; Rossi, V.; Humbert, S. *Methodological Guidelines for the Life Cycle Inventory of Agricultural Products*; Version 2.0; World Food LCA Database (WFLDB), July 2015; Available online: https://quantis-intl.com/wp-content/uploads/2017/02/wfldb_methodologicalguidelines_v3.0.pdf (accessed on 26 April 2019).
27. Guinée, J.B.; Gorrée, M.; Heijungs, R.; Huppes, G.; Kleijn, R.; Koning, A.; de Oers, L.; van Wegener Sleeswijk, A.; Suh, S.; Udo de Haes, H.A.; et al. Handbook on life cycle assessment. In *Operational Guide to the ISO Standards*; I: LCA in Perspective, IIa: Guide. IIb: Operational Annex, III: Scientific Background; Kluwer Academic Publishers: Dordrecht, The Netherland, 2002; p. 692. ISBN 1-4020-0228-9.
28. Goedkoop, M.J.; Heijungs, R.; Huijbregts, M.; De Schryver, A.; Struijs, J.; Van Zelm, R. *ReCiPe: A Life Cycle Impact Assessment Method Which Comprises Harmonised Category Indicators at the Midpoint and the Endpoint Level*, 1st ed. Report I: Characterisation. 2008. Available online: http://www.lcia-recipe.net (accessed on 26 April 2019).
29. Knudsen, M.T.; Meyer-Aurich, A.; Olesen, J.E.; Chirinda, N.; Hermansen, J.E. Carbon footprints of crop from organic and conventional arable crop rotations-using a life cycle assessment approach. *J. Clean. Prod.* **2014**, *64*, 609–618. [CrossRef]

© 2020 by the authors. Licensee MDPI, Basel, Switzerland. This article is an open access article distributed under the terms and conditions of the Creative Commons Attribution (CC BY) license (http://creativecommons.org/licenses/by/4.0/).

Article

Environmental Sustainability of Bioenergy Strategies in Western Kenya to Address Household Air Pollution

Ricardo Luís Carvalho [1,2,*], Pooja Yadav [3], Natxo García-López [1], Robert Lindgren [1], Gert Nyberg [4], Rocio Diaz-Chavez [5], Venkata Krishna Kumar Upadhyayula [6], Christoffer Boman [1] and Dimitris Athanassiadis [3]

1. Thermochemical Energy Conversion Laboratory, Department of Applied Physics and Electronics, Umeå University, 90187 Umeå, Sweden; natxo.garcia@umu.se (N.G.-L.); robert.lindgren@umu.se (R.L.); christoffer.boman@umu.se (C.B.)
2. Centre for Environmental and Marine Studies, Department of Environment and Planning, University of Aveiro, 3810-193 Aveiro, Portugal
3. Department of Forest Biomaterials and Technology, Swedish University of Agricultural Sciences, 90183 Umeå, Sweden; pooja.yadav@slu.se (P.Y.); dimitris.athanassiadis@slu.se (D.A.)
4. Department of Forest Ecology and Management, Swedish University of Agricultural Sciences, 90183 Umeå, Sweden; gert.nyberg@slu.se
5. Stockholm Environment Institute, Africa Centre, World Agroforestry Centre, Nairobi 30677, Kenya; rocio.diaz-chavez@sei.org
6. Department of Chemistry, Umeå University, 90187 Umeå, Sweden; krishna.upadhyayula@umu.se
* Correspondence: ricardo.teles@ua.pt; Tel.: +46-907-866-756

Received: 31 December 2019; Accepted: 5 February 2020; Published: 7 February 2020

Abstract: Over 640 million people in Africa are expected to rely on solid-fuels for cooking by 2040. In Western Kenya, cooking inefficiently persists as a major cause of burden of disease due to household air pollution. Efficient biomass cooking is a local-based renewable energy solution to address this issue. The Life-Cycle Assessment tool Simapro 8.5 is applied for analyzing the environmental impact of four biomass cooking strategies for the Kisumu County, with analysis based on a previous energy modelling study, and literature and background data from the Ecoinvent and Agrifootprint databases applied to the region. A Business-As-Usual scenario (BAU) considers the trends in energy use until 2035. Transition scenarios to Improved Cookstoves (ICS), Pellet-fired Gasifier Stoves (PGS) and Biogas Stoves (BGS) consider the transition to wood-logs, biomass pellets and biogas, respectively. An Integrated (INT) scenario evaluates a mix of the ICS, PGS and BGS. In the BGS, the available biomass waste is sufficient to be upcycled and fulfill cooking demands by 2035. This scenario has the lowest impact on all impact categories analyzed followed by the PGS and INT. Further work should address a detailed socio-economic analysis of the analyzed scenarios.

Keywords: agroforestry; waste valorization; sustainable development goals; renewable energy; bioenergy transitions; circular bioeconomy; clean cooking; life-cycle assessment; energy policy

1. Introduction

It is estimated that over 40% of the world's population is currently relying on solid-fuels for cooking and heating [1]. According to the World Health Organization (WHO), the inefficient utilization of biomass and coal for these purposes constitutes today's largest global environmental health risk [2,3]. Worldwide, over 4 million deaths occur per year from illnesses related to the smoke from solid-fuel combustion indoors, which mainly affects women and children [4,5]. In developing countries, wood and charcoal continue to play a vital role in meeting household energy demands, where it remains easily accessible and affordable [6–8]. Thus, the transition to cleaner cooking fuels in advanced cookstoves constitutes an important way to address several of the 17 Sustainable Development Goals (SDGs) [9,10], contributing to address at least five of the SDGs, including the: (1) Good health and well-being (SDG 3); (2) Gender equality (SDG 5); (3) Affordable and clean energy (SDG 7); (4) Climate action (SDG 13); and (5) Life on land (SDG 15).

In Sub-Saharan Africa (SSA), only 35% of the population have access to electricity and 80% of the people rely on traditional firewood, charcoal, animal dung and agricultural residues for cooking [11,12]. Negative environmental impacts such as global warming due to the emission of carbon dioxide (CO_2), nitrous oxide (N_2O) and methane (CH_4), eutrophication related with the emission of nitrogen oxides (NO_x) and N_2O, acidification associated with the emission of sulphur dioxide (SO_2) and NO_x, and toxicological effects on humans related with the emission of NO_x and particulate matter (PM) are potentially intensified by traditional cooking. In Kenya, the cooking sector emissions are driven by rapid population and economic growth [13], and household air pollution (HAP) causes 15,600 deaths with direct impacts on the health of around 15 million people [14–17].

In the Kenyan context, the transition to advanced cookstoves using upgraded biomass fuels (e.g., wood pellets and biogas) produced via renewable energy strategies constitutes a relevant way to mitigate HAP. Such locally produced biofuels based on waste valorization and agroforestry strategies have a great potential to enhance land restoration and livelihoods, as agroforestry is a powerful tool to enhance multiple ecosystem services [18]. Famers in Kenya are responsible for producing a substantial part of the food consumed in the region. In this context, an integrated management of crop-residues at the farm scale can tackle challenges of food security, poverty and climate change [18,19]. Furthermore, the utilization of crop-residues remains an opportunity for sustainable bioenergy production in rural and peri-urban communities [20]. In such a perspective, the Kenyan government has worked on designing proper strategies towards the deployment of efficient bioenergy systems [21,22], including the use of both agricultural crop [23] and industrial biomass residues [21] for the production of biogas and densified biomass fuels. The Kenya Country Action Plan (CAP) for Clean Cookstoves and Fuels has established a target to promote the installation of cleaner cookstoves in 5 million households by 2020 [24]. With a thermal efficiency over two times higher than that achieved by improved cookstoves using wood-logs and sticks [25,26], advanced biomass cooking solutions such as micro-gasifier and biogas cookstoves can be key in technologies to address such clean cooking systems.

Despite the existing initiatives, there is currently a knowledge gap on how to design sustainable cooking strategies in the context of emerging circular bioeconomies. Few studies have analyzed the life-cycle environmental impact of integrated biomass fuel/cookstove strategies on the mitigation of HAP at the sub-national level. Various studies have been analyzed the impact of different energy transition options [8,27,28]. In Western Kenya, Carvalho et al. [26] have applied the Long-Range Energy Alternative Planning (LEAP) software to analyze the energy savings and emissions caused by distinct bioenergy strategies on HAP in Kisumu County. The study was applied for the time span between 2015 and 2035, showing part of the environmental benefits of such strategies. Although the previous energy modelling study [26] shows the HAP mitigation potential at the sub-national level, there is currently a limited number of studies assessing the overall environmental performance of these transitions in a life-cycle assessment (LCA) perspective.

As defined by the International Standard Organization (ISO) in the ISO 14040 [29], LCA is a technique used to quantify the environmental impacts of a product system like a cooking fuel over its whole life cycle, from raw material acquisition through production, use, end of life, treatment, recycling, and disposal [29]. A previous LCA study conducted in Kenya observed that biogas from animal dung and ethanol from wood as cooking fuels had the best environmental performance in almost all environmental impact categories while charcoal briquettes from wood exhibited poor environmental performance due to emissions resulting from kiln operation [30]. Okoko et al. [31] confirmed the higher carbon footprint of unimproved charcoal value-chain in relation to alternative biomass energy solutions for cooking in Kenya and Tanzania. Lansche and Müller performed a comparative LCA on traditional biomass and biogas household cooking systems in Ethiopia, demonstrating the high potential for environmental improvements when adopting biogas systems. Although the previous LCA studies analyzed the environmental impacts of alternative cooking value-chains in a certain static moment in time, they did not include a LCA considering regional energy forecasts, i.e., a dynamic LCA approach [32]. The dynamic LCA method can be applied to analyze environmental impacts in different time spans. Pehnt [33], for instance, considered the time-variation of electricity mix in the assessment of GHGs and acidification impacts. García-Gusano et al. [34] combined energy forecasts using LEAP with a dynamic LCA approach to evaluate the effects of various coal power plant retrofits over time. Although various dynamic LCA studies conducted, there is currently a lack of approaches dedicated to analyzing household energy transitions in the developing region context.

In the present research, a dynamic LCA based data provided by a previous energy forecast study [26] is conducted to determine the environmental sustainability of emerging biomass cooking strategies in the Kisumu County (Western Kenya). Although most biomass cookstoves have not progressed to the point that they are equivalent to Liquefied Petroleum Gas in terms of efficiency and cleanliness in the household indoor environments [35], this study is confined to evaluate the advances in the local use of biomass resources in the context of renewable energy and bioeconomy transitions. WIth this background, the present work constitutes not only an LCA of alternate biomass value-chains, but also a methodological development in the integration of LEAP/LCA tools for the analysis of the environmental sustainability of distinct energy policies. Despite the fact that several studies have conducted life-cycle analysis of products and systems at the country [34,36] and city levels [37,38], few have combined integrated energy models with LCA to evaluate the environmental impact of energy policy scenarios at the sub-national level. The Kisumu County in Western Kenya was selected for the case study as it presents a large availability of endogenous biomass resources, including agricultural and industrial residues that can potentially be used for the production of cooking biofuels. The region presents an important area of fertile agricultural land that may serve for the establishment of local-based agroforestry systems [39,40], which can be used to produce wood fuel in a more sustainable and resilient manner.

2. Materials and Methods

The present study is focused on analyzing the environmental performance of four bioenergy transition strategies for household cooking in the Kisumu County (Kenya), considering the importance of biomass local resources for addressing the resilience of communities in the developing country context. The study uses some of the results from a previous energy modelling study [26] as input data to conduct a dynamic LCA. Although the study does not measure the economic and social implications of the different transition scenarios, the design of the different scenarios for the environmental assessment is based on the premise that the efficient valorization and use of local biomass resources is an important waste management solution [31]. Additionally, this study includes the application of woody agroforestry biomass systems, which are powerful tools to enhance access to energy and food, land restoration and sustainable livelihoods [18,39]. In Western Kenya, the sustainable use of biomass for cooking has an important role to enhance various ecosystem services through the creation of local businesses. Considering the limitations observed to collect social and economic data, the present environmental assessment serves as a first step for conducting a full sustainability assessment on advanced local biomass-based cooking transition options. The environmental assessment is focused on the relevance of introducing renewable energy systems in a developing region of Kenya in the context of the SDGs [10,41].

Considering the limited number of dynamic LCA studies applied to the developing region context, the present study focuses on conducting an environmental impact assessment of the bioenergy transition strategies previously analyzed in an energy modelling study conducted by Carvalho et al. [26]. In line with the previous study, the present research analyzes the environmental sustainability of the energy transition options in relation to a business-as-usual scenario for the time span between 2015 and 2035. The dynamic LCA approach considers the evolution of the household energy mix computed in the LEAP model for that time span. The LCA results are presented for both the baseline and end years, considering the projections in the evolution of energy demand and supply until the year 2035, taking into account historical changes in the economy and demographic conditions in the Kisumu County. As in the LEAP study [26], in the present LCA study, the BAU scenario also considers that no policies will be introduced to mitigate environmental impacts of traditional cooking systems. The present LCA also considers the evolution of the household energy demands according to the following biomass/cookstove transition scenarios (Figure 1):

(i) Business as usual, considering the evolution according to historical trends on population growth and urbanization, whereas the use of liquefied petroleum gas (LPG) and electricity will continue the same and the share of traditional cooking systems will be reduced according to historical trends (no transition policies adopted);
(ii) Improved cookstoves combusting wood-logs produced via locally sustainable agroforestry systems (ICS), and biomass briquettes produced with residues with crop residues such as maize cobs;
(iii) Micro-gasifying cookstoves using pelletized fuels made of sugarcane bagasse and woody biomass produced in agroforestry systems (PGS);
(iv) Biogas stoves with household organic waste and animal manure being anaerobically digested to a gaseous mixture including methane (CH_4) in local biodigestors (BGS);
(v) An Integrated (INT) scenario evaluated a mix of the ICS, PGS and BGS scenarios (Figure 1).

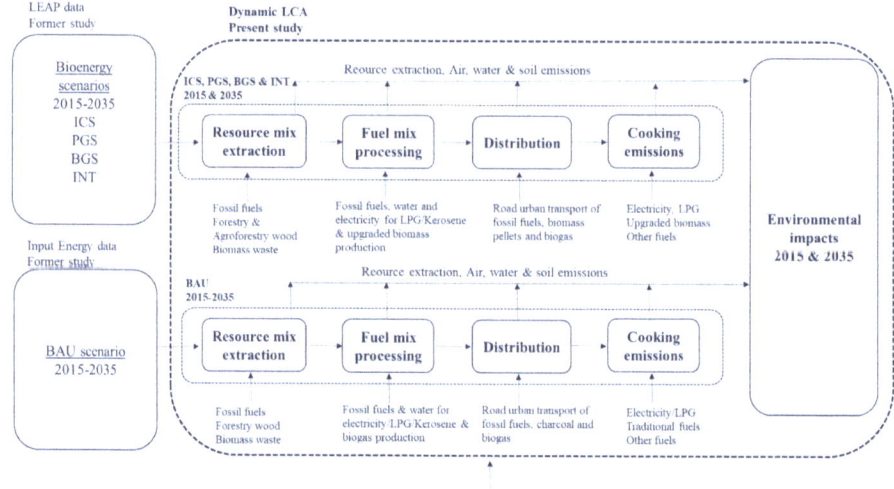

Figure 1. Overall description of the dynamic LCA method applied based on input data generated in the previous LEAP study (Original figure designed by the author).

The LEAP software results are used to conduct the LCA analysis based on the previous energy modelling task, i.e., demand driven (bottom-up) energy model, previously applied for providing energy and emission projections for policy analysis at the regional level [26,42]. The LEAP modelled results used in the LCA are based on technical data provided by the Technology and Environmental Database (TED) coupled to the LEAP software. In this case, the TED is used to calculate emissions from various types of energy systems in Kenyan households [42–44]. The energy demand results used consider the evolution in the household energy use and demographic conditions in the Kisumu County, including different uses of fuel/cookstove systems in the time span between 2015 and 2035 [26]. Considering that the present LCA study is based on the same scenarios as those projected in the previous LEAP study conducted by Carvalho et al. [26], this work also considers that policy measures should be implemented before 2035 to promote cleaner and sustainable solutions for renewable energy cooking.

2.1. Demographic Data

The main demographic information used to generate the input projection data for the dynamic LCA is the population of the Kisumu County, per capita income in both urban areas and county, the typical household size, number of households in both urban and rural areas, as well as the useful energy demand for cooking and lightning (Table 1).

Table 1. Demographic and energy data of Kisumu County for the baseline year 2015 used by Carvalho et al. [26] to generate the LEAP data to the LCA model [26].

Type of Data	Amount	Units	Reference
Population	1155	1000 units	[45]
Income/capita	1.440	USD	[46,47]
GDP growth/capita	4.100	Percent	[41]
Household size	4.400	Nr. of people	[48]
Nr. of houses	227.0	1000 units	[48]
Rural houses	118.0	1000 units	[48]
Urban houses	109.0	1000 units	[48]
Peri-urban/total nr.	60.00	Percent	[49]
Useful cooking energy demand [a]	0.1200	TOE	[50–52]
Useful lightning energy demand [b]	2.500	MWh	[53]

[a] Based on the average annual useful energy demand for cooking; [b] Based on the average annual useful energy demand for lightning in a mid-income urban household in Kenya.

This study is based on the fact that the population in the Kisumu County will increase by 99% in the period between 2015 and 2035. By 2035, the population income per capita is expected to grow up to 2900 USD per year, whereas the income per capita in the city of Kisumu (urban income), is expected to be 1.5 times higher than in both the urban and rural areas of the Kisumu County.

According to the LEAP modelling results [26] used in the LCA model, in the baseline year, the household energy use was over 9 million Gigajoule (GJ), with wood-logs and charcoal being the main cooking fuels used in the year 2015. In the LCA study, it is also considered that, despite the trends in urbanization, a significant part of the peri-urban population is expected to continue living in informal settings with limited access to electricity and LPG.

Thus, the present study stresses the importance of potential life-cycle based environmental improvements associated with the introduction of alternative biomass cookstove strategies in the region. In line with the previous LEAP study, the present LCA study also explores the fact that, by 2035, bottle biogas and biomass pellet cooking systems might be an affordable way for a substantial part of the Kisumu County's population to mitigate environmental impacts related to current traditional cooking practices [26].

2.2. Resource and Energy Data

In line with the previous LEAP study [26], the projected use of natural resources and final energy used for cooking and lightning used to model the LCA inputs is computed considering the household energy use patterns in the historical years between 2010 and 2014 [26]. The calculation of the energy use is performed according to each type of cooking and lightning fuel/technology system, in order to model the household energy demands between 2015 and 2035. In the ICS scenario, all the woody biomass used in 2035 is expected to be produced via agroforestry systems considering the available agricultural land in both the Kisumu County and the nearby county of Siaya. As there is no sufficient amount of agricultural land available today that can be converted to agroforestry land systems, in this study it is considered that part of the woody biomass produced via agroforestry was sourced by Siaya County. In the PGS, a full replacement of traditional cookstoves by pellet micro-gasifying stoves is expected to occur in the year 2035. In urban areas, the biomass pellets are expected to be produced by a mixture of woody biomass from agroforestry (50%) and sugarcane bagasse (50%), considering that sugarcane bagasse is the most important crop-residue produced in the industrial sugar belt around the city of Kisumu. In rural areas, biomass pellets are assumed to be fully produced via the densification of woody biomass produced in agroforestry systems. Finally, in the BGS scenario, a full replacement of traditional cookstoves by biogas stoves is projected to occur for the year 2035. Here, half of the biogas is expected to be produced through the anaerobic digestion of animal manure and the other half through the digestion of municipal household waste produced in the Kisumu County. The LEAP functions used to compute the energy data that serves as input data to the LCA model are described in the Eq. 1-3 and Appendix A of the previous research published by Carvalho et al. [26].

2.3. Life-Cycle Assessment

After modelling the supply of energy in the various scenarios, the LCA is applied according to the international ISO14040 standard, which defines three main steps to conduct a LCA: (i) Aim and Scope; (ii) Inventory analysis; (iii) Interpretation. This study aims to analyze material and energy flows, and quantify the environmental impacts of distinct bioenergy value-chains for cooking across the life-cycle stages of the different systems, including feedstock collection, processing, distribution and use, as described in Table 2. This study has the final goal of identifying opportunities for environmental improvement, supporting decision makers to understand the sources and magnitudes of impacts throughout the life cycle of each system [31]. Although this study does not include any social and economic assessment of the proposed bioenergy strategies, the application of LCA in this study can be strategic to support environmentally sustainable energy planning policies in the Kisumu County.

Table 2. Bioenergy value-chains for cooking, life cycle stages, assumptions, and source of data based on a previous scenarios established by Carvalho et al. [26,40].

Life Cycle Stages		Policy Option and Bioenergy Value Chains			
		BAU	ICS	PGS	BGS
Feedstock collection	Assumption	Unsustainable harvesting no regrowth—Manual [54]	Sustainable harvesting agroforestry—Manual [39,54]	Sustainable harvesting of bagasse and agroforestry wood—Manual [39,54]	Sustainable waste & manure harvesting—Manual
	Ecoinvent processes adapted [1,2]	Residual wood, dry| market	Residual wood, dry| market	bagasse straw (50%)/wood pellet (50%) prod| market [c]	Biogas, from grass| market
Feedstock processing	Technology	Inefficient charcoal kilns	Efficient char. kilns	Efficient pelleting engines	Efficient municipal digestors
	Efficiency/source	10–15%	20%	99%	50–60%
Distribution	Transport/distance	Rural: None. Urban: Bicycle (30 km)	Rural: None. Urban: Bicyc. (30 km)	Rural: None. Urban: Bicyc. (30 km)	Rural: None. Urban: NA
Use	Cooking efficiency [3]	<20%	>20–30%	30–45%	50–60%

[1] Liquified Petroleum Gas (LPG) was considered according to the Ecoinvent process Liquefied petroleum gas {RoW}\| market for\| APOS, U; [2] Electricity was considered to be produced according to the Ecoinvent process Market for electricity, medium voltage KE; [3] The stoves with the indicated range of cooking efficiencies dominate the mix of fuel/cookstove systems; [c] Two Ecoinvent processes, respectively one applied for biomass residues and the other for woody biomass.

2.3.1. System Boundaries

In this study, an LCA is performed in function of the cumulative energy demand for the various energy cooking scenarios analyzed in LEAP. The LCA is based on a regionalized inventory and impact assessment from raw material extraction to energy end-use for cooking. The calculations are performed by using the software Simapro 8.5 by applying the ReCiPe method [55].

The study is conducted taking into account the ISO standard 14040 [29,56], according to the following life-cycle stages (Figure 1):

- Production and transportation of the mix of the cookstove fuel feedstock from the production site to the processing location;
- Processing of the mix of feedstock into a form of ready to be used fuels in cookstoves;
- Distribution of the different types of fuels from the processing locations to the respective retailers or consumers;
- Use of the fuels via their combustion or use of electricity in a cookstove, according to the mix of cooking systems;

2.3.2. Inventory Analysis

The inventory analysis is based on data from provided by the LEAP modelling study conducted by Carvalho et al. [26], and information collected from the literature and background data from version 3.1 of the ecoinvent [57] and Agrifootprint databases [57]. In order to compensate for the unavailability of regional Life Cycle Inventory (LCI) data, we permitted the use of direct proxies for certain processes [58]. Some fuel categories (e.g., charcoal, briquettes, wood from unsustainable forest management) are not included in the Ecoinvent database and values have to be approximated (e.g., for charcoal it is assumed that to produce 1 ton of charcoal, 5 tons of wood are necessary). For other fuel categories, e.g., kerosene, biogas and wood pellets only general values (global averages) on raw material extraction and production (upstream processes) are used. The Ecoinvent data processes used in the study are presented in Appendix A (Tables A1–A3).

2.3.3. Impact Assessment

The environmental performance of the different transition policy options in the Kimusu County are assessed taking into account the following environmental impact categories: (i) Global warming; (ii) Ozone formation, human health; (iii) Particulate matter formation; (iv) Terrestrial acidification; (v) Water consumption; (vi) Freshwater eutrophication; (vii) Marine eutrophication; (viii) Mineral resource scarcity; (vix) Fossil resource scarcity and (x) Land use. These are considered to reflect the most relevant/critical environmental impacts which can be associated with the combustion of solid-fuels in cookstoves. In this work, the first category is mostly associated with the combustion of fossil and biomass fuels and its effect on the emission of atmospheric pollutants with a global warming potential such as CH_4, N_2O, carbon monoxide (CO) and particulate matter. The second category is associated with the formation of ozone, which is related to processes that release NOx compounds into the environment. The third process is directly associated with atmospheric emissions from combustion processes, as referred to in the first category. The forth category is associated with the acidification of the soil due to emissions of certain nitrogen compounds such as ammonia (NH_3) and NOx, which can be associated with the use of chemicals and fertilizers that cause emissions to the soil. The fifth category is related with the consumption of water consumption and availability in the ecosystems and can be associated with its incorporation in industrial processes, which can be specifically used for the production of some chemicals and other materials. The sixth and seventh categories are associated with the excessive presence of certain nutrients in water systems, due to the release of phosphorus and nitrogen compounds, respectively, which can be associated with the use of certain chemicals and fertilizers. The eighth and ninth categories are mostly associated with the extraction of raw materials from nature and the scarcity of these resources. The last category can be associated with relative species loss due to the use of land, related to the processes of land transformation, land occupation and land relaxation.

3. Results

3.1. Biomass Resources and Energy Demands

The Kisumu County has an agro-industrial and municipal organic waste feedstock of over 1.8 million t biomass (Table 3). The present work assumed that all the mass of biomass feedstock analyzed are suitable for an efficient and sustainable conversion to upgraded cooking fuels such as biomass pellets and biogas.

Table 3. Biomass available in BAU, 2015 (t) and use in the BAU, ICS, PGS, BGS and INT, 2035 (t).

	Availability BAU, 2015	Demand BAU, 2035	Demand ICS, 2035	Demand PGS, 2035	Demand BGS, 2035	Demand INT, 2035
Wood-logs (forest) [1]	1.05×10^4	9.28×10^5	-	-	-	-
Wood-logs (agrofor.) [2]	5.65×10^5	-	1.63×10^5	1.34×10^5	-	5.60×10^4
Maize cobs	9.50×10^5	-	1.88×10^5	-	-	1.40×10^4
S. bagasse	2.60×10^5	-	-	1.05×10^5	-	3.10×10^4
Org. waste	2.35×10^5	-	-	-	2.04×10^5	2.00×10^4
Manure	4.26×10^5	-	-	-	2.04×10^5	2.00×10^4
Total	1.94×10^6	-	3.51×10^5	2.39×10^5	4.08×10^5	1.41×10^5

[1] Considering wood harvesting manual practices unsustainable with no regrowth. [2] Considering the demand for woody biomass will be produced via sustainable agroforestry systems in the available agricultural land in both the Kisumu (35%) and Siaya (65%) Counties. [3] Potential in the year 2015 to produce wood-logs with the available agricultural land, considering that 10% of the agricultural land can be converted to agroforestry land and the productivity of 10 t of woody biomass per ha of agroforestry land.

In the BAU scenario, in the year 2035, the demands for wood-logs in the Kisumu County would exceed the forest production capacity and potential agroforestry systems in that scenario and year. In the BAU scenario, in the year 2035, it will be necessary to import around 861 thousand ton of wood-logs from other regions, an amount that corresponds to around 92.8% of the total use.

According to the ICS scenario, the current capacity for producing woody biomass via sustainable agroforestry systems is able to satisfy 35% of the future demand of wood-logs. In this scenario, the current availability of feedstocks of maize cobs is sufficient to fulfill 100% of the demands for briquetting this biomass residue and transform it to a usable cooking fuel (Table 3). In the PGS scenario, the current capacity of agricultural land for the production of woody biomass in sustainable agroforestry systems can satisfy 42% of the future demands for wood-logs. According to the assumptions and projections made for this scenario, the current availability of sugarcane bagasse is sufficient to satisfy 100% of the demands for this biomass feedstock to produce sufficient biomass pellets for satisfying the energy cooking demands in 2035. Finally, in both the BGS and INT scenarios, the biomass feedstocks available in 2015 will be sufficient to address 100% of the demands for the production of wood-logs and upgraded fuels made of various biomass feedstocks considered in those two scenarios.

3.2. Fuel Energy Inventory Data

In the BAU scenario, due to population growth and the rate of urbanization, the amount of fuel in terms of energy consumption is expected to increase by over 50% until 2035 in relation to the baseline year from around 9 to approximately 14 million GJ. This condition is related to the fact that there will be more people with energy needs and a higher fraction of the population consuming charcoal in urban areas, in case no policy is adopted to change the patterns of energy usage for cooking. In the ICS scenario, the projected increase in the energy consumption is expected to smoothen in relation to the

BAU, since the energy consumption by 2035 will be in the order of 13 million GJ (Figure 2). In the PGS scenario, the energy consumption is expected to increase by 26% in the period between 2015 and 2035, reaching a value in the order of 12 million GJ. The smallest increase in the energy consumption is predicted for 2035 in the BGS scenario, being this in the order of 8%. In the INT scenario, the energy consumption is expected to increase by 20% to 11 million GJ in 2035.

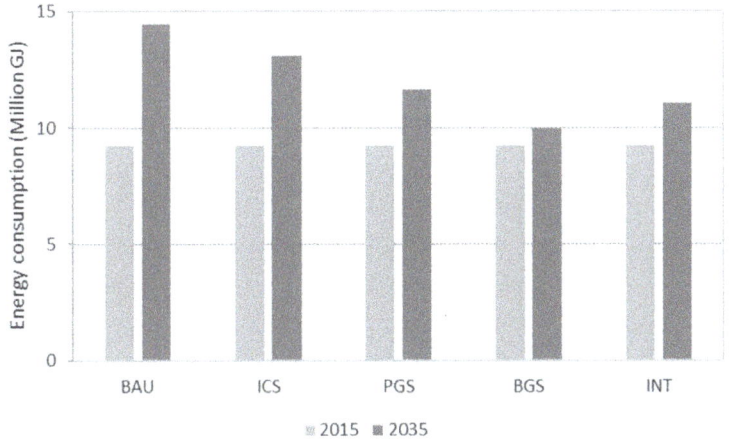

Figure 2. Fuel use in terms of energy consumption as input for the LCA model in the BAU, ICS, PGS, BGS and INT scenarios in 2015 and 2035.

In the BAU scenario, the greenhouse gas (GHG) emissions are expected to increase by 10%, ranging from around 220 thousand t of CO_{2e} in 2015 to around 245 thousand t of CO_{2e} in 2035. In the ICS scenario, the GHG emissions are predicted to be reduced by 18% to a value around 180 thousand t of CO_{2e} in 2035 (Figure 2). The amount of fuel input flows to the LCA model regarding t fuel production and cooking life-cycle stages by type of fuel/cooking system in each scenario, are respectively described in detail in the Appendix A (Tables A2 and A4).

3.3. Environmental Impact Assessment

In the BAU scenario, in the year 2015, approximately 62% of the use of household energy for cooking was associated with the use of open fires, whereas the use of charcoal represents 27% of the residential energy consumption for cooking, and the use of kerosene stoves represents 6% of the energy use for cooking (Figure 3).

In the BAU scenario, the LCA results also reflect the effects of population growth and increasing urbanization in Western Kenya between 2015 and 2035 with an increase of 96% in all impact categories (Figure 3; Table 4). In this scenario, with the increased urbanization, an increased number of households is expected to have access to charcoal for cooking, reducing the direct exposure to certain household air pollutants. However, the LCA results also reflect that this improvement is not reflected in an overall reduction of environmental impacts, a fact that can be associated with the low thermochemical conversion performance of charcoal kilns.

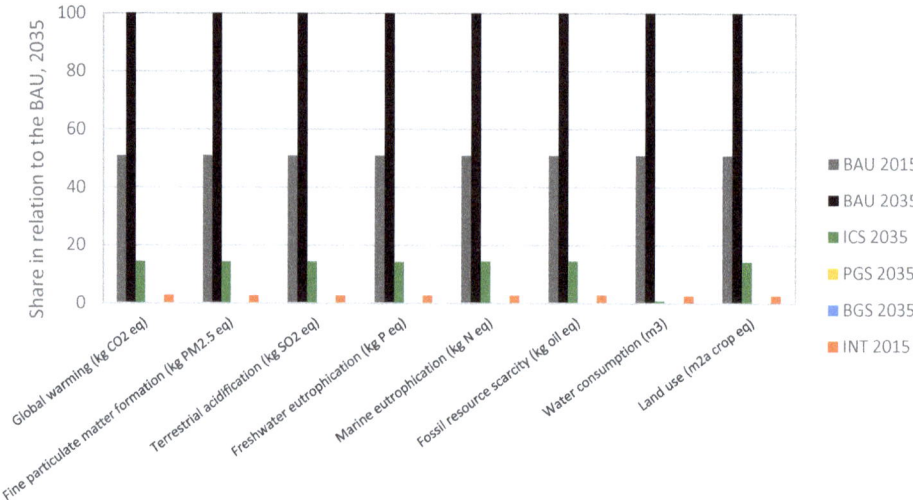

Figure 3. Environmental impacts in the BAU 2015, ICS 2035, PGS 2035, BGS 2035 and INT 2035 scenarios in relation to the BAU scenario, according to the percentage determined in the BAU 2035 scenario.

Table 4. Environmental impacts based on critical ReCiPe categories which are associated with biomass cooking activities in the Kisumu County in the BAU scenario in the years 2015 and 2035.

Impact Categories [1]	BAU 2015	BAU 2035
Global warming (kg CO_{2eq})	3.4×10^{10}	6.7×10^{10}
Fine particulate matter formation (kg $PM_{2.5eq}$)	1.5×10^{8}	3.0×10^{8}
Terrestrial acidification (kg SO_{2eq})	1.5×10^{8}	3.0×10^{8}
Freshwater eutrophication (kg P_{eq})	9.5×10^{6}	1.9×10^{7}
Marine eutrophication (kg N_{eq})	8.9×10^{5}	1.7×10^{6}
Fossil resource scarcity (kg oil_{eq})	1.1×10^{10}	2.1×10^{10}
Water consumption (m^3)	6.7×10^{8}	1.3×10^{9}
Land use (m^{2}a $crop_{eq}$)	9.4×10^{10}	1.9×10^{11}

[1] Calculated according to the ReCiPe method [55].

The impact of the introduction of improved cookstoves in the ICS scenario is predicted to contribute to the reduction of the effect of household energy activities on "Global warming", "Fine particulate matter formation", "Terrestrial acidification", "Freshwater eutrophication", "Marine water eutrophication", "Fossil resource scarcity", "Water consumption" and "Land use" between 80 and 90% in relation to the BAU scenario in the year 2035. Additionally, the introduction of advanced gasifier cookstoves combusting pelletized fuels is expected to reduce the environmental impacts of all the referred categories by over 98% in relation to the BAU 2035 scenario (Figure 3; Table 5). Furthermore, the replacement of traditional cookstoves by biogas stoves is predicted to reduce those impacts by the same order of magnitude as in the previous scenario. In the INT scenario, the environmental impacts associated with the analyzed impact categories are expected to be around 97% lower than those observed in the BAU scenario in 2035 (Figure 3; Table 5).

In the year 2035, and for the ICS scenario, the introduction of ICSs using wood-logs produced in agroforestry systems will contribute to an over 80% reduction in all the environmental impacts analyzed in this study in relation to the BAU scenario. Such improvements are partly associated with the fact that the improved cookstoves are around twice more efficient than traditional cooking systems, which results in significant fuel savings. Additionally, the use of woody biomass produced in a more sustainable way is expected to reduce the pressure on natural ecosystems. However, despite reflecting the reduction in the harvesting of woody biomass from Kenyan forests, the life cycle inventory in this study does not account for the positive environmental impacts related to the introduction of agroforestry systems, i.e., those impacts associated with the preservation of biodiversity and possible land-use changes (LUCs), due to limitations in accessing to real datasets that can reflect certain localities. Such environmental aspects are not accounted in the life-cycle inventory for scenarios of biodiversity losses and LUCs enhanced by the rate of deforestation associated with higher consumption of woody biomass from the forests. In the same year, for both the PGS and BGS scenarios, and for all analyzed impact categories, the introduction of the advanced cookstove systems in 2035 will contribute to reducing the environmental impacts by over 90% in relation to the BAU scenario in 2035. In a similar way, the technology advancements allow the achievement of even higher thermal energy conversion efficiencies in relation to traditional systems. Beyond that, these scenarios result from the utilization of biomass residues in micro-gasifier and biogas cookstoves, which are going to either be disposed of in natural environment or inefficiently burned in open fields if no measures are introduced. Furthermore, in this study, no environmental impacts are allocated to the upcycled biomass residues (i.e., sugarcane bagasse, animal manure and organic waste), which explains the fact that these two biomass/fuel systems have a higher environmental performance than the improved cookstoves. Finally, in the INT scenario, the adoption of the mixed biomass cookstove strategies is expected to contribute to reduce the environmental impacts by a value between 94% and 95% (Table 5).

Table 5. Environmental impacts based on critical ReCiPe categories which are associated with biomass cooking activities in the Kisumu County in the ICS, PGS, BGS and INT scenarios in 2035.

Impact Categories [1]	ICS 2035	PGS 2035	BGS 2035	INT 2035
Global warming (kg CO_{2eq})	9.7×10^9	2.1×10^8	1.6×10^8	1.9×10^9
Fine particulate matter format. (kg $PM_{2.5eq}$)	4.3×10^7	4.3×10^5	2.7×10^5	8.2×10^6
Terrestrial acidification (kg SO_{2eq})	4.3×10^7	5.1×10^5	3.8×10^5	8.2×10^6
Freshwater eutrophication (kg P_{eq})	2.7×10^6	2.6×10^4	6.1×10^3	5.0×10^5
Marine eutrophication (kg N_{eq})	2.5×10^5	4.1×10^3	5.6×10^2	4.8×10^4
Fossil resource scarcity (kg oil_{eq})	3.0×10^9	5.8×10^7	4.8×10^7	5.9×10^8
Water consumption (m^3)	1.9×10^7	7.9×10^5	3.3×10^5	3.5×10^7
Land use (m^2a $crop_{eq}$)	2.6×10^{10}	4.0×10^8	4.5×10^5	5.0×10^9

[1] Calculated according to the ReCiPe method [55].

In this study, a small difference is observed for the variations in the environmental impacts of the different biomass cookstove transition measures in relation to the BAU scenario in the year 2035. However, it is possible to observe that the BGS and PGS scenarios are the ones with higher reduction of the environmental impacts in relation to the analyzed categories. This result follows the trend observed for the evolution of the GHG emissions as shown in a previous study conducted by Carvalho et al. [26]. These environmental improvements can mostly be explained by the high efficiency of the thermochemical conversion processes associated with the combustion of biogas and pelletized fuels, which has a significant potential effect on reducing impacts such as on those associated with "Global warming", "Fine particulate matter formation", "Water consumption" and "Land use".

4. Conclusions

Considering the premise that local-based and renewable energy solutions are needed in the short to medium term to address HAP in developing countries, this study shows that the transition to biogas and micro-gasifying cookstove systems combusting upgraded biofuels, i.e., biogas and biomass pellets, greatly contributes to reducing critical environmental impacts beyond deforestation and HAP associated with a potential future phase out of traditional cooking practices in Western Kenya. In the BGS scenario, the availability of animal manure and organic waste in the year 2015 (661 thousand t) is 4% higher than the expected consumption of this feedstock in the year 2035 (408 thousand t). In the PGS scenario, the availability of agricultural land to produce agroforestry woody biomass in the

Kisumu County in 2015 (around 57 thousand t) is not sufficient to satisfy the projected demands for this feedstock in the year 2035 (around 134 t); imports from the neighboring county of Siaya should be considered to adjust future supply and demand. Additionally, in this scenario, the amount of sugarcane bagasse available in 2015 (260 thousand t) is more than enough to satisfy 100% of the demands for this feedstock in the projected PGS scenario in the year 2035 (around 105 thousand t).

Overall, the LCA confirms the hypothesis of increasing the environmental sustainability of current cooking systems and related fuel value-chains in Western Kenya through the introduction of advanced cookstoves combusting either biogas (BGS scenario) or pelletized fuels (PGS scenario). Such measures are expected to reduce by over 80% the environmental impacts for all the referred categories analyzed in relation to the BAU scenario in the year 2035. Although this study has demonstrated the great environmental benefits of the suggested measures, further studies are required to consolidate this sustainability assessment through the performance of both social and economic assessments of the proposed valorization alternatives.

By showing that biogas and pellet fuel/advanced cookstove value-chains are viable in terms of the mitigation of the environmental impacts, this study may orient the scientific community and energy sector stakeholders about the environmental benefits of such valorization and energy efficiency alternatives at the sub-national level. As the suggested bioenergy cooking strategies mitigate a wide range of life-cycle environmental impacts beyond HAP, an interesting and specifically relevant area for future research is related to the analysis of the environmental externalities (e.g., on human health) associated with the mitigation of greenhouse gas and PM emissions in Western Kenya. In such a context, this study does not yet fully provide consolidated recommendations to decision-makers responsible for designing energy policies in the Kisumu County, which will only be possible after a solid socio-economic assessment is carried out.

Future work in this area is highly relevant in order to provide key recommendations not only for policy makers, but also for local biomass industries, which could be interested in capitalizing investments in alternate value-chains. Furthermore, government incentives for the deployment of innovative logistical systems and tax credits—including the deployment of advanced biomass valuation alternatives and cookstove via local community-based projects—present a potential to enhance sustainable livelihoods through bottom-up policies design with the participation of the population.

Author Contributions: R.L.C., corresponding author, worked on the conceptualization, methodology, formal analysis, data collection and writing of the original draft. P.Y., V.K.K.U. and D.A. collaborated with R.C. in the modelling work with the software LEAP and Simapro. They also collaborated in the validation of the results in collaboration with the other co-authors. C.B., G.N., N.G.-L., R.L. and R.D.-C. contributed to the conceptualization of the work and writing review and editing. C.B. and D.A. were responsible for the project administration and supervised the work conducted in collaboration with the other co-authors. They were also involved in funding acquisition for the achievement of the goals of this research. All authors have read and agreed to the published version of the manuscript.

Funding: This research was funded by the Swedish Research Council FORMAS through the project Sustainable Biomass Utilization in Sub-Saharan Africa for an Improved Environment and Health, Dnr. 942-2015-1385. The work conducted by the researcher Ricardo Carvalho was financed by the Postdoctoral grant number JCK-1516, funded by the Kempe Foundation, and thanks are also due to FCT/MCTES for the financial support to CESAM (UID/AMB/50017/2019), through Portuguese national funds. The LCA work was financed by Bio4Energy, a strategic research environment appointed by the Swedish government.

Acknowledgments: Part of the work conducted by Ricardo Carvalho was funded with the technical support of the Group of Energy and Environment of the Federal Institute of Education, Science and Technology of Ceará, where the researcher is integrated.

Conflicts of Interest: The authors declare no conflict of interest.

Appendix A

Information about the inventory data for the LCA are presented in the tables below.

Table A1. Dataset used from the Ecoinvent database related to the LCA study in every fuel.

System	Ecoinvent Database Unit Process Used
Openfire wood	Residual wood, dry {GLO}| market for| APOS, U
ICS wood	Residual wood, dry {GLO}| market for| APOS, U
Kerosene	Kerosene {RoW}| market for| APOS, U
Trad charcoal [1]	Residual wood, dry {GLO}| market for| APOS, U
Impro charcoal [1]	Residual wood, dry {GLO}| market for| APOS, U
Electric	Market for electricity, medium voltage KE
LPG	Liquefied petroleum gas {RoW}| market for| APOS, U
ICS briquette residue	Wood pellet, measured as dry mass {RoW}| market for wood pellet| APOS, U
ND-gasifier biomass pellets (50/50) [1]	Straw, stand-alone production {GLO}| market for| APOS, U (50%)/Wood pellet, measured as dry mass {RoW}| market for wood pellet| APOS, U (50%
FD-gasifier biomass pellets (50/50) [1]	Straw, stand-alone production {GLO}| market for| APOS, U (50%)/Wood pellet, measured as dry mass {RoW}| market for wood pellet| APOS, U (50%)
ND-gasifier wood pellets	Wood pellet, measured as dry mass {RoW}| market for wood pellet| APOS, U
FD-gasifier wood pellets	Wood pellet, measured as dry mass {RoW}| market for wood pellet| APOS, U
Biogas digester	Biogas, from grass {GLO}| market for| APOS, U
Biogas stove	Biogas, from grass {GLO}| market for| APOS, U

[1] Based on LHVwood = 18MJ/drykg. To produce 1 kg of charcoal, 5 kg of wood are needed.

Table A2. Energy inputs in the fuel production life-cycle stage (TJ), considering each fuel system.

Year	2015	End Year 2035				
System	BAU	BAU	ICS	PGS	BGS	INT
Unsustainable wood [1]	5280	4462	-	-	-	-
Sustainable wood (agroforestry) [2]	-	-	2927	-	-	565.0
Kerosene	2185	1624	1624	1624	1624	1624
Charcoal	1146	3152	-	-	-	-
Electricity	470.1	2955	2955	2955	2955	2955
LPG	90.30	1039	1039	1039	1039	1039
Maize cobs	-	-	3391	-	-	569.6
Sugarcane Bagasse	-	-	-	3781	-	1121
Wood-chips (agroforestry) [3]	-	4462	-	-	-	-
Animal manure	23.10	-	2927	-	-	565.0
Municipal waste	-	1624	1624	1624	1624	1624

[1] Roundwood produced from manual harvesting practices with no regrowth. [2] Roundwood produced with manual harvesting with regrowth of nitrogen fixing trees [39,40]. [3] Wood-chips produced in wood chipping engine after sustainable production of wood in agroforestry systems.

Table A3. Material/distance inputs in the fuel transport life-cycle stage (tkm [1]) for each fuel system in urban areas.

Year	2015	End Year 2035				
System	BAU	BAU	ICS	PGS	BGS	INT
Wood	6.43×10^6	3.86×10^6	2.53×10^6	-	-	3.33×10^2
Briquettes	2.97×10^5	-	5.35×10^6	-	-	8.49×10^5
Charcoal	4.20×10^4	2.30×10^6	-	-	-	-
Biomass pellets	9.00×10^3	-	-	6.30×10^6	-	1.87×10^6
Biogas cylinder	2.40×10^4	-	-	-	3.06×10^6	1.22×10^6
LPG	6.43×10^6	6.18×10^5	6.18×10^5	6.18×10^5	6.18×10^5	6.18×10^5
Kerosene	2.97×10^5	4.92×10^5	4.92×10^5	4.92×10^5	4.92×10^5	4.92×10^5

[1] Considering the average distribution distance of 30 km from the production areas to the households.

Table A4. Energy inputs in the cooking life-cycle stage (TJ), considering each cooking system.

Year	2015	End Year 2035				
System	BAU	BAU	ICS	PGS	BGS	INT
Openfire wood	5067	4178	-	-	-	-
ICS wood	203.4	284.0	2895	-	-	532.6
Kerosene	237.6	725.7	725.7	725.7	725.7	725.7
Trad charcoal	433.8	1320	-	-	-	-
Impro charcoal	575.8	1832	-	-	-	-
Electric	19.30	524.7	524.7	524.7	524.7	524.7
LPG	74.10	532.8	532.8	532.8	532.8	532.8
ICS briquette residue	-	-	3391	-	-	569.6
ND-gasifier biomass pellets (50/50) [1]	-	-	-	968.0	-	399.3
FD-gasifier biomass pellets (50/50) [1]	-	-	-	2813	-	721.6
ND-gasifier wood pellets	-	-	-	879.9	-	333.5
FD-gasifier wood pellets	-	-	-	158.4	-	52.80
Biogas digester	23.10	79.80	79.80	79.80	1299	930.4
Biogas stove	5067	-	-	-	1939	768.8

[1] Biomass pellets are a mixture of biomass residues with 50% bagasse and 50% woody biomass produced in agroforestry systems.

References

1. Bonjour, S.; Adair-Rohani, H.; Wolf, J.; Bruce, N.G.; Mehta, S.; Prüss-Ustün, A.; Lahiff, M.; Rehfuess, E.A.; Mishra, V.; Smith, K.R. Solid fuel use for household cooking: Country and regional estimates for 1980–2010. *Environ. Health Perspect.* **2013**, *121*, 784–790. [CrossRef] [PubMed]
2. Lim, S.S.; Vos, T.; Flaxman, A.D.; Danaei, G.; Shibuya, K.; Adair-Rohani, H.; Amann, M.; Anderson, H.R.; Andrews, K.G.; Aryee, M.; et al. A comparative risk assessment of burden of disease and injury attributable to 67 risk factors and risk factor clusters in 21 regions, 1990–2010: A systematic analysis for the Global Burden of Disease Study 2010. *Lancet* **2012**, *380*, 2224–2260. [CrossRef]
3. Chafe, Z.; Brauer, M.; Héroux, M.; Klimont, Z.; Lanki, T.; Salonen, R.O.; Smith, K.R. *Residential Heating with Wood and Coal: Health Impacts and Policy Options in Europe and North America*; World Health Organization: Copenhagen, Denmark, 2015.
4. Rehfuess, E. *Fuel for Life: Household Energy and Health*; Geneva World Health Organization: Geneva, Switzerland, 2006; pp. 1–23. Available online: http://www.who.int/indoorair/publications/fuelforlife/en/ (accessed on 6 February 2020).
5. UNICEF. *Clear the Air for Children—The Impact of Air Pollution on Children*; UNICEF: New York, NY, USA, 2016; ISBN 1904097243.
6. Sanches-Pereira, A.; Tudeschini, L.G.; Coelho, S.T. Evolution of the Brazilian residential carbon footprint based on direct energy consumption. *Renew. Sustain. Energy Rev.* **2016**, *54*, 184–201. [CrossRef]

7. Felix, M.; Gheewala, S.H. A review of biomass energy dependency in Tanzania. *Energy Procedia* **2011**, *9*, 338–343. [CrossRef]
8. Clough, L. *The Improved Cookstove Sector in East Africa: Experience from the Developing Energy Enterprise Programme (DEEP)*; GVEP-Global Village Energy Partnership International: London, UK, 2012; p. 108.
9. Rosenthal, J.; Quinn, A.; Grieshop, A.P.; Pillarisetti, A.; Glass, R.I. Clean cooking and the SDGs: Integrated analytical approaches to guide energy interventions for health and environment goals. *Energy Sustain. Dev.* **2018**, *42*, 152–159. [CrossRef] [PubMed]
10. United Nations. *Transforming Our World: The 2030 Agenda for Sustainable Development*; United Nations: New York, NY, USA, 2015.
11. Johnson, O.; Wanjiru, H.; Ogeya, M.; Johnsson, F.; van Klaveren, M.; Dalla Longa, F. *Energising Kenya's Future: Reducing Greenhouse Gas Emissions and Achieving Development Aspirations*; Stockholm Environment Institute: Nairobi, Kenya, 2017.
12. Ouedraogo, N.S. Africa energy future: Alternative scenarios and their implications for sustainable development strategies. *Energy Policy* **2017**, *106*, 457–471. [CrossRef]
13. Dalla Longa, F.; van der Zwaan, B. Do Kenya's climate change mitigation ambitions necessitate large-scale renewable energy deployment and dedicated low-carbon energy policy? *Renew. Energy* **2017**, *113*, 1559–1568. [CrossRef]
14. Government of Kenya (Ministry of Energy and Petroleum). *Ministry of Energy and Petroleum Draft National Energy and Petroleum Policy*; Government of Kenya: Nairobi, Kenya, 2015; pp. 1–130.
15. World Health Organization (WHO). *Kenya: Country Profile of Environmental Burden of Disease*; WHO: Geneva, Switzerland, 2009; p. 1.
16. Global Alliance for Clean Cookstoves. *Kenya Country Action Plan*; Global Alliance for Clean Cookstoves: Washington, DC, USA, 2013; pp. 1–30.
17. Institute for Health Metrics and Evaluation. Global Burden Disease Compare, Seattle, WA. Available online: https://vizhub.healthdata.org/gbd-compare/ (accessed on 6 February 2020).
18. Food and Agriculture Organization of the United Nations. *Agroforestry for Landscape Restoration*; Food and Agriculture Organization of the United Nations: Rome, Italy, 2017.
19. African Smallholder Farmers Group (ASFG). *Supporting Smallholder Farmers in Africa: A Framework for an Enabling Environment*; African Smallholder Farmers Group: London, UK, 2013.
20. Berazneva, J.; Lee, D.; Place, F.; Jakubson, G. Allocation and Valuation of Non-Marketed Crop Residues in Smallholder Agriculture: The Case of Maize Residues in Western Kenya. *Ecol. Econ.* **2017**, *152*, 1–21.
21. Ministry of Energy and Petroleum. *Draft Strategy and Action Plan for Bioenergy and LPG Development in Kenya*; Ministry of Energy and Petroleum: Nairobi, Kenya, 2015; pp. 1–8.
22. Dietz, T.; Börner, J.; Förster, J.J.; von Braun, J. Governance of the bioeconomy: A global comparative study of national bioeconomy strategies. *Sustainability* **2018**, *10*, 3190. [CrossRef]
23. Murphy, J.; Bochmann, G. *IEA Bioenergy—Task 37: Biogas from Crop Digestion*; International Energy Agency: Paris, France, 2011; Volume 37, pp. 1–23.
24. SE4ALL. *Sustainable Energy For All—Kenya Action Agenda*; 2016; pp. i–64.
25. Carvalho, R.L.; Jensen, O.M.; Tarelho, L.A.C. Mapping the performance of wood-burning stoves by installations worldwide. *Energy Build.* **2016**, *127*. [CrossRef]
26. Carvalho, R.L.; Lindgren, R.; García-López, N.; Nyambane, A.; Nyberg, G.; Diaz-Chavez, R.; Boman, C. Household air pollution mitigation with integrated biomass/cookstove strategies in Western Kenya. *Energy Policy* **2019**, *131*, 168–186. [CrossRef]
27. Njenga, M.; Iiyama, M.; Jamnadass, R.; Helander, H.; Larsson, L.; de Leeuw, J.; Neufeldt, H.; Röing de Nowina, K.; Sundberg, C. Gasifier as a cleaner cooking system in rural Kenya. *J. Clean. Prod.* **2015**, *121*, 208–217. [CrossRef]
28. Torres-Rojas, D.; Lehmann, J.; Hobbs, P.; Joseph, S.; Neufeldt, H. Biomass availability, energy consumption and biochar production in rural households of Western Kenya. *Biomass Bioenergy* **2011**, *35*, 3537–3546. [CrossRef]
29. International Standard Organization ISO 14040:2006—Environmental Management—Life Cycle Assessment—Principles and Framework. Available online: https://www.iso.org/standard/37456.html (accessed on 24 February 2019).

30. Global Alliance for Clean Cookstoves. *Comparative Analysis of Fuels for Cooking*; Global Alliance for Clean Cookstoves: New York, NY, USA, 2016.
31. Okoko, A.; Reinhard, J.; von Dach, S.W.; Zah, R.; Kiteme, B.; Owuor, S.; Ehrensperger, A. The carbon footprints of alternative value chains for biomass energy for cooking in Kenya and Tanzania. *Sustain. Energy Technol. Assess.* **2016**, *22*, 124–133. [CrossRef]
32. Su, S.; Li, X.; Zhu, Y.; Lin, B. Dynamic LCA framework for environmental impact assessment of buildings. *Energy Build.* **2017**, *149*, 310–320. [CrossRef]
33. García-Gusano, D.; Iribarren, D.; Dufour, J. Is coal extension a sensible option for energy planning? A combined energy systems modelling and life cycle assessment approach. *Energy Policy* **2018**, *114*, 413–421. [CrossRef]
34. Smith, K.R. Cleaning up fuels in two fronts. *Nature* **2014**, *511*, 2014. [CrossRef]
35. Hossain, N.; Zaini, J.; Meurah, T.; Mahlia, I. Life cycle assessment, energy balance and sensitivity analysis of bioethanol production from microalgae in a tropical country. *Renew. Sustain. Energy Rev.* **2019**, *115*, 1–14. [CrossRef]
36. García-sánchez, M.; Güereca, L.P. Environmental and social life cycle assessment of urban water systems: The case of Mexico City. *Sci. Total Environ.* **2019**, *693*, 1–15. [CrossRef]
37. Albertí, J.; Brodhag, C.; Fullana-i-palmer, P. First steps in life cycle assessments of cities with a sustainability perspective: A proposal for goal, function, functional unit, and reference flow. *Sci. Total Environ.* **2019**, *646*, 1516–1527. [CrossRef]
38. Ståhl, L.; Nyberg, G.; Högberg, P.; Buresh, R.J. Effects of planted tree fallows on soil nitrogen dynamics, above-ground and root biomass, N2-fixation and subsequent maize crop productivity in Kenya. *Plant Soil* **2002**, *243*, 103–117. [CrossRef]
39. Carvalho, R.; Yadav, P.; Lindgren, R.; García-López, N.; Nyberg, G.; Diaz-Chavez, R.; Upadhyayula, V.K.; Boman, C.; Athanassiadis, D. Bioenergy strategies to address deforestation and household air pollution in Western Kenya. In Proceedings of the 27th European Biomass Conference and Exibihition 2019, Lisbon, Portugal, 27 May 2019.
40. Peša, I. Sawdust pellets micro gasifying cook stoves and charcoal in urban Zambia: Understanding the value chain dynamics of improved cook stove initiatives. *Sustain. Energy Technol. Assess.* **2017**, *22*, 171–176. [CrossRef]
41. Ministry of Environment and Natural Resources. *Kenya's Intended Nationally Determined Contribution (INDC)*; Ministry of Environment and Natural Resources: Nairobi, Kenya, 2015.
42. Wernet, G.; Bauer, C.; Steubing, B.; Reinhard, J.; Moreno-Ruiz, E.; Weidema, B. The ecoinvent database version 3 (part I): Overview and methodology. *Int. J. Life Cycle Assess.* **2016**, *21*, 1218–1230. [CrossRef]
43. Kadian, R.; Dahiya, R.P.; Garg, H.P. Energy-related emissions and mitigation opportunities from the household sector in Delhi. *Energy Policy* **2007**, *35*, 6195–6211. [CrossRef]
44. Emodi, N.V.; Emodi, C.C.; Murthy, G.P.; Emodi, A.S.A. Energy policy for low carbon development in Nigeria: A LEAP model application. *Renew. Sustain. Energy Rev.* **2017**, *68*, 247–261. [CrossRef]
45. Limmeechokchai, B.; Chawana, S. Sustainable energy development strategies in the rural Thailand: The case of the improved cooking stove and the small biogas digester. *Renew. Sustain. Energy Rev.* **2007**, *11*, 818–837. [CrossRef]
46. KNBS. *Kenya Census Population Statistics 2019*; KNBS: Nairobi, Kenya, 2019.
47. IMF. *World Economic Outlook (WEO) Database*; IMF: Washington, DC, USA, April 2018; p. 2015.
48. Trading Economics. Kenya GDP Per Capita 1960–2018. Available online: https://tradingeconomics.com/kenya/gdp-per-capita (accessed on 17 July 2018).
49. Government of Kenya. *Kisumu County Fact Sheet-1. 2015*; Government of Kenya: Nairobi, Kenya, 2015; pp. 1–15.
50. UN-Habitat. *Kisumu City Development Strategies (2004–2009)*; United Nations: Nairobi, Kenya, 2004.
51. Smith, K.R.; Uma, R.; Kishore, V.V.N.; Lata, K.; Joshi, V.; Zhang, J.; Rasmussen, R.A.; Khalil, M.A.K. *Greenhouse Gases from Small-Scale Combustion Devices in Developing Countries, Phase II a, Household Stoves in India*; United States Environmental Protection Agency: Washington, DC, USA, 2000.
52. Jetter, J.; Zhao, Y.; Smith, K.R.; Khan, B.; Yelverton, T.; Decarlo, P.; Hays, M.D. Pollutant emissions and energy efficiency under controlled conditions for household biomass cookstoves and implications for metrics useful in setting international test standards. *Environ. Sci. Technol.* **2012**, *46*, 10827–10834. [CrossRef]

53. MacCarty, N.; Still, D.; Ogle, D. Fuel use and emissions performance of fifty cooking stoves in the laboratory and related benchmarks of performance. *Energy Sustain. Dev.* **2010**, *14*, 161–171. [CrossRef]
54. Magambo, B.; Kiremu, C. A Study of Household Eletricity Demand and Consumption Patterns in Nairobi. Master Thesis, University of Nairobi, Nairobi, Kenya, 2010.
55. KNBS. *County Statistical Abstract Kisumu County 2015*; Kenya National Bureau of Statistics: Nairobi, Kenya, 2015; pp. 1–19.
56. Huijbregts, M.A.J.; Steinmann, Z.J.N.; Elshout, P.M.F.; Stam, G. *ReCiPe 2016: A Harmonized Life Cycle Impact Assessment Method at Midpoint and Endpoint Level Report I: Characterization*; National Institute for Public Health and the Environment: Bilthoven, The Netherlands, 2016.
57. Morelli, B.; Cashman, S.; Rodgers, M. *Life Cycle Assessment of Cookstoves and Fuels in India, China, Kenya and Ghana*; United States Environmental Protection Agency: Washington, DC, USA, 2017.
58. Milà i Canals, A.; Azapagic, A.; Doka, G.; Jefferies, D.; King, H.; Mutel, C.; Nemecek, T.; Roches, A.; Sim, S.; Stichnothe, H.; et al. Approaches for Addressing Life Cycle Assessment Data Gaps for Bio-based Products. *J. Ind. Ecol.* **2011**, *15*, 707–725. [CrossRef]

© 2020 by the authors. Licensee MDPI, Basel, Switzerland. This article is an open access article distributed under the terms and conditions of the Creative Commons Attribution (CC BY) license (http://creativecommons.org/licenses/by/4.0/).

MDPI
St. Alban-Anlage 66
4052 Basel
Switzerland
Tel. +41 61 683 77 34
Fax +41 61 302 89 18
www.mdpi.com

Energies Editorial Office
E-mail: energies@mdpi.com
www.mdpi.com/journal/energies

www.ingramcontent.com/pod-product-compliance
Lightning Source LLC
LaVergne TN
LVHW070457100526
838202LV00014B/1741